Captain Rock

JAMES S. DONNELLY, JR., is professor emeritus of history at the University of Wisconsin–Madison. He is the author of *The Great Irish Potato Famine, The Land and the People of Nineteenth-Century Cork,* and *Landlord and Tenant in Nineteenth-Century Ireland.* He is also co-editor of *Irish Popular Culture, 1650–1850,* and of *Irish Peasants: Violence and Political Unrest, 1780–1914.*

Captain Rock

The Irish Agrarian Rebellion
of 1821–1824

JAMES S. DONNELLY, JR.

The Collins Press

FIRST PUBLISHED IN IRELAND IN 2009 BY
The Collins Press
West Link Park
Doughcloyne
Wilton
Cork

First published in the USA in 2009 by The University of Wisconsin Press
www.wisc.edu/wisconsinpress/

British Library Cataloguing in Publication Data

Donnelly, James S.
  Captain Rock : the Irish agrarian rebellion of 1821-1824.
  1. Agriculture and state—Ireland—Munster—History—19th
  century. 2. Farm tenancy—Economic aspects—Ireland—
  Munster—History—19th century. 3. Tenant farmers—Ireland—
  Munster—Economic conditions—19th century. 4. Munster
  (Ireland)—Social conditions—19th century.
  I. Title
  338.1'8419'09034-dc22
      ISBN-13: 9781848890107

Typesetting by Judy Gilats
Typeset in Escrow
Printed in Ireland by ColourBooks Ltd

The cover image is a copy of a lithograph by P. Ronan of a hand-colored etching *c.* 1824. M. Cleary of Nassau Street, Dublin, printed the lithograph. Depicted in the illustration are four adherents to the cause of "Captain Rock". Seated around a table with tankards and guns, three of them have come together to "swear in" a new member of the movement on a book. The title given to the lithograph is "Captain Rock's banditti: swearing in a new member". (Nicholas K. Robinson Collection of Caricature, held within the Early Printed Books Department of Trinity College Library, Dublin) *for*

JENNIFER *and* JOHN

EILEEN *and* BRAD

ELIZABETH *and* JEFF

Captain Rock

Acknowledgments

My first thanks must go to those who made possible the completion of a project begun long ago. The extraordinary generosity of the Centre for Irish Studies at the National University of Ireland Galway (NUIG), which hosted my stays in 2006 and 2009, enabled me to finish the writing of this volume and to shepherd it through the production process this past winter and spring. I am enormously grateful to Louis de Paor, Centre director, and to Tadhg Foley, chair of the Centre's board, for providing me with ideal scholarly accommodations on both occasions and for their unstinting hospitality to my wife Joan and myself. Samantha Williams, Centre administrator, made daily life there an unalloyed pleasure by her thoughtful concern in a wide range of practical matters. I would have been unable to occupy these Centre posts without a fellowship funded in 2006 by the Faculty of Arts at NUIG and without vital financial support furnished by a fellowship cosponsored in 2009 by NUIG and the Irish American Cultural Institute. The confidence placed in me by the Irish American Cultural Institute, extending far beyond its fellowship, has sustained my scholarly career over a much longer period and has earned my endless gratitude. The stimulating intellectual company of scholars from different disciplines at the Centre furnished me with an added boost in getting the job fully done. I am particularly grateful to Clare Carroll, Timothy Collins, Michelle Comber, Verena Commins, Nessa Cronin, Seán Crossan, David Doyle, John Eastlake, Joyce Flynn, Leo Keohane, Jenny McCarthy, Méabh Ní Fhuartháin, and Philip O'Leary for their comradeship while I shared the Centre with them. David Doyle was selflessly dedicated in resolving certain key questions relating to relevant manuscript materials at the National Archives of Ireland in Dublin.

Also providing goads to hard work and models of scholarly accomplishment were numerous friends in the History Department in Galway, who lent books, asked good questions, gave sound advice, opened up new perspectives, and generally buttressed in ways big and small my efforts to reach the finsh line. I wish to thank Caitríona Clear, John Cunningham, Enrico Dal Lago, Mary Noelle Harris, Gerard Moran, Niall Ó Ciosáin, Dáibhí Ó Cróinín, and Ciarán Ó Murchadha (now of St. Flannan's College, Ennis). I am especially grateful to my Galway colleagues Nicholas Canny, Steven Ellis, and above all Gearóid Ó Tuathaigh, who smoothed the path of a visiting scholar in crucial ways and made my wife Joan and me feel very welcome indeed. Gerry Moran and Ciarán Ó Murchadha never failed to provide immediate research assistance whenever asked, and John Cunningham was extremely generous in the same way. Among other NUIG colleagues who provided encouragement and practical assistance, I must express my gratitude to Michael Kavanagh, Nollaig Ó Muraíle, and Anthony Varley. Both of our stays in Galway were greatly enhanced by the hospitality and special friendship extended by Josephine Griffin and Colm Luibhéid, Fionnuala and Alf Mac Lochlainn, Maura and Michael O'Connell, Maureen O'Connor and Tadhg Foley, Máirín Ní Dhonnchadha, and Hilary and David Harkness.

At different times over the years numerous institutions (besides those already named) have generously supported research on this project. Leading the way have been different units of the University of Wisconsin-Madison. The Graduate School and the History Department have never lagged in funding my research. The Anonymous Fund of the UW-Madison has made an extremely generous grant in aid of the publication of this book. I am also deeply indebted to those other institutions that have supported the research on which this study is based: the Guggenheim Foundation, the Institute for Advanced Study in Princeton, N.J., the Woodrow Wilson International Center for Scholars in Washington, D.C., and the Institute of Irish Studies at the Queen's University of Belfast.

A number of institutions and individuals provided illustrations for this book or granted permission to use material in copyright, and their helpfulness and generosity must be acknowledged. Archivist Felicity

Harper of Powderham Castle in Devon and Librarian Jane Cunning-ham of the Courtauld Institute of Art in London secured permission for the use of a portrait of the third Viscount Courtenay; special thanks are due to the Earl of Devon in this matter. Director Peter Murray and Exhibitions Officer Dawn Williams of the Crawford Art Gallery in Cork elicited the permission of an anonymous owner for the reproduction of a well-known painting by Irish artist Daniel Maclise. Charles Benson and Timothy Keefe were both instrumental in making available three illustrations from the Nicholas Robinson Collection in Trinity College Dublin, including the cover image. And my special friend John Cussen of Newcastle West, Co. Limerick, was very gen-erous in furnishing two other illustrations drawn from his invaluable collection of documents relating to the Courtenay estates.

The professional advice and assistance of numerous archivists and librarians has enabled me to access and excavate the sources essential to the completion of this study. When I worked at the National Library of Ireland in Dublin, Alf Mac Lochlainn, Patricia Donlon, Noel Kissane, and Gerard Lyne kindly facilitated my research in manifold ways. The staff of the former State Paper Office at Dublin Castle opened many a box of the State of the Country Papers to my wide-eyed inspection. (This vast collection has since been transferred to the National Archives of Ireland.) And the staff of Memorial Library in Madison have continually worked to advance my research on Irish agrarian rebellion and to furnish technical assistance in the repro-duction of images. I am deeply grateful to two former bibliographers in European history and social studies—Erwin Welsch and Barbara Walden—for their persistent efforts to enhance the impressive holdings of Memorial Library in modern Irish history in general. I am extremely grateful once more to the staff of the Micro-Imaging Laboratory in Memorial Library for their expertise in producing illustrations of the highest quality. For granting access to relevant materials in their collections, I also wish to thank the heads of the Cork Archives Institute, the Local Studies Department of the Tipperary County Library in Thurles, the Historical Records Centre in Ennis, the Public Record Office of Northern Ireland in Belfast, the Public Record Office, Kew, England (now the National Archives), and the Public Archives of Canada in Ottawa.

On the lengthy journey through the historical records that undergird this book, many friends have helped to direct my steps and to open fruitful new lines of inquiry. At the beginning of this journey Timothy and Carmel O'Neill furnished the academic platform, the social networks, and the mental stimulation that set the course. Sharing the company of Paul E.W. Roberts as we both burrowed deeply into the State of the Country Papers was always exhilarating. My close collaboration with the historical sociologist Samuel Clark broadened my horizons and rekindled my concern with Irish agrarian rebellion. Long friendships with Liam Kennedy and Kerby A. Miller have shaped many aspects of this project; both of them have made essential contributions to this study that stem from their extraordinary knowledge of prefamine Ireland and from their unfailing kindness in sharing that knowledge. Other close friends have given unstintingly of their own expertise. I have never stopped learning from Laurence Geary, David W. Miller, and Gary Owens, all of whom have assisted me in important ways. I am especially grateful to Larry Geary for his painstaking reading of the whole manuscript and for his many useful suggestions and corrections. I owe a heavy debt as well to Anita Olson, who executed with remarkable efficiency the daunting task of bringing an earlier draft of large portions of this book into the computer age.

Members of the Madison History Department have unfailingly supported my scholarly work. At the head of this group stands Tom Archdeacon, who, along with his wife Marilyn Lavin, has been a major prop in my family's life for more than four decades. A trip with Tom and Marilyn to Tom's Irish relatives in the counties of Waterford, Cork, and Kerry in the early 1990s led me almost accidently to the "birthplace" of the Rockites in Newcastle West. The encouragement provided by my colleagues Ted Hamerow and Rich Leffler, and their wives Diane and Joan, will always be remembered with great appreciation.

I have incurred obligations to a numerous group of other scholars who have shaped my thinking about subjects treated in this book or have given me practical assistance that resolved thorny issues. For such help I wish to express my gratitude to Thomas Bartlett, Maurice Bric, the late Galen Broeker, Sean J. Connolly, L. Perry Curtis, Jr., David Dickson, Terence Dooley, the late Michael Dore, Tom Dunne, David Fitzpatrick, Tom Garvin, Raymond Gillespie, J.J. Lee, Ian

McBride, the late Oliver MacDonagh, Donald MacKay, W.A. Maguire, John A. Murphy, Liam Ó Duibhir, Fergus O'Ferrall, Cormac Ó Gráda, Pádraig Ó Macháin, Stanley H. Palmer, James G. Patterson, Bill Power, W.E. Vaughan, Brian Walker, and Kevin Whelan.

Special words of thanks are due to John Cussen, a solicitor by training and a historian by instinct and by dint of full immersion in the history of Newcastle West and County Limerick. Without the rare printed works and manuscripts that he provided so freely, the origins of the Rockite movement on Viscount Courtenay's estates would have remained largely unknown to me. He pointed me in the direction of helpful new publications that opened additional windows on this subject. And he was unflagging in his encouragement and endlessly tolerant of my demands on his time and energy amid the everyday pressures of a modern solicitor's office.

At every stage of the publication process I have benefited from the skill and proficiency of staff members at the University of Wisconsin Press and of others beyond the Press. Acquisitions Editor Gwen Walker patiently and ingeniously resolved a host of issues going far beyond the duties normally associated with her formal title. Her constant support for this book has been an extraordinary blessing. Press Director Sheila Leary showed a degree of courage in embracing such a big book that I hope will be handsomely rewarded; her backing for the entire series on the History of Ireland and the Irish Diaspora has exceeded all reasonable expectations of the coeditors Tom Archdeacon and myself. My manuscipt has been greatly improved by the searching eyes and ceaseless care of two wonderful editors—Mary Magray and Adam Mehring—who have saved me from innumerable errors. For the excellent maps I wish to thank Tanya Buckingham of the Cartography Laboratory. For the impressive index I am indebted to Margie Towery. For the fine appearance of the book in general, I am deeply obliged to Production Manager Terry Emmrich and to my longtime friend, the compositor and designer Judy Gilats. It has been a distinct pleasure to work with The Collins Press in Cork, on the overseas edition of this book.

I must also pay a brief tribute to my former graduate students over the years in Madison. I have learned more from them and their own research and publications than they will ever realize. The listing of the dissertations of many of them in the bibliography of this book hardly begins to acknowledge the debt that I owe to their flair for archival investigation and penetrating analysis. Nor does this listing capture the deep impact already made by a significant number of them on the historiography of modern Ireland.

With the encouragement of my wife Joan, to whom in everything I owe the most, I wish to dedicate this book to our three daughters and our three sons-in-law. The six of them have enriched my life beyond measure, and this is a fitting opportunity to make partial recompense for all that they have done to brighten my existence. The dedication will also put them under some pressure to read this book to the end.

Captain Rock

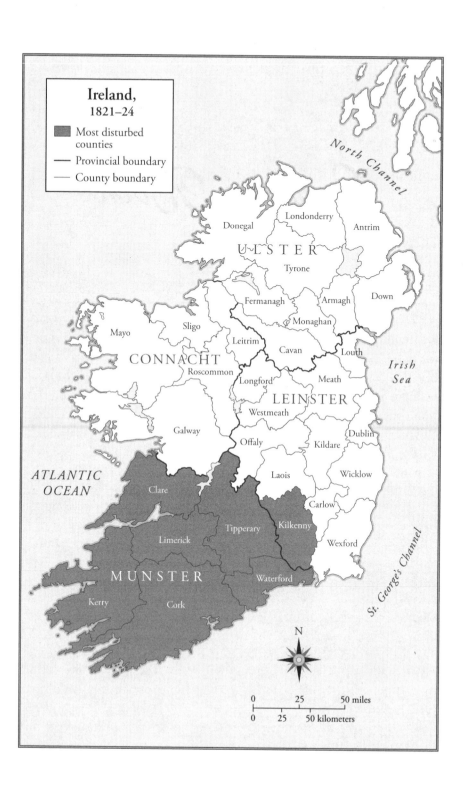

Ireland,
1821–24

■ Most disturbed
counties
— Provincial boundary
— County boundary

North Channel

Donegal Londonderry Antrim

ULSTER

Tyrone

Fermanagh Armagh Down

Monaghan

Mayo Sligo

CONNACHT

Leitrim Cavan Louth

Roscommon Longford Irish
Sea

Meath

Galway Westmeath LEINSTER

Dublin

Offaly Kildare

Laois Wicklow

ATLANTIC
OCEAN

Clare

Carlow

Tipperary Kilkenny

Limerick Wexford

MUNSTER Waterford

Kerry Cork

St. George's Channel

N

0   25   50 miles

0   25   50 kilometers

Introduction

On the night of 29–30 October 1816 a large party of about a hundred men surrounded Wildgoose Lodge, the dwelling of Edward Lynch in the townland of Arthurstown in County Louth, about ten miles from Dundalk on the east coast of Ireland. Earlier that year, Lynch had successfully prosecuted three men under the Whiteboy Act of 1776 for breaking into the lodge in search of arms—a trial culminating in the hanging of those convicted. Incensed by Lynch's role as "informer," some members of the group outside his house that night set it on fire and prevented the eight men, women, and children inside from escaping the flames. In response to this horrendous crime the authorities strenuously exerted themselves to apprehend and capitally convict those responsible, and in the end "the majesty of the law" exceeded even the original atrocity in its violence. As many as eighteen men were executed for their alleged involvement in the burning and murders at Wildgoose Lodge. Patrick Devan, the reputed leader, a schoolmaster and the clerk of a local Catholic chapel, was executed at the scene of the crime in July 1817; he was gibbeted and his carcass allowed to hang in chains for almost two years at Corcreeghagh, his birthplace. Most of the others' bodies were also gibbeted in groups of three or four at conspicuous places in County Louth. The sheer number of executions, the grisly gibbeting of numerous bodies, and the dissection of others by surgeons (intended as public ignominy) served to create a strongly adverse popular reaction, particularly as many local people believed that innocent men had suffered alongside the guilty. Helping to mold public opinion was the widespread conviction that those who took it upon themselves to administer unofficial justice to informers had morality on their side. Attuned to the sympathies of Irish country people in such matters, Carleton would later conclude his fictionalized account of the event with an emphatic passage intended to distill popular attitudes to Devan's fate. "The peasantry," Carleton asserted in reference to Devan's gibbeted body, "frequently exclaimed on seeing him, 'Poor Paddy!' A gloomy fact that speaks volumes!"[1]

This crime in County Louth punctuated the end of a much wider wave of Irish agrarian turmoil in the years 1813–16, and though almost unique as a "multitudinous murder" (in Carleton's phrase), Wildgoose Lodge was itself only one hillock in a whole mountain range of agrarian murders in Ireland during the period 1800–45. Indeed, the last executions at Dundalk for the Wildgoose Lodge killings were barely over when troubles began around Newcastle West in County Limerick that spawned the great Rockite movement of 1821–24, an exceptionally violent agrarian rebellion fueled by a combination of economic, sectarian, and political motives.

Named for their mythical leader "Captain Rock,"[2] the Rockites were especially remarkable for the frequency of their resort to murder and incendiarism as weapons of warfare. They garnered support extending far beyond the ranks of the poor; their movement eventually embraced many of the better-off farmers in the southern region where they exercised an extraordinary sway—most prominently in Limerick and Cork, but also in portions of Kerry, Clare, Tipperary, Waterford, and Kilkenny. The intensity of their grievances, the frequency of their resort to sensational violence, and their appeal on key issues—especially rents and tithes—across a broad front presented a nightmarish challenge to Dublin Castle. This challenge prompted a major reorganization of the police, a purging of the local magistracy, and the introduction of large military reinforcements. Adding fuel to the conflagration was a great upsurge in sectarianism and millenarianism that accompanied this agrarian rebellion. Prophecies of imminent Protestant doom gained a firm hold at the popular level among Catholics in the affected region—indeed, far beyond the seven aforementioned counties. In short, the Rockites showed little inclination to submit to control by Catholic priests or members of the landed elite and presented a stiff test to the Protestant Ascendancy in Ireland.[3]

The Rockite movement involved, then, a sustained outburst of agrarian violence greater than any that had previously occurred in Ireland. This is no small claim, for there had already been a long line of organized and widespread rural protests stretching back to the first "Whiteboy" movement of the early 1760s—a movement that gave rise to common use of the term "Whiteboyism" to denote the seemingly endless succession of agrarian disturbances for which Ireland became notorious in the late eighteenth and early nineteenth centuries.[4] The Whiteboys of the early 1760s were not the very first manifestation of regional agrarian protest in Ireland; that distinction apparently belongs to the "Houghers," who inflicted considerable damage on stock-rearing gentlemen and other graziers in certain western counties in the years 1711–12. By killing or cutting the hamstrings of large numbers of cattle and sheep belonging to flockmasters and graziers in parts of Connacht and a few adjacent counties, these rebels sought to keep the large-scale rearers of cattle and sheep from monopolizing ever larger tracts of pasture land and to preserve or extend tillage farms and small-scale livestock operations. Feeling themselves oppressed by rich proprietors and wealthy landholders, "the poor" and those of higher social status (including some paternalistic gentlemen) in these western counties sought through violence and intimidation to impose strict limits on the expansion of what would later be called grazing "ranches."[5] Houghing on an extensive scale was not unknown in the western region at certain points later in the century; there were outbreaks in 1779, in the mid-1790s, and in the aftermath of the 1798 rebellion.[6]

But if one takes a long-term view of Irish agrarian rebellion in the eighteenth and the early nineteenth centuries, it was not the west but rather the south that saw the most persistent outbreaks of agrarian rebellion. (Often attached to the rebellious "south" were the southeastern county of Kilkenny and the southwestern county of Clare.) It was in this southern region of the country that the phenomenon of Whiteboyism, with its mobilizing oaths of secrecy and loyalty, became most deeply entrenched between 1760 and 1845.[7] It must be acknowledged that in the revolutionary decade of the 1790s the midland counties (stretching from the hinterlands of Dublin in the east to County Galway in the west) witnessed extraordinary levels of agrarian protest (as well as political violence) that exceeded whatever disturbances occurred in the southern province of Munster or adjacent districts.[8]

Nevertheless, apart from the 1790s, collective agrarian violence between 1760 and the Great Famine was most closely identified with the Munster counties and adjoining portions of Leinster. ("Ribbonism" had a different geographical profile, along with other distinctive features, and should not be conflated with Whiteboyism.)[9] The dominance of Captain Rock and his followers in the southern region in the early 1820s thus constituted a recurrence of patterns of collective behavior with a long pedigree.[10] But the Rockites carried agrarian rebellion in Ireland to a higher plane of intensity than it had ever before reached; their extreme violence invites and repays close study.

Why Economic Explanations Are Insufficient

It was once fashionable to insist that Whiteboyism was a phenomenon that could be explained more or less exclusively in economic terms.[11] There seemed much to be said in favor of this view. The grievances that the agrarian rebels expressed were for the most part economic in nature—excessive tithes, exorbitant rents, evictions, escalating food prices, not enough potato ground (or conacre land), high taxes, and heavy priests' dues. The most frequent precipitants of collective action stemming from such complaints were the recurrent economic events of acute price depressions or serious subsistence crises, though there were some outbreaks of collective violence over tithes that had less to do with low prices or food scarcity than with the extreme distaste of tillage farmers for a tithe system that increased the burden of these clerical taxes whenever they expanded their production of grain crops and potatoes. Admittedly, for Catholics and Presbyterians there were religious objections to tithes as involuntary but legally required payments that went for the upkeep of the Anglican church establishment in Ireland. But the fact that Catholic and Presbyterian protesters usually campaigned for the reduction of tithes rather than for their outright abolition seemed to indicate the primacy of economic over religious motives among the opponents. The case of the Rightboy movement of 1785–88 appeared to offer further evidence in the same direction. Some Anglican landed gentlemen—"the gentry Whiteboys"—provided part of the leadership for this great agitation, while many members of the rank-and-file manifested almost as much hostility toward priests' dues as they did toward tithes.[12] And when historians sought to explain the timing of the three greatest agrarian rebellions of the early nineteenth century, they did not have to look far to see that they each coincided with a severe downturn in the agricultural and indeed wider economy of the country.[13]

But research carried out on agrarian and popular political movements since the early 1980s has clearly established that at certain important junctures millenarian and sectarian currents flowed strongly into Irish protest and revolt. What has been discovered with respect to Ireland is reasonably congruent with what scholars of millenarianism in other historical eras and other areas of the world have also found.[14] Among the antecedent predisposing or causative circumstances identified by various scholars, three are of particular relevance to the Irish millenarian phenomena considered in this book. First, disaster or fear of catastrophe, such as famine, epidemic disease, or massacre, has been a recurrent cause of millennial cults that have mobilized masses of people. Second, millenarian movements are likely to emerge in those agrarian societies where peasants are effectively excluded from the political apparatus of the state and lack secular political organizations of their own to defend or advance their common interests. And third, activist millennial cults have tended to flourish in colonial countries, especially during the early stages of the imposition of imperial regimes or after the indigenous peoples have incurred repeated defeats in their efforts to resist foreign domination through secular military or political means.[15] "Central to this form of millenarianism," one historian has declared, "is the belief that the oppressors are about to be cast down, even annihilated, with the help of supernatural beings."[16]

British imperial domination of Ireland was far from being in its initial phases in the late eighteenth or early nineteenth centuries. But in this period Irish millenarianism was in some respects closely akin to that often found in colonial contexts. It provided much of the ideological fuel for the interconnected Defender and United Irish movements of the 1790s, both of which aimed at the overthrow of British rule, predicated on expectations of military assistance from revolutionary France combined with native insurrection. In addition, the savage suppression of the 1798 rebellion and, later, the fall of Napoleon—who was surely a messianic figure in Ireland—made British rule seem impervious to overthrow by human effort alone.

In a provocative and often penetrating article, however, Patrick O'Farrell attributed what he considered the weakness of millennialism in Ireland before and after 1800 to the strength of anti-English feeling among the Catholic population. In O'Farrell's opinion hatred of England and of the English (or Anglo-Irish) in Ireland was a secular substitute for millennial visions in popular Irish Catholic culture.[17] Apparently because his general notion of what constitutes millenarianism was far too specific and restrictive, O'Farrell failed to see that the very hostility he recognized as perdurable was at times capable of becoming the keystone of an Irish Catholic millennialism. The hatred felt by many Catholics for Protestants and "the English" entered heavily into the millenarian dreams widespread in Ireland during the revolutionary 1790s.[18] Very similar sentiments took firm hold again under different circumstances and in a novel form around 1820. At that time the prophecy on which popular Catholic attention chiefly focused did not herald an imminent second advent. Instead, it was postmillennial; Christ's second coming was not to occur until long after the arrival of a millennium conceived as the obliteration of Protestant heresy and, by extension, the destruction of the Protestant church and state in Ireland.[19] Of course, Irish Catholic country people associated the prospective fulfillment of this prophecy, and similar ones current at the same time, with abolishing economic injustices and taking revenge for historic wrongs. But O'Farrell was misleading when he claimed that the destruction of Protestantism "was envisaged more as a time of retribution, an end to quite concrete grievances, than the dawning of a new era."[20] In fact, retribution, deliverance from grievances, and the new epoch went hand in hand, all being viewed as integral to the millennium long overdue.

Was there persistent millenarian content to the long succession of regional agrarian revolts in Ireland during the late eighteenth and early nineteenth centuries? Scholars once gave a resounding negative to this question. O'Farrell dismissed millennialism as seemingly irrelevant to agrarian agitation before the Great Famine: "In the century following the 1760s the popular rebellious elements in Ireland were too remote from their nominal religion, too ignorant of its content, and too consistently opposed by its ministers to entertain any idea of setting their protests in a religious context."[21] Much more guardedly, Joseph Lee has observed, "No genuine millenarian movements, endemic in most peasant societies, swept the Irish countryside, despite attempts to disseminate Pastorini's prophecies in the 1820s."[22] With good reason, as we have seen, both O'Farrell and Lee insisted on the rational attitudes and behavior of Irish agrarian rebels: O'Farrell referred to "the reactionary pragmatic realism of the Whiteboy phase of Irish turbulence—the 1760s to the 1830s," while Lee stressed the rebels' "limited, concrete, pragmatic programme," the "relentless realism" of the Whiteboy mind.[23] Indeed, in O'Farrell's view the very pervasiveness and effectiveness of agrarian secret societies in prefamine Ireland was one important reason for the lack of resort to millennial ideologies or cults.[24] And as a general explanation of the absence of genuine millenarian movements, O'Farrell's contention is highly persuasive. It constitutes a significant contribution to the comparative study of millennialism in peasant societies.

Yet if the capacity for effective self-defense helps to explain why millenarianism appealed less to Irish country people between 1760 and 1840 than it has to peasants in other countries at various times, the contrast should not be drawn too sharply. Lee and O'Farrell both assumed that activist millennial movements, or revolts strongly influenced by millennial ideas, have arisen more often than the facts indicate. It is extremely doubtful that millenarian movements are "endemic in most peasant societies," as Lee once alleged.[25] Norman Cohn and other scholars have shown that the medieval West produced numerous millennial cults in both rural and urban settings.[26] In a relatively small number of instances, now perhaps almost too well known, the millennial ideas of the cults spilled over into social protest and rebellion. But as Sylvia Thrupp has written, "In the vast majority of the many hundreds of medieval peasant revolts and urban revolutions on record there is no evidence of any millennial influence."[27]

If it is incorrect to assume that protest movements in peasant societies have regularly been infused with millenarian beliefs, there is also no certainty that when millennial ideas do penetrate the consciousness of those engaged in revolt, the apocalyptic notions displace whatever limited and realistic goals the rebels may earlier have had. At times revolutionary millenarians have no doubt succeeded in substituting their boundless salvationist dreams for the restrained objectives with which social movements have commenced, but on other occasions millennial doctrines, though present, have not become the actual charter of collective action.[28] A particular millenarian vision permeated the Rockite movement of 1821–24, but the immediate goals of the Rockites were not boundless or wildly unrealistic. Nor did the unusual prevalence of apocalyptic notions diminish the practical effectiveness of these agrarian rebels. The realization of their pragmatic aims depended on human agency, while the destruction of Protestantism was seen to hinge mostly on supernatural intervention—certainly a sensible division of labor under the circumstances.

Though millennial ideas did not provide the Rockite movement with a program of action, they served other important purposes. Such beliefs helped to rally Catholic country people to the Rockite cause and to confirm them in their allegiance in the face of severe government repression by presenting the Protestant establishment in church and state as a doomed edifice on the verge of annihilation. Millennial ideas also performed an additional function. They assisted in integrating within the same movement Catholics whose material interests frequently clashed, namely, landless laborers and cottiers on the one hand and the larger farmers on the other.[29] Acceptance of the prophesied ruin of Protestantism was concentrated among the lowest strata of Catholic rural society, but many middling and some substantial farmers also gave credence to this millennial vision. Though a shared belief in Protestant oppression and its impending destruction was not enough to obliterate class conflict between the Catholic poor and the larger Catholic farmers, anti-Protestant millennialism almost certainly reduced the sharpness of such antagonism.

Class Conflict below the Landed Elite

By the early 1820s the need to bridge class antagonisms below the level of the landed elite had become more acute than ever. Class conflict of this type had been a persistent feature of Irish agrarian society at least since the economic recovery that followed the terrible famine of 1740–41, in which as many as three to four hundred thousand people are estimated to have perished.[30] That disaster, measured in proportion to the population existing on its eve, was almost certainly an even greater calamity than the Great Famine a century later; it inflicted the worst mortality on the southern province of Munster, killing at least one-fifth of its inhabitants. As David Dickson has observed, the crisis of the early 1740s "can be seen as hitting hardest the new class of near-landless cottiers created by the land-engrossing cattlemen and sheep-masters of lowland Munster."[31]

Economic modernization and the commercialization of Irish agriculture had thus begun prior to the famine of 1740–41, but these processes accelerated over the remainder of the eighteenth century and especially during the French revolutionary and Napoleonic wars from 1793 to 1813. Of course, this long stretch was not a time of uninterrupted prosperity; the period was punctuated by some very painful years owing to bad weather, food scarcity, or temporarily low agricultural prices. In addition, the great inflation during the wartime boom after 1793 certainly victimized the poor of countryside and town, whose wages persistently fell behind the increased costs of food and the ever-rising rents of potato ground.[32] But for Irish landowners and the better-off members of the farming classes, the period extending from the middle of the eighteenth century to the end of the Napoleonic wars saw much more of prosperity than of its absence. There was a long and dramatic secular rise in rental incomes that supported a conspicuously lavish lifestyle among the landed elite (the building of the numerous townhouses in Georgian Dublin and of some great Palladian country mansions were the most obvious examples).[33] And as Kevin Whelan has demonstrated, this period also saw the rise of a Catholic rural middle class of "strong farmers" and middlemen whose frugality and lack of ostentation were matched by their acquisitiveness and industry.[34] This latter group of Catholic graziers and large farmers was especially prominent in Leinster and the lowlands of Munster. There was all too clearly a pronounced process of social differentiation in train, fueled by the ongoing commercialization of Irish agriculture and creating a highly unbalanced class structure. In this new Irish rural world the landless and the land-poor were falling badly behind a distinct minority of middling and large farmers who were coming to hold an ever-expanding proportion of the land available for cultivation or grazing.

By the eve of the Great Famine gross inequality of income had become characteristic of Irish rural society. At the top of the structure in 1841 there were approximately 50,000 rich and 100,000 "snug" farmers, the mean size of whose holdings was 80 and 50 acres, respectively; together with the 10,000 or so landowners, they cultivated roughly half the land and controlled access to most of the rest. In the middle of the structure were 250,000 family farmers; the mean size of their holdings was about 20 acres, and they usually met their labor needs from within the family. At the bottom—the depressingly broad base of the social pyramid—were an estimated 1.3 million poor peasants and laborers. These poor peasants, or cottiers (300,000, with a mean holding of 5 acres), and the laborers (1 million, with a mean holding of 1 acre) both worked for wages and purchased food. In addition, most of the laborers hired conacre plots each year from the larger farmers or the landowners, who were also the chief employers.[35]

Increasingly, then, cottiers and small tenants found themselves confined to marginal holdings of little value so that cattlemen, flockmasters, and large tillage farmers might have fuller scope for their profitable enterprises. Inevitably, the masses of landless laborers and cottiers felt trapped within an economic system that condemned them to a life of frequent and often perpetual privation and misery. And yet such was the nature of this system that it offered just enough food, employment, and land to the hordes of the rural poor that the birth and fertility rates among them kept outstripping those of the rural Catholic middle class, at least until the "demographic adjustment" finally began to take hold once the wartime boom was reversed beginning in 1813.[36] Even afterward, in some parts of Ireland this adjustment was scarcely evident. For example, Clare in the southwest, despite its limited resources of good land, experienced perhaps the highest rate of population increase of any county in Ireland right up until the Great Famine.[37] This huge imbalance in the rural social structure (especially evident in the south and the west), fueled class conflict of the kind that Joseph Lee presented as the characteristic form of agrarian violence and rebellion in early nineteenth-century Ireland.

Alternative Explanations

Lee's interpretation was challenged, however, by Michael Beames in an article appearing in *Past and Present* in 1978.[38] From a detailed examination of twenty-seven agrarian murders in County Tipperary during the late 1830s and the 1840s, Beames concluded that class antagonism between laborers or cottiers and farmers was a much less potent source of violent conflict than Lee had maintained. Instead, Beames stressed that disputes over the occupation of land were the leading cause of "peasant assassinations," and in his 1983 book he presented evidence purporting to show that such disputes gave rise to more agrarian crime than any other single factor or motive.[39] The aggrieved parties in these disputes, in his view, belonged especially to the class of small landholders, and their enemies were landlords— usually "improving" landlords bent on the consolidation of farms— and their agents or servants.[40] These would-be modernizers of Irish agriculture, Beames seemed to be saying, were threatening the survival of a distinct and solidary peasant community.

For different reasons David Fitzpatrick also found fault with Lee's class-conflict thesis.[41] Fitzpatrick closely studied an extraordinarily violent parish (Cloone) in County Leitrim during the 1830s and 1840s; his chief conclusion was that "conflict *within* social strata was probably still more pervasive than conflict *between* strata."[42] Speaking generally, he argued that the extreme scarcity of such resources as land and employment in prefamine Ireland regularly pitted farmers against farmers, and laborers against laborers, either as individuals or more often in rival groups or factions. This emphasis on conflict within social strata is related to Fitzpatrick's perception of prefamine social structure, which he saw as much less static than either Lee or Beames—more like "a ladder which one could climb up or slip down," rather than "a pyramid on which each man felt he had been assigned (perhaps unfairly) his proper station."[43] Against Beames's notion of a solidary peasant community enforcing its ethical standards through Whiteboyism, Fitzpatrick contended that rival factions repeatedly masqueraded as communities.[44]

Despite Fitzpatrick's healthy iconoclasm, and despite Beames's unintentional resurrection of something approaching the traditional view, Lee's revisionist position was long dominant. With the important exception of the tithe war of the early 1830s, Samuel Clark basically accepted Lee's thesis when skillfully treating rural collective action before the famine in his important book of 1979, *Social origins of the Irish land war.*[45] Even more supportive of Lee's interpretation was a carefully researched, highly original, and wide-ranging essay published by Paul Roberts in the 1983 collection entitled *Irish peasants.* In this essay Roberts argued that the far-famed but hitherto ill-understood feud between the Caravats and Shanavests in east Munster in the early decades of the nineteenth century was deeply rooted in class conflict over access to land, employment, and food. As Roberts portrayed this feud, and his account remains generally persuasive for

13

the years 1806–11, the Caravats were lower-class Whiteboys, comprising within their ranks agricultural laborers, industrial workers, and smallholders. The Shanavests, by contrast, though they included some farm workers bound by economic ties to their employers, were drawn from the rural middle class of farmers, publicans, and shopkeepers, with farmers usually in the ascendant. In response to the violent efforts of the Caravats to impose Whiteboy "laws," the farmer-led Shanavests frequently formed vigilante organizations to keep the Caravat poor in check; they also established factions that fought pitched battles with their lower-class enemies at fairs, patterns, and other large public gatherings. Roberts strongly suggested that the element of class conflict inherent in the Caravat-Shanavest rivalry between 1806 and 1811 was recapitulated in faction-feuding and agrarian rebellion during the rest of the early nineteenth century.[46]

Formulating a Different Explanation

All of these interpretations possess merit in varying degrees, but none does sufficient justice to the diversity and complexity of rural violence before the Great Famine. The principal weakness of earlier scholarly efforts at explanation in this area may be simply stated: it is that the interplay between economic fluctuations and the social composition of agrarian rebellions has not been adequately appreciated. Neither Beames nor Fitzpatrick shows any sustained concern with this issue. Both chose *not* to make particular agrarian movements or upheavals the focus of analysis, even though these waves of unrest accounted for by far the greater portion of the collective violence that took place in early nineteenth-century Ireland. Both are too ready to generalize from limited sets of data: Beames (in his article at least) from the single crime of murder in only one county, and in a period when there was no major wave of unrest; Fitzpatrick largely from the events in a single parish, albeit a very violent one. As a result, Fitzpatrick and especially Beames seem oblivious to the fact that a whole string of early nineteenth-century movements were essentially conflicts between the organized poor and substantial farmers or graziers. Into this category fall not only the Caravat/Shanavest rivalry illuminated by Roberts but also the Threshers of 1806–7, the so-called Ribbonmen of 1819–20, and above all the Terry Alts of 1829–31.[47]

While Beames and Fitzpatrick attached too little importance to class conflict, Roberts and Lee attached too much. Roberts tended to view the Caravat/Shanavest rivalry of 1806–11, a feud that erupted in a period of agricultural prosperity, as the paradigm of the outbreaks of Whiteboyism that marked the early nineteenth century.[48] Lee, on the other hand, put forward the agrarian upheavals of 1813–16, 1821–24, and the early 1830s—all three fueled by agricultural depression—as exemplifying the dominance of class conflict.[49]

Elsewhere I have advanced a different argument that I seek to expand in this book on the Rockites.[50] A simplified version of the argument would run as follows. Movements that arose in periods of prosperity, when farm prices were buoyant and land values were rising sharply, were usually dominated by the landless and the land-poor, whose fortunes were adversely affected by the prevailing economic winds. Their participation in rebellion at such times is to be explained by their determination to restrain the inflation of conacre rents and food prices, to boost wages, and to frustrate the land-acquisitive tendencies of large farmers and graziers—tendencies that were especially pronounced during periods of strong economic expansion.[51] On the other hand, agrarian upheavals (including those of the Rockites) that were fueled by a drastic decline in agricultural prices were generally marked by a distinct widening in the social composition of the rebellious groups. In the context of depression, to be sure, the poor had many reasons to maintain their tradition of activism, but they were now less often preeminent among the agrarian rebels. Conditions of acute depression did indeed create more fertile ground for the forging of alliances of various kinds across the traditional lines of social division.

The Influence of the Tithe Grievance

One grievance that perennially shaped the social composition of Irish agrarian protest was that of tithes, which agrarian rebels had always placed at or near the top of their list of complaints ever since the first Whiteboys in the early 1760s.[52] This particular grievance had the capacity to bring together in rebellious alliance a wide range of social groups and to multiply the number of participants from the lowest levels of the rural and urban social scale in Munster and parts of Leinster.[53] There were three anomalous features of the Irish tithe system that need to be made clear at the outset, as two of them are essential to understanding why Whiteboyism became so deeply entrenched in the southern counties and why this issue was the most persistent and widespread of all agrarian complaints in that region through the period of the Rockites and beyond. The first anomaly was the curious fact that the tithe of potatoes was generally restricted to the six Munster counties and to adjacent portions of Leinster; outside of this region, though it was not completely unknown, the tithe of potatoes was very rare. The second anomaly was that, of the various tithes due from the produce of the soil, the highest level of taxation by the Anglican church and its ministers generally fell on potatoes; the rates of tithe or tax, in other words, were invariably heavier for an acre of potatoes than for an acre of wheat, oats, barley, or hay.[54] As the chief staple in the diet of the poor, the potato appears to have established a position of dominance earlier in Munster (gradually during the period 1750–1800) than in the other provinces, and cultivation of this root was probably still expanding there until the Great Famine.[55] For most farmers by the early nineteenth century, the potato had become an important adjunct of their diet, but for the cottiers and laborers of Munster and much of the rest of Ireland, it was almost their only food, along with "sky-blue"—milk from which the cream had been removed.[56] In short, it could fairly be said that the heaviest single burden of the tithe system in Munster and a certain portion of Leinster fell on members of the rural community least able to bear it—those at the bottom of the socioeconomic scale.

If these two anomalies help to explain the prominence of cottiers and laborers in the Whiteboy movements of Munster and some parts of Leinster between the 1760s and the 1830s, there was a third anomaly that affected—negatively or positively—landholders or tenants all over the country. This was the general exemption that pastureland had long enjoyed from liability to tithes; with minor exceptions, tillage farmers—from the biggest to the smallest—bore almost the whole brunt of supporting the Anglican church establishment in Ireland.[57] The peculiar workings of the law of tithes therefore meant that if economic trends favored an extension of tillage, as was certainly the case in the 1780s, farmers shifting into tillage from pasture became subject to a new and significant burden that they immediately came to resent. In such a shift on an extensive scale lies much of the explanation for the rise of the great Rightboy movement of 1785–88.[58] But practically every tillage farmer, whether new to the cultivation of crops or not, was aggrieved about tithes. How could tillage farmers not be aggrieved about a system that left the much wealthier cattlemen and sheepmasters almost entirely free from a set of burdens that only those growing crops had to shoulder? Considered together, these anomalies of the Irish tithe system laid the foundations for collaboration in rebellious activity—cooperation across the lines of a social divide that had the potential under certain common circumstances to set subsistence cultivators and commercial tillage farmers at serious odds.[59]

Social Composition as Shaped by Rents and Evictions

The other principal grievance of the Rockites—rents considered grossly exorbitant and the eviction of tenants who fell too far behind in paying them—influenced the social composition of the movement in quite complicated ways. In the gestation of the Rockite campaign on Viscount Courtenay's estate in west Limerick, as chapter 1 will show, there occurred the surprising spectacle of middlemen tenants of gentry and Protestant background mobilizing subtenants to resist the exactions of a newly installed land agent on a property where his lax predecessor had allowed enormous arrears of rent to accumulate. It was far more usual, however, to find grasping or needy middlemen as the targets of Rockite agrarian violence, for most middlemen in this period faced a double peril—repudiation of debts by struggling or bankrupt smallholders, who lashed out against pressing middlemen, and the widespread desire of head landlords (proprietors in the strict sense) to terminate the leases of middlemen whose own indebtedness made them much less attractive as intermediate landlords, responsible for the paltry and now often uncollectible rents of hordes of cottiers and small farmers.[60]

Normally, most large farmers, to whom the tags of "snug" or "comfortable" were commonly applied in this period, would have considered Whiteboyism as the resort of country people much less respectable than themselves. But the punishing economic downturn that commenced late in 1819 left not even the wealthiest farmers unscathed, and they too joined the general clamor for sweeping reductions of rent. Indulgence, however, did not come naturally or easily to Irish landowners. Since they too had creditors demanding to be paid, the result was a steep increase in evictions. No doubt the poorest tenants generally took the brunt of this landlord response, but because the depression menaced even formerly well-off farmers with insolvency or the inability to pay nearly as much in rent as they had promised, such landholders now assumed a different attitude toward Rockite violence aimed at frustrating the collection of "normal" rents or at punishing those landed proprietors, middlemen, and other rich farmers who resorted to the dispossession of defaulting tenants. Under certain circumstances, therefore, the Rockites claimed the allegiance not only of cottiers, laborers, and artisans but also of large numbers of middling and large farmers who feared eviction or who wanted to see their rents steeply reduced.[61]

Local Protests or Regional Movements?

There was once a tendency in writing about Irish agrarian rebellion to see the different outbreaks of protest and violence as arising mainly from local grievances and as led by local leaders whose authority was confined to narrowly circumscribed areas. Still one of the best-known writers on Whiteboyism in the early nineteenth century is Sir George Cornewall Lewis, whose book entitled *On local disturbances* was published in 1836.[62] Lewis embraced the concept of Whiteboyism as a type of agrarian trade unionism, but one that lacked any recognizable national or even regional organization. Instead, Lewis preferred to portray the phenomenon as involving a multiplicity of local bodies with complaints or grievances that were held in common in many different places.[63] Some professional historians have been inclined to view Irish agrarian secret societies of the late eighteenth and early nineteenth centuries as primarily local in nature, or they have asserted that other historians have exaggerated the degree to which meaningful supralocal cohesion prevailed among the agrarian rebels.[64] Taking strong issue with the localist interpretation in this book, I argue that the Rockites, like quite a number of their predecessors and successors, constituted a regional agrarian revolt that exhibited many supralocal features. While not altogether discounting the significance of contagion, chapter 2 of this book seeks to demonstrate that the spread of the Rockite movement from its origins on Viscount Courtenay's estate to other portions of west Limerick and to neighboring counties was the result of deliberate and carefully orchestrated activity designed to extend the sway of Captain Rock and his "laws" or "regulations" over a very wide geographical area. The insurrectionary phase of the movement at the end of 1821 and the very beginning of 1822 displayed a considerable degree of organization, even if the thousands who turned out in northwest Cork were not formidable militarily. The debilitating impact of the near famine in the spring and summer of 1822 was in its own way a testament to the character of the movement: the subsistence crisis effectively shut down the campaign everywhere in the south, but once this crisis had passed, the movement resurrected itself and remained vital over much of Munster and in County Kilkenny until at least the early months of 1824. In this later period too the Rockites frequently manifested supralocal organization in the conduct of their operations. Admittedly, most Rockite actions (especially if all types are taken into account) involved small parties of a dozen or fewer activists, but there were a considerable number of major rebel enterprises before and after the spring and summer of 1822 in which scores or even hundreds of Rockites were engaged. The rationale for depicting the Rockite rebellion as a regional movement derives its force even more strongly from the widely shared consciousness among the rebels that they were seeking adherence to a body of laws or a code of regulations that were designed to have a general rather than simply a local applicability. For this widely shared consciousness many of the large number of surviving threatening letters and notices produced by the Rockites provide abundant documentation.[65]

Political Ideas and Popular Mentalité

Another tendency in historical writing about Irish agrarian rebellions of the late eighteenth and early nineteenth centuries has been to minimize their political content and even at times to stress their economic motivations almost to the point of excluding political ones altogether. Of course, this observation does not apply to the 1790s. There is general agreement among historians that radical and revolutionary ideas (including much millennialism) often penetrated to the lower reaches of the social scale in town and country during that tumultuous and transformative decade. The mutually antagonistic sectarianisms of Catholics and Protestants were infused with huge amounts of new venom in the 1790s—before, during, and after the 1798 rebellion itself. Indeed, the bloody Protestant-led repression that followed in the wake of the 1798 rebellion occurred on a scale that made it impossible for Catholics to forget or forgive. And while millenarian ideas were promoted or accepted by numerous Presbyterian radicals in the 1790s, in the latter part of that decade millennial beliefs became much more widely diffused among rank-and-file Defenders and United Irishmen outside the northeast.[66] It would be no exaggeration to claim that the political and agrarian rebels of the 1790s bequeathed a substantial legacy of sectarianized political ideas to ordinary Catholics over much of Ireland in the early nineteenth century. As this book on the agrarian revolt of 1821–24 will repeatedly show, the rhetoric of numerous Rockite notices and threatening letters as well as the language of Rockite oaths, verbal signs, and passwords frequently bore the telltale marks of their origins in the 1790s.

Ribbonism and the Rockite Movement

One of the specific political legacies of the sectarianized 1790s was the development of Ribbon lodges or societies in many Irish cities and towns as well as in certain parts of rural Ireland during the early nineteenth century. This was a shadowy world on which historians have so far managed to throw only a limited amount of light. Tom Garvin and Michael Beames have made the most significant contributions to our knowledge of this subject.[67] Ideologically, Ribbonism was a form of nationalism combining popular political radicalism or republicanism with anti-Orange Catholic sectarianism.[68] There is some disagreement about the geographical distribution of this phenomenon. According to Garvin, the heartlands of Ribbonism in the early nineteenth century were Dublin city, north Leinster, north Connacht, and south Ulster; Ribbonism was also "reported from all over north Ulster at various times during the 1815–45 period."[69] Beames, on the other hand, has given a somewhat different picture of its geographical distribution: "It was strongest in Dublin, the counties of the eastern seaboard, and parts of Ulster. It could also be found in various inland towns through which the Royal and Grand Canals passed, as well as in the industrial areas of Great Britain."[70] Nevertheless, both Garvin and Beames agree that the Munster counties were virtually free of organized Ribbonism throughout the early nineteenth century. Garvin stated that in his extensive survey of the surviving evidence he had "only come across one reference to Ribbonism proper in Munster—

a letter from the Dublin leaders, picked up by the police in the Munster county of Tipperary in 1839."[71] Similarly, Beames almost completely dissociated Ribbonism from agrarian unrest in the southern counties: "Ribbonism was in fact strongest in those areas least affected by peasant disturbances in the pre-famine period."[72]

This study of the Rockite movement, by contrast, will suggest that Ribbonism was not absent from Munster in the early 1820s and did play a role of modest significance in the agrarian rebellion of those years. It did so mainly by furnishing points of contact between radicalized urban workers or artisans and aggrieved country dwellers. Beames has rightly called attention to the "distinct, if not overwhelming, urban emphasis" of Ribbonism.[73] Generalizing about the social composition of the Ribbon lodges after a close investigation of the scanty detailed information that has survived, Beames observed: "It is the lower urban trading and carrying classes which seem to have provided the core of the membership, with the public house operating as the focal point of their activities."[74] Ribbon societies with a similar membership and focus appear to have come into existence in some urban centers of Munster and south Leinster by the early 1820s.[75] They not only served as vehicles of sociability and popular recreation but also provided the organizational means through which numerous lower-class urban Catholics expressed their desire to throw off Protestant politico-religious domination and economic oppression. There is also considerable evidence that *rural* Catholics of the lower orders in the south often came into contact with organized or informal Ribbonism when they had reason to visit nearby cities or towns. It seems likely that this contact increased in times of economic crisis such as 1813–16 and 1821–24. It may also have expanded during the fever epidemic of 1816–19 (the peak years of the epidemic were 1817 and 1818),[76] when popular panic about the spread of the often-lethal disease prompted the revival of anti-Protestant millennial beliefs.[77] What is not in doubt is that the Ribbon lodges of some southern towns helped to shape the ways in which the organization and ideology of the Rockite movement developed in their hinterlands. For this reason too the tendency to view agrarian rebellion as apolitical is not helpful in coming to a fully rounded understanding of the Rockites and their world.

Captain Rock and Nationalism

While this book certainly does not argue that the Rockite movement was predominantly political, it does contend that there were numerous political features of this agrarian rebellion, even beyond its sectarian and millenarian aspects. The violent struggle waged against the tithe system was designed not only to deprive tithe owners of most or all of their income but also in part to extract under great popular pressure the kind of legislative concessions that the government had been resisting for many decades. Something similar might also be said about exorbitant rents and evictions, though on these matters the general Rockite aim was not so much to change the behavior of the government as it was to alter the conduct of landowners, middlemen, and noncompliant large farmers.

In their quest for a more just economic order, at least some Rockites took inspiration from a remembered Jacobite or Jacobin past or from news that filtered into the Irish countryside about radical or revolutionary movements taking place in other lands. Their understanding of foreign political events, of Jacobitism or Jacobinism, and of Irish Catholic rule in earlier centuries was no doubt sketchy at best, but it was knowledge of the mere fact of revolution—at home or elsewhere, past or present—that kindled hope of far-reaching political and economic change in Ireland, or more precisely, in those parts of Ireland that fell under the sway of Captain Rock in the early 1820s. For example, the writer of one threatening notice posted on a gate in Fermoy, Co. Cork, in January 1822 invoked both the Jacobite past of the eighteenth century and recent events in Greece to express his desire for a kind of Catholic revolutionary nationalism: "I [John Rock] am aided by the king of France, Spain, and the rest of the Chatolick [*sic*] powers of Europe. Yea, . . . the Greeks ar[e] overturning their infidel masters, and is it not a shame for us, where we are so powerful, that [we] don't crush the serpents of this nation. There is thirty thousand men coming from Spain and France [who] will maintain war with England."[78] This kind of revolutionary millenarianism will be discussed at length in chapter 4, but it should not be thought that the Rockites, or the Catholic schoolmasters who regularly put their ideas in written form, had little conception of Ireland as a "nation" or of England as the historic enemy of Ireland. On the contrary, these ideas occurred repeatedly in Rockite notices and threatening letters. The writer of one

22

such notice directed to "the farmers of the county of Tipperary" in December 1821 complained bitterly: "We are oppressed by the tyrannical laws of the English government, which we cannot endure no longer; . . . they crowd us up with rents, tythes, and taxes, which we daily sustain the burthen of, without the least abatement of rents. We [will] have the Irish parliament and king crowned in Ireland, as these formerly was."[79] The writer of another notice posted in County Limerick early in January 1822 sounded the tocsin of Catholic sectarianism and rebellion against the Protestant "heretics" from England whose coming had destroyed what was remembered as a nation ruled by Catholics, but a Catholic restoration was declared to be at hand: "This nation was once a seat of holiness but is now the throne of wickedness, and that since the coming of those devils into our holy nation. Arise and make yourselves strong for the battle, for there is not a man that falls in defence of his religion and liberty that his soul won't be received into heaven the moment it is out of his body."[80] And closely linked to the idea of Ireland the holy nation was Ireland the sorely oppressed nation: "Now, brethren," began yet another County Limerick document of the same month (headed "Captain Rock's speech to the men of Ireland"), "from James the 2nd's time to this period you have borne the greatest calamity under affliction, tyranny, slavery, and persecution that ever a nation suffered." But again, according to this notice, the Protestant oppressors of the Irish nation were about to be cast down by divine retribution: "God will revenge it upon them. This year is the year of liberty when the seed of Luther will be locked by hell's inexorable doors."[81] It would probably be wrong to think that abstract notions such as an Irish nation or England as the source of its ills sprang as readily to the minds of ordinary Rockites as did the much more concrete "evils" embodied by those local agents of oppression against whom the rebels now came to direct so much violence. But in the face of the large surviving body of material that sheds a flood of light on popular *mentalité* at this juncture, it would be equally misleading to believe that Irish country people, however illiterate, were not politicized in significant ways or were incapable of ever thinking in abstract political terms.

An Economy Shattered

That the Rockites exhibited an acute sense of suffering from economic injustice is not at all surprising, for Ireland in the early 1820s did indeed seem to be a "most distressful nation," a country crushed by in-supportable financial burdens. Part of the difficulty was the set of huge adjustments made necessary beginning in 1813 by the bursting of the wartime inflationary bubble. From the peak of the boom in 1812 to the bottom of the first postwar trough in the harvest year 1815 (September 1815–August 1816), the prices of oats and barley plum-meted by almost 60 percent, and wheat prices plunged by as much as 43 percent. Butter producers remarkably escaped any serious dam-age, but those who raised livestock were much less fortunate, as beef prices fell by 22 percent between 1812 and 1815, and those of bacon and hams declined by no less than 55 percent.[82]

Worse was yet to come. The grain, hay, and potato harvests of 1816 were seriously deficient owing to an extremely wet summer and to more rain and widespread flooding in the autumn.[83] Food prices soared. In a report from Dublin in November 1816 the price of the quartern loaf was said already to have doubled since a year earlier, and the price of a stone of potatoes was stated to have tripled over the same twelve months. The potato crop was described as "almost destroyed," and most of the new flour from the last harvest was certain to be "found unfit for bakers' use."[84] Starvation in Dublin (and no doubt elsewhere) had begun. The spring and summer of 1817 necessarily saw scarcity worsening in both rural and urban areas. An editorial writer in the *Dublin Evening Post* bewailed the general immiseration in August 1817: "Dublin, with some exceptions, . . . we look upon as a city more than half ruined. Cork, we have been informed, is almost entirely so."[85]

Fortunately, the yields of grain and potatoes in 1817 were good or even better than average, and the three harvest years 1816–18 (Sep-tember 1816 through August 1819) saw a strong recovery in grain prices, along with reasonably good prices for butter and beef as well as high prices for pig meat.[86] But prices alone do not tell the whole story. Besides the subsistence crisis of 1816–17 resulting from dam-aged crops, soaring food prices, and heavy unemployment, the sum-mer of 1818 saw a major drought. Though first-quality Cork butter reached the "enormous price" of £7 per hundredweight in early Sep-tember of that year, dairy farmers, deprived of grass, brought many fewer firkins of butter to market.[87] The story with respect to cattle in the summer and autumn of 1818 was similar: prices were extremely high, but livestock numbers had been declining, and the drought meant that cattle and sheep were slow to mature and often had to be retained on sparse pastures. Reporting on sales at the heavily attended Smithfield market in Dublin, one expert complained, "Cattle are scarcer this year than we ever remember them, which is solely owing to the unusual dryness of the summer and the consequent scarcity of feeding."[88]

Moreover, any upturn in the performance of the agricultural economy was overshadowed by the great fever epidemic of 1816–19, which was one of the worst outbreaks of disease since the famine of 1740–41 and which caused a nationwide panic as its ravages spread over most of the country. Two contemporary investigators of the fever epidemic of 1816–19 estimated that about a quarter of the Irish population, or 1.5 million people, had contracted the disease and that perhaps as many as 65,000 people had died. Even if these figures significantly exaggerate the mortality, this epidemic entailed a national health crisis of enormous magnitude.[89] Hardly had the scourge of epidemic disease abated by early 1819 than in the latter part of that year there began an economic downturn that was even more debilitating than the first postwar trough in 1813–16.

When hard times returned beginning in the fall of 1819, the toll was now worse than before. Over the four harvest years from 1819 to 1822 (September 1819–August 1823), the prices of the three main grain crops were halved; beef prices fell by nearly 30 percent, and those of bacon and hams by nearly 60 percent. This time dairy farmers did not escape the general punishment, as butter prices plummeted by 41 percent between 1818 and 1822, and the improvement in 1823 was meager. In general, the second postwar depression was much worse than the first, partly because the pastoral sector suffered as much as that of tillage and partly because the potato harvests of 1821 and 1823 were especially poor.[90] More will be said about the acute economic crisis of 1819–23 in chapter 1 of this book, but the summary provided here is intended to show that the rise of the Rockite movement was aligned with and largely fueled by both an immediate economic crisis of far-reaching proportions and by a sense that the country had been dealt a series of hammer blows since the defeat of Napoleon, a legendary hero for Irish Catholics whose overthrow in 1815 seemed in retrospect to portend their extraordinary subsequent tribulations.

One

Origins of the Movement

Contemporaries were practically unanimous in assigning responsibility for the origins of the Rockite movement to the harsh behavior of a single man—Alexander Hoskins, the chief agent of Viscount Courtenay's 34,000-acre estates centered around the small town of Newcastle West, about ten miles from the Kerry border in County Limerick.[1] A longtime resident of Newcastle who authored a now extremely scarce pamphlet entitled *Old Bailey solicitor*, which was a relentless indictment of the Hoskins regime, declared that the agent's administration was "a reign of tyranny and oppression, of meanness and artifice, that defies the page of history to produce a parallel."[2] The chorus of bitter complaints against Hoskins arose partly from his abuses and irregularities as a magistrate, usually in connection with his management of the Courtenay estates. Eventually, as many as fourteen out of sixteen local magistrates told a high-ranking police official that the government should strike Hoskins's name out of the commission of the peace.[3] Echoing this strong opinion among the local landed elite, William Gregory, the civil undersecretary at Dublin Castle, declared flatly in November 1820, "I believe nothing can be more oppressive than the conduct of Lord Courtenay's agent."[4]

Courtenay himself was the antithesis of a "good landlord." The Courtenays were an old established family in County Limerick. Ancestors of the subsequent Earls of Devon, their original grant went back to the late sixteenth century, when Sir William Courtenay "became possessor of the great estates carved out of the earl of Desmond's principality."[5] In addition to their Irish property, the family owned a large estate in Devonshire, with a seat at Powderham Castle. The current representative of the family, William Courtenay, the third viscount, was a perpetual absentee from his Irish and English estates because his open homosexuality made an ordinary life as a resident landlord in either country nearly impossible.

Courtenay was merely a boy when he became a central figure in one of the most infamous scandals of the late eighteenth century. At the age of ten he became the object of the affections of his older cousin William Beckford, whose merchant father had been lord mayor of London and was generally considered the wealthiest man in England until his death in 1770. From their base in Jamaica several generations of the Beckfords had accumulated a vast fortune from the ownership of West Indian sugar plantations. As the only legitimate son of the lord mayor, William Beckford inherited a fortune of £1 million at the age of twenty-one in 1780, and he initially commanded an annual income of about £100,000. Love letters subsequently written by Beckford to William Courtenay, later the third viscount, were maliciously printed in the newspapers and created a huge scandal beginning in the summer of 1784. The charges of sexual misconduct soon drove Beckford from England to the continent for an extended period.[6]

William Courtenay, who became the third viscount on his father's death in 1788, was, like Beckford, flamboyantly homosexual. He ran through much of his inheritance by high living and ostentatious displays of wealth. The widespread homophobia of the time compelled him to reside abroad permanently, first in America, where he purchased property along the Hudson River in New York State, and later in France; he was to die in Paris in 1835.[7] Wherever he lived, he never demonstrated the least financial restraint. Among his many costly diversions was that of "making aquatic excursions in the [English] Channel in a yacht most incomparably fitted out" at a personal expense of £20,000.[8] Constantly living beyond his means, Lord Courtenay was forced to sell land in order to meet the demands of his creditors. According to a report in the *Leinster Journal* in January 1806, property belonging to him had recently been sold for £206,000. Continuing extravagance helped to prompt further sales, as in De-

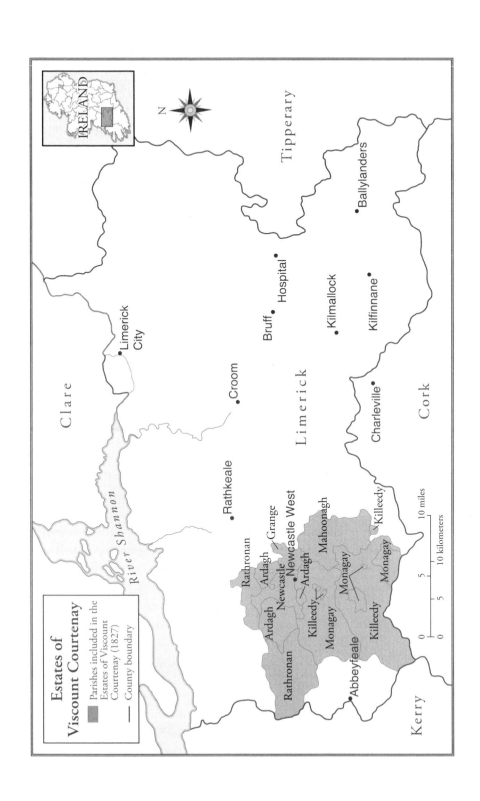

Estates of
Viscount Courtenay

Parishes included in the
Estates of Viscount
Courtenay (1827)

— County boundary

IRELAND

N

Clare

River Shannon

Limerick
City

Croom

Rathkeale

Rathronan

Ardagh Grange
Newcastle
Newcastle West
Ardagh
Ardagh

Killeedy
Monagay

Rathronan

Abbeyfeale

Killeedy

Mahoonagh
Monagay

Monagay

Killeedy

Monagay

Killeedy

Limerick

Bruff

Hospital

Kilmallock

Kilfinnane

Charleville

Cork

Tipperary

Ballylanders

Kerry

0   5   10 miles

0   5   10 kilometers

cember 1817, when 11,000 acres of "excellent fertile land" in the districts of Newcastle, Rathkeale, and Charleville were put on the market. Purchasers at the auction in Dublin paid a total of nearly £124,000 for their cheap acquisitions. Well before this auction, Courtenay's financial profligacy had led to the appointment of trustees to oversee the management of his property.[9]

Lax Management under Edward Carte

Supervision seemed essential for another reason. Laxity in the estate office at Newcastle and the impact of the severe agricultural depression of 1813–16 had led to the accumulation of vast arrears of rent, amounting to £63,500 in September 1818, or four and a half times the annual rental of £14,000.[10] This alarming development occurred during the lenient stewardship of Edward Carte, who had become agent in 1811, when the economic climate was radically different, at the height of the agricultural prosperity associated with the Napoleonic wars.[11] The new agent was popular and apparently sought to dampen any local tendency toward religious discord between Catholics and Protestants. At a meeting of "the tenantry" of the estate at Newcastle West in January 1813, resolutions were passed lauding Carte for resisting recent attempts to sow religious antagonism among inhabitants of the town; these resolutions were reportedly signed by almost three hundred "highly respectable Protestant and Catholic gentlemen."[12] Carte apparently possessed some experience as a land surveyor; he conducted a survey of the roads on the estate in 1813,[13] thus showing concern with one of the most serious economic problems of this largely mountainous area of the southwest. (The lack of good roads was to make it very difficult for the government to repress the Rockite movement militarily in its earliest phase; the great road-building program executed by Richard Griffith in the early 1820s went far to remedy this deficiency.)[14] Carte and his predecessors were reportedly liberal spenders on improvements to the estate and the town of Newcastle, mainly with a view to providing extensive employment to laborers and cottiers. This was allegedly in accord with Viscount Courtenay's wishes, "it being the benevolent desire of the lord of the soil to grant a large portion of his income yearly towards the employment of such labourers as were able to discharge the rents of their small cabins and gardens."[15]

This portrait of the third Viscount Courtenay (1768–1835) by Richard Cosway, who flourished in the Regency era and painted the future George IV in 1780, shows the ostentatious dress to which members of the British and Irish aristocracy had long been partial. Lord Courtenay's mother produced numerous daughters before his birth in her life-threatening effort to produce a male heir. Because of his openly gay lifestyle, Lord Courtenay (called "Kitty" by his relatives and friends) was the target of such hostility and prejudice from heterosexuals that he was compelled to live abroad for most of his adult life—a life spent in grand extravagance of many kinds. Though more than once he teetered on the brink of bankruptcy, he salvaged the reputation of the Courtenay family in 1831, when the House of Lords granted his petition for the restoration of the long-dormant Earldom of Devon. Upon his death in 1835 the English tenants of the ninth Earl of Devon (as he had become) reportedly welcomed his return in a stately funeral to the family seat at Powderham Castle in Devonshire. His Irish tenants would never have been so hospitable. (Reproduction of the portrait of the third Viscount Courtenay by the kind permission of the Earl of Devon and the Photographic Survey, Courtauld Institute of Art, London)

In addition, not long after the wartime boom gave way to depression, steep abatements were conceded in March 1814 to middlemen and other agricultural tenants. These were abatements of the "war rents" in force from 1811 to 1814—"the period at which the estate was last let."[16] In other words, Lord Courtenay's property had been largely relet to the tenants at the very height of the Napoleonic boom, and those rents simply became unsustainable once prices plunged downward. But in spite of the large and timely abatements, arrears accumulated remorselessly until, as previously noted, they had come to exceed well over £60,000 by September 1818. The enormity of the arrears, it seems, finally prompted the trustees to cashier Edward Carte as agent in August of that year and to install Hoskins in his place.[17]

An Oppressive New Regime

An Englishman lacking any previous land-agency experience in Ireland, Hoskins had spent the previous twenty years in London as a chancery solicitor before coming to live at Newcastle with his teenage son Thomas and a marriageable daughter. He did have a long-standing connection with Lord Courtenay's Irish estates. He had been a partner in the firm of Smith and Hoskins of Lincoln's Inn, which presided over land sales on the County Limerick property during the first two decades of the century, with the last such sale occurring in 1817.[18] But otherwise his knowledge of Ireland and its customs (the culture of this district of Limerick was still generally Gaelic) was apparently of the most rudimentary kind. Boding still worse were certain deep traits of personality and character that, if the author of *Old Bailey solicitor* and other observers are to be trusted, included vindictiveness, duplicity, extreme stubbornness, a taste for cruelty, and a strong preference for physical force over firm but quiet measures.[19]

The assigned task of Alexander Hoskins was evidently to raise the current income of the property and to lower the appallingly large arrears that had piled up while Carte was in charge. Hoskins set to work with a single-mindedness and tenacity that did credit to his dedication to duty but reflected very poorly indeed on his judgment and humanity. First, he increased the annual rental from £14,000 to £19,650, or by as much as 40 percent, on the basis of a new assessment of the worth of the estates by "experienced valuators."[20] Included in this operation was a doubling of the rents of the town fields and gardens adjacent to Newcastle, which were "the principal means of support to the inhabitants of a town perfectly insulated and totally destitute of trade or business."[21] Second, Hoskins announced to the tenantry that the extraordinary abatements in effect from March 1814 to March 1818, said to be equivalent to 60 percent of the rents due, were to be replaced by smaller reductions in the rents payable for the two years ending in March 1820; after that date the abatements were to cease altogether and the full reserved rents were to be paid.[22] Lastly, Hoskins initiated attempts to collect arrears from defaulting tenants, though it was decided that a substantial portion of the sum outstanding in September 1818 was to be written off.[23]

Up to a point, these efforts were certainly defensible. The grain and potato harvests of 1818 and 1819 were abundant; corn prices had recently improved somewhat (they did not slump badly again until after August 1819); butter was in reasonably strong demand; and livestock prices had recovered a degree of their wartime buoyancy and remained remunerative until the summer of 1820.[24] In short, the economic context at the outset was favorable to Hoskins's designs and not demonstrably unfair to the tenants in general. Furthermore, parts of the Courtenay estates were in the hands of middlemen possessing sizeable beneficial interests under old prewar leases, and to them abatements could reasonably be denied. Other portions of the property, where leases taken before 1790 had recently expired, could legitimately be saddled with some increase of rent upon renewal.[25]

On the other hand, the magnificent indulgence shown by the previous agent Edward Carte would long live in the tenants' memory, and the sharp tightening of the screws by his successor, even if it had been done without unusual severity, was bound to be resented. In addition, it was not only the middlemen and other agricultural tenants who were made to suffer. The town tenants of Newcastle, whose garden and field rents were doubled, firmly believed that they had been guaranteed no rise during the continuance of their town-tenement leases, to which their fields and gardens were mere appendages with fixed rents of 40s. an acre.[26] Moreover, Hoskins was said to have abruptly ended the traditional policy of spending heavily on estate and town improvements that furnished large-scale employment to laborers and artisans. Indeed, the author of *Old Bailey solicitor* made the no doubt exaggerated claim that during the three years of Hoskins's agency, out of a rental of nearly £20,000, not even sixpence had been spent on work-creating improvements.[27] Nor is there any evidence to show that Hoskins was prepared to modify his policies once the prices of livestock and grain began their precipitate decline.

Courting Extreme Unpopularity

At the same time that Hoskins was forcing pain and sacrifice on others, he himself spiritedly pursued an extravagant gentry lifestyle—extravagant at least in the eyes of Newcastle observers. So that he could properly entertain his social equals, he imported from London a butler, a waiting-maid, and a pastry cook. He purchased "costly furniture and hangings" to ornament the Castle.[28] He hired a sportsman whose job it was to kill a wide variety of game for his table and to fill the Castle pond with fish netted in local rivers, which Hoskins then fed "for his nice appetite."[29] Indeed, declared a hostile commentator, "so keen" was "the stomach of this modern Dives that at his table he was known to partake of beef, mutton, turkey, duck, hare, rabbit, grouse, partridge, woodcock, and snipe at one time, besides a variety of sweetmeats."[30] Much "delicious animal food" was "flung to dogs and pigs" at the Castle when it would have sufficed to feed "a great many poor housekeepers and other poor objects that were generally starving in the town of Newcastle." And there was no hint of paternalism: "not a person [who was hungry] dare enter the gate or ask for a morsel of food."[31]

Ostentation showed itself in other ways as well. Hoskins set out to become a notable foxhunter. He purchased a stud of horses at a cost of £300 and filled new kennels with "greyhounds, spaniels, lurchers, and starters" at a further expense of £400. He acquired a handsome carriage and gig, and he reportedly hired a staff of servants "suitable for a nobleman." Through all this display Hoskins became "what we call in Ireland a damn good fellow." The purposes of this pomp included the advancement of his local electioneering interests and the making of a worthy match for his daughter. Success attended this latter goal, with the Protestant bishop of Limerick coming to Newcastle to solemnize the marriage. Local Protestants were not a little surprised and annoyed, since no bishop of the established church had shown his face in the town for over thirty years, not even "to perform the duty of Confirmation conformable to the 17th Canon of the Church of England."[32]

Alexander Hoskins lived in this mansion called "The Castle" at Newcastle West from 1818 until his dismissal in October 1821. Nearby were the remains (two great halls) of the medieval castle once belonging to the Earls of Desmond, whose principality in and around Newcastle was confiscated by Elizabeth I and granted to Sir William Courtenay of Powderham in Devonshire in 1591. His descendant the third Viscount Courtenay succeeded to the former Desmond domain and other lands in County Limerick as an only son (he had thirteen sisters) on his father's death in 1788. Though he never resided in "The Castle," he was represented there by a succession of gentlemen land agents whose high social status helped them to secure the perquisites of Big House life. "The Castle" (seen here in a photograph of ca. 1900) was an unusually long house of two stories with nine bays and an attic—sited within the precincts of the ancient Desmond castle. During the Civil War of 1922–23 the house was gutted by fire and demolished soon thereafter. (Courtesy of John Cussen)

Only too aware of Hoskins's high living, the tenants (especially the larger ones, who were often Protestant) did not meekly submit to the financial sacrifices he demanded of them. Soon after Hoskins publicized his more stringent measures concerning abatements, concerted resistance began to be organized. This had as its ultimate object the continuation of the rent reductions in force since 1814. Participating in this resistance were numerous middlemen on the Courtenay estates, at least some of whom appear to have mobilized their undertenants for action. Admittedly, the well-informed author of *Old Bailey solicitor* claimed that the agrarian combination against Hoskins was the work of the occupying peasantry, with the role of the "respectable tenants" limited to a posture of benevolent inaction:

> The physical strength of the estate, the body of the tenantry, who could not upon this property be said to rank much higher in the scale of society than a mere peasant, it being mostly let out in very small divisions, and all sort of redress denied them, entered into a combination and conspiracies against him; the more respectable tenants, who by their advice or influence might have prevented the operation of these plots, looked on with perfect indifference, well knowing that any opposition they should offer would only tend to root the common enemy more firmly in his situation.[33]

But it appears highly probable that at least some middlemen verbally encouraged or actively promoted the struggle against Hoskins, rather than simply looking on approvingly.

Especially prominent in the group of protesting middlemen were three of the former agent's relatives: his brother Robert Carte, Sr., and his nephew Robert Carte, Jr., both of Woodlawn, and Robert Parker of Glenquin, the husband[34] of his niece. Apparently because they were in receipt of profit rents, Hoskins refused any abatement at all to two of these men, and he no doubt adopted the same stern line with other holders of substantial intermediate interests. When Edward Carte's three relatives declined to pay without the old abatements, Hoskins was quick to exercise his legal rights in the most rigorous manner. He had Robert Carte, Sr., arrested and jailed, though much to his disgust, Carte was soon let out on bail; he had Robert Carte, Jr., served with a *latitat*, a writ that assumed that the defendant was in hiding and summoned him to respond in the Court of King's Bench; and he took steps to have Parker's goods distrained.[35] To treat

The roll of murdered landlords and land agents in nineteenth-century Ireland includes a significant number whose unconscionable behavior clearly outraged the feelings of their tenants. As depicted in this sketch from the Land War of the early 1880s, the Mayo landowner Walter Bourke acquired infamy partly by personally serving writs of eviction on some of his own tenants at gunpoint. He was shot and killed in June 1882. Over sixty years earlier, Alexander Hoskins engaged in a whole series of extreme actions that outraged the feelings of many of Viscount Courtenay's tenants. Among Hoskins's bitterest enemies were middlemen of gentry status and close relatives of the former agent whom Hoskins severely persecuted. His conduct incited some of Lord Courtenay's tenants to conspire to murder him. (*Illustrated London News*, 14 May 1881)

respectable gentlemen in this high-handed fashion was to invite a ri-
poste. With volleys of stones, an "armed multitude" partly composed
of Parker's undertenants drove off the bailiff's party sent by Hoskins
in December 1819 to demand possession of Parker's livestock and
other effects under a distress warrant. This embattled middleman,
who owed £415 in arrears, reportedly gathered a second armed crowd
to frustrate a sheriff's sale of his livestock. Only after special arrangements had been made for military assistance to prevent a rescue did
Hoskins succeed in impounding Parker's cattle.[36]

In his protracted struggle with Parker, Hoskins inflicted such a variety of injuries and indignities as to attract widespread sympathy and
support for his victim. At one point, acting under an information al-
leging that Parker and his then pregnant wife "had rescued some articles of furniture" from Hoskins's drivers, a party of armed police and
estate bailiffs roused the Parkers from their beds at five in the morning, arrested them, marched them five miles on foot into Newcastle,
and paraded them through the streets before handing them over to a
magistrate. From this harsh treatment Mrs. Parker soon afterward
went into premature labor and was delivered of a stillborn child.[37] For
as long as six months, from June to December 1819, Hoskins placed no
fewer than fifteen keepers on Parker's farm to prevent him from removing his stock or crops. Eventually, "every single atom" of Parker's
effects, along with those of Robert Carte, Sr., "were sold by auction in
the square of Newcastle." Finally, Hoskins served Parker with ejectments from all his lands, and any redemption became impossible when
soon afterward Parker's house, along with "all the timber and valuable improvements," was completely destroyed by a fire that had been
deliberately set. People blamed Hoskins for putting someone up to
this malicious act "for the purpose of finally exterminating Mr.
P[arker] from off the estate by rendering him incapable of redeeming
the farm."[38]

The First Captain Rock

It was from Parker's lands and in the context of this bitter conflict with
Hoskins that the original Captain Rock emerged. This man was
Patrick ("Paddy") Dillane, the son of a blacksmith in the Shanagolden
district and himself bred to the same trade. Dillane had first come to
Newcastle for work, and then at some point during Hoskins's tenure as

agent he became an assistant to Walter ("Watty") Fitzmaurice, also a blacksmith, who lived at a place called the Strand, located very near Robert Parker's farm at Glenquin. Besides being a substantial middleman on Lord Courtenay's estate, with numerous tenants holding under him, Parker was also a considerable employer of labor as a road contractor, and in 1821 had a county presentment for an extensive road in his own locality. Hoskins, however, had been harrying Parker in his capacity as a road builder too and had earlier seized and canted a number of horses and carts being used in the construction. With the work almost finished, Hoskins sent a party of his own laborers to complete the job. What happened next is described by the author of *Old Bailey solicitor*, the only known source for this event:

> The peasantry about the hills, viewing the transaction of sending men to repair the road with an eye of jealousy, particularly as none of themselves were to be employed, rose up against the labourers [sent by Hoskins] and actually drove them from off the road with stones and sticks, and it was by the activity and exertions of Paddy Dillane on that day that he was called Carrigah [*sic*] or Rock, he being so expert in flinging stones from a gravel quarry at the invaders. . . .[39]

On the very next day, from a roadside ditch at Lissurland, he "put several slugs" into Hoskins's steward as he and another of the agent's employees passed by. And just a few days later still, it was again Dillane who was the key figure in the fatal wounding of Thomas Hoskins, the agent's son, an event to be discussed in detail below. At a subsequent trial of others for this crime Dillane boasted that "he was regularly christened Captain Rock by a schoolmaster named Morgan from Glenmore Hill, by pouring a glass of wine on his head."[40]

Another middleman whose undertenants apparently sprang to his defense, and their own, when Hoskins attacked him was William Brown of Rathcahill. Brown held a large mountain farm near Newcastle under a lease for his own life granted in 1795, and the same land had been in his family "for ages." Brown and his ancestors were "of very high respectability" and widely popular; their "unbounded hospitality and charitable disposition secured them the respect and confidence of the peasantry and gentry." It did not take Hoskins long to begin a campaign against Brown and his undertenants, with the stringent demand "for all rents and arrears to the day," followed by the seizure and impounding of "every beast" on Brown's farm, and cul-

The land agent and two associates portrayed in this sketch, protected as they were by a mounted police escort, escaped harm from the two men crouching behind the stone wall who had intended to shoot the agent. But small parties of agrarian rebels carried out many successful ambushes of their adversaries during the Rockite era. Among the earliest instances was the nonfatal shooting of Alexander Hoskins's steward, followed in May 1821 by the fatal wounding of the agent's son Thomas Hoskins. Taking the lead in both events was the blacksmith Patrick (Paddy) Dillane—reputedly the first "Captain Rock." (William Steuart Trench, *Realities of Irish life* [London, 1868], facing p. 215)

minating in the service of ejectments on the title. Loud complaints to Lord Courtenay's trustees forestalled further action by Hoskins, as the trustees ordered him to desist. But the damage had already been done. Hoskins's treatment of Brown was considered outrageous, and "if there were any cause," declared the author of *Old Bailey solicitor*, "more than another calculated to disturb the repose of the county of Limerick in that period," this was the one. For the undertenants were again ready to provide violent support for the beleaguered middleman: "Here was another step to foment disturbances. The wrath of the peasantry residing on that mountain farm was excited against such cruel injustice. [There was] a wild, uncultivated people, with numerous offspring, naturally prone to revenge and being led to believe they were to be exiled from their homes, and their children to starve if Alexander [Hoskins] succeeded."[41]

Hoskins saw himself and the estate trustees as targets of a conspiracy directed by the relatives and adherents of the former agent Edward Carte as well as by the clerks and retainers who had been dismissed along with Carte. The Carte party allegedly kept in its pay "men of bad character" ready to do the bidding of its leaders and social superiors in resisting the collection of rents and arrears. Carte's partisans were even accused of fomenting riots between the two chief factions into which the tenants of the Courtenay estate, or at least a portion of them, were divided.[42] But the same charge was leveled by the other side at Hoskins. He was accused of arming one faction "to fight against another faction obnoxious" to his views, of deliberately ignoring faction fights at Newcastle and elsewhere, thus failing in his duty as a magistrate, and of following the old English maxim of *Divide et impera*. Indeed, for some time he was considered "the captain general of one party" or faction. But Hoskins was said to have eventually lost influence with the faction that was initially swayed by him. When the two contending factions came to appreciate that Hoskins strove to keep them divided so that he might "more easily obtain dominion over them, they made friends and came to an adjustment of all disputes."[43] Whether the Carte party worked to bring about this reconciliation and then exploited it for its own purposes is unclear. What may safely be concluded is that the most active members of the Carte group were bound together by a bruising sense of grievance over their loss of place and favor at the estate office and by deep chagrin at the discontinuance of the old abatements, and that they had economic and social connections sufficient to attract a wide following capable of violence.

But Hoskins himself saw displays of physical force and the practice of espionage as essential to his efforts to intimidate his numerous opponents and to carry out what he conceived to be his obligations to his employers. As part of his establishment during the last two years of his agency, he maintained a body of "twenty or thirty armed men, who were regularly drilled by a pensioner then in his employment, all of whom kept watch day and night in and about his castle." In addition, he assembled "a band of informers and spies whom he kept in regular pay"; they seem to have been especially busy at night listening to conversations at the different public houses scattered through the town of Newcastle.[44] Hoskins's enemies also claimed that he had organized and armed a team of bailiffs, drivers, and other retainers who terrorized the "peaceable inhabitants" of the district. The continual task of seizing the livestock and other effects of defaulting tenants must itself have required a sizeable staff. On one notable occasion, in May 1820, Hoskins's bailiffs and their assistants drove over fifteen hundred head of cattle into pounds in or near Newcastle.[45] On a different occasion Hoskins was said to have treated a crowd of partisans at Newcastle with meat and drink and then to have sent this "furious drunken mob" to attack the house of the former agent's nephew Robert Carte, Jr.[46] Another observer who drew attention to Hoskins's forceful tactics was the Reverend William Ashe, a tenant on the Courtenay property for over forty years and until recently the Anglican curate of Newcastle. Hoskins, complained Ashe in November 1820, "sends out armed men unexpectedly to distrain for rent; these are men of bad character." Even if a desire for revenge did motivate certain members of the Carte party, Hoskins fully reciprocated by what Ashe described as his ceaseless efforts to "power down the whole weight of his vengeance on anyone that assists, goes bail for, or in any manner attempts to show any kindness to those he thinks fit to fall out with."[47] On top of the continual distraining of defaulters, Hoskins actually proceeded to the point of permanent eviction in no fewer than forty cases.[48]

Failed Tactics and Violent Reprisals

Despite the all too apparent vigor of his tactics, Hoskins was unable to secure much rent from the tenants, rich or poor. By his own account in November 1820 he had managed in the two years since his appointment to reduce the arrears by less than £6,500, and his books showed that as much as £57,000 then remained due.[49] He had earlier complained bitterly to Dublin Castle that the lack of military aid had frustrated all his exertions, only to be politely informed that the use of troops in sorties to recover rent by means of distraint was incompatible with the general orders of February 1817.[50] Given this unhelpful response, and given the near unanimity of local-gentry opinion against Hoskins by the end of 1820, his prospects of defeating the combination among the tenants were small indeed. Yet Hoskins was temperamentally incapable of retreat. Because of his obstinate severity the Newcastle district was seething with discontent.[51]

Hoskins had long been a marked man. He admitted in January 1820 that his life had often been threatened, and in December of the same year an attempt was made to murder him as he walked early one morning in the nursery of his demesne.[52] In spite of the armed guards who supposedly patrolled the castle grounds, a would-be assassin with a pistol fired at Hoskins at close range, perforating the shoulder of his coat with a pistol ball. For a minute or so Hoskins came to grips with the man who had nearly killed him, but then his nerve apparently failed. Only a short time earlier, Hoskins "had exclaimed he would not be in dread of *ten*, nay *eighty* Irishmen opposed to him," but "in the very moment of danger [he] abandoned his hold and ran off to his castle, bawling and roaring like a bull, altho' not receiving the slightest hurt." If Hoskins escaped physical harm, he could not avoid the psychological scars. For a while at least, "he shut himself up in his castle and suffered none to approach his person but his most confidential vassals"; he also fortified the castle "with double bars of iron" and put himself "into every posture of defence."[53]

Increasing Hoskins's terror of assassination in 1821 was his perusal of "a book entitled the *History of the Christian church*, written by a fanatic of the name of Signior Pastorini, signifying [that] killing a heretic [is] no murder, in which it was suggested to be just to destroy all [Protestant] reformers."[54] Discussion of the work and influence of "Pastorini" forms a substantial part of this book at a later stage (see

OUTRAGE
AND
REWARD

AT a Meeting of the undersigned Magistrates, held at NEWCASTLE, in the County of Limerick, on the 26th day of December, 1820, to take into consideration the present state of the Country, and particularly the barbarous attempt of assassination on ALEXANDER HOSKINS, Esq, on the 16th day of December instant.

BRUDENELL PLUMMER, ESQ. IN THE CHAIR.

RESOLVED—That we lament to find that this part of the County has become much disturbed

RESOLVED—That holding the system of Assassination in the utmost abhorrence, we will use every exertion in our power to bring the Perpetrators thereof, as well as the Conspirators of that diabolical act, to condign punishment ; and for the better discovery thereof, we will pay the sums annexed to our names for such private Information as shall be given to us, or any of us, whereby the said Conspiracy shall be discovered and the Perpetrators convicted, within Six Months from the date hereof ; and if any person concerned in said outrage, shall give us information against his accomplices, (except the person who actually made said attempt on said ALEXANDER HOSKINS'S Life,) so as the rest of his accomplices shall be convicted, he shall not only be entitled to said Reward, but application shall be made by us to Government for his Pardon.

The attempted murder of Alexander Hoskins on 16 December 1820 prompted a meeting of magistrates and others at Newcastle ten days later. The resolutions adopted at the meeting led to this poster condemning the "barbarous" outrage, attributing the crime to a set of unknown "conspirators," and offering a substantial reward for information leading to the conviction of those responsible. Among the subscribers to the reward were over sixty of Lord Courtenay's tenants, but apart from Hoskins himself, who pledged £50, and the estate trustees, who promised £200, the subscribers in general did not distinguish themselves by generosity. The chairman of the meeting, Brudenell Plummer of Mount Plummer, set a rather poor example by offering only £5 13s. 6d.—the Sterling equivalent of 5 guineas. (Courtesy of John Cussen)

chapter 4); here we are concerned with the impact that the notorious treatise made on Hoskins after it was given to him by "one of his trusty guards." According to the author of *Old Bailey solicitor*, the experience could scarcely have been more unsettling:

> Hoskins after a review of this book betook to a concealment almost of his person. He wore a belt under his clothes and always kept loaded pistols in his pocket. His aspect became gloomy and he regarded every stranger with a glance of timid suspicion. He travelled very seldom, but when he did, it was with hurry and precipitation, and [he] never slept two nights successively in the same apartment.[55]

The millenarian tract that so unnerved Hoskins had an entirely different effect among his adversaries, helping to stimulate them to rebellion. In the Shanagolden district a schoolmaster named Hall "set himself up as a prophet." He "declared that the king would never be crowned, and all degrees of respectability should be leveled, and equality universally established in titles and estates." Though Hall did not derive these particular notions from the *History of the Christian church*, he did insist "on the authority of Signior Pastorini's writings" and succeeded in trading on "the credulity and ignorance of the poor peasantry." Indeed, he was said by this hostile witness to have "carried his insolence to an immeasurable pitch."[56] It must have been Hoskins's awareness of the spread of millenarian ideas among Lord Courtenay's tenants that prompted him to take an interest in Pastorini's book in the first place.

But if Hoskins now went in dread of assassination, and took all sorts of precautions to guard against it, the one precaution that he did not take was to change his policies. On the contrary, in the aftermath of the attempt to kill him in December 1820, he "then on the double repeated his acts of cruelty and oppression. Writs, executions, processes, driving, and ejectments became the order of the day."[57] This intensified campaign elicited a dramatic but again abortive attack. In June 1821 "a large body" of armed men went so far as to mount a nocturnal assault on the Castle itself, marching right up to the gates before being driven off by the guards.[58] Though Hoskins and his men succeeded in repelling this attack and in wounding some of the raiders, he was reported soon afterward to be "under hourly apprehension of assassination" and to be fortifying the castle against a second assault.[59]

The vengeance that Hoskins himself always managed somehow to escape was eventually visited, however, on his nineteen-year-old son Thomas. A party of seven men, hired to perform the deed, shot him and fractured his skull at Barna on 27 July 1821 as he rode along a country road in company with his sportsman and a servant boy. (The young Hoskins died of his wounds five days later.)[60] The ringleader of the assassination party was the original Captain Rock, Paddy Dillane, who later described how Thomas Hoskins, wounded and begging for his life, was cruelly dispatched.[61] His killers showed anything but remorse at the scene: they "danced and played on a fife for about an hour," according to one witness at a trial for the murder two months later.[62] The elder Hoskins was still not ready to change course. In fact, he "declared publicly [that] altho' everyone of his family should be murdered," he was determined to remain as agent and to collect rent from the tenants as usual. He would, he boasted dangerously, "oblige them to bring it in upon their knees, and if they resisted, he would hang them on the square of the town in numbers (these were his expressions), which language it was natural to suppose only irritated the people."[63]

But if Hoskins shouted defiance at his enemies, they turned it back on him with every bit as much vehemence. The tenants were said to have "made a vow" never to attend any auction of distrained cattle or to buy such cattle under any circumstances, no matter how low the price. They also solemnly agreed to withhold "whatever money they had from the year's produce until Hoskins was removed and their farms fixed at rents suitable to the peace."[64] Giving a sharp edge to this concerted plan was a campaign of violence. A number of Hoskins's drivers were shot; several informers were "murdered in the noon day" and "houses consumed by fire." Any person attending a fair or market within forty miles of Newcastle who was pointed out as an employee of Hoskins was sure to be beaten. And in his last six months it was virtually impossible to find men willing to go "out upon the estate, either to ask for rent or to prevent dilapidation." In a few cases where evictions had occurred, "bodies of men in open day with perfect impunity" carried off large quantities of valuable timber by the cartload, leaving uncut "not so much as would make a handstick."[65] Clearly, Hoskins's authority had largely collapsed. Applying for the dispatch of troops in late August, a group of neighboring magistrates asserted

that the Newcastle district had recently experienced "dreadful outrages which have no parallel save in times of open rebellion"; they
maintained that large bodies of men were assembling, "many hundreds
together," for illegal purposes.[66]

Belated Exit in Disgrace

These events soon rang down the curtain on the stormy career of
Alexander Hoskins, who was now so despised that his assassination
would have been widely considered a meritorious act.[67] In a desperate
attempt to retain the agency that was fast slipping from his grasp,
Hoskins made overtures to "many of the respectable tenantry," sug-
gesting that if they would support him before the trustees, he was
ready to reinstate the dismissed estate officials, to grant an abatement
of 50 percent, and in general to "become their most particular
friend."[68] Reportedly, he even had an interview with, and gave some
money to, "the very person who it was confidently asserted fired the
pistol at him" in December 1820.[69] But it was much too late for this
discreditable gambit to succeed. In October 1821 the trustees of the
Courtenay estates announced that three gentlemen had been appointed to receive and investigate claims for abatements of rent from
the tenants.[70] All legal proceedings against them were stayed.[71] While
this was good news to the tenants, the best must have been the removal
of Hoskins and his replacement by Alfred Furlong. An attorney and
County Limerick native, Furlong was described as knowledgeable, presumably in Irish agrarian ways, and prudent.[72] When Furlong and several estate trustees arrived in Newcastle from London, "the air was
rent with acclamations of joy from an innumerable multitude of people."[73] Furlong's reinstitution of the old abatements of 1814–18 and his
demolition of Hoskins's "despotic maxims" immediately earned him
"the universal approbation of all orders of the Courtenay tenantry."[74]

Hoskins was slow to clear out. Before he would agree to give Furlong
possession of the Castle, he seems to have demanded £800 for improvements allegedly made to the orchards and other portions of the
demesne.[75] His enemies claimed that instead of having made improvements, he was actually "committing the most disgraceful outrages to
the house, garden, and demesne" during the last few months of his stay.
He was said to have shot all the pigeons, drained the pond of its fish,
cut up the demesne with horses, and felled and burned "beautiful tim-

ber."[76] Nevertheless, so as to be finally rid of him, Furlong apparently met his demand. On 16 December Hoskins departed at last, to the great satisfaction of the author of *Old Bailey solicitor*, who recorded what "a curious sight" it was "to view this usurper stealing out of Newcastle in an old hackney carriage, guarded by a military escort of dragoons and police, amidst the groans and hisses of the people."[77]

The utterly imprudent Alexander Hoskins certainly left behind an inauspicious legacy. At the very time of his dismissal the government sent Major General Sir John Lambert to take command of a greatly increased number of troops from the headquarters established at Newcastle.[78] That town quickly became "a strong garrison and, as such, exhibited a warlike appearance. The streets perpetually resounded with the martial drum, the ear-piercing fife, or the shrill twanging or warbling notes of the hoarse trumpet and bugle."[79] By the end of October some four to five thousand soldiers had been stationed in County Limerick or on its borders in an effort to cordon off the disturbed districts.[80] Nearly two hundred of these troops were garrisoned, appropriately enough, at Lord Courtenay's castle.[81]

The Widening Campaign

News of the exploits and successes of the agrarian activists on the Courtenay estate around Newcastle enjoyed wide circulation and helped to provide inspiration and incentive for the broad offensive that soon developed. Those who were discontented elsewhere traded on the violence that had occurred in the Newcastle district. The agent of Limerick landowner and former M.P. W.H. Wyndham-Quin received a threatening notice in October 1821 declaring that unless he complied with its commands, he would be shot like Thomas Hoskins.[82] Farther afield, in Kilflyn parish in north Kerry, a notice posted on the church gate in November menaced all drivers of livestock and collectors of tithes with the fate of young Hoskins.[83] Other public notices demanded rent reductions of 60 percent, and though this figure may have been chosen independently, it coincided with the abatement for which the tenants of the Courtenay property were reportedly contending.[84] Their struggle gained added notoriety from the publicity surrounding the Limerick assizes in September. During these proceedings one prisoner was acquitted of the attempted murder of Alexander Hoskins and another was found not guilty in the killing of his son. Both of these defendants were blessed in having as their counsel the illustrious Daniel O'Connell,[85] who had conducted a withering cross-examination of Alexander Hoskins himself in an earlier suit that achieved folkloric status.[86] Two more prisoners, charged with shooting at a foreman employed by Hoskins, were also acquitted.[87] Immediately following the assizes, a number of witnesses for the prosecution in cases stemming from the disturbances on the Courtenay estate, including several women, were waylaid and assaulted as they were returning home.[88] Indeed, one of these witnesses, James Buckley, was actually stoned to death at Ardagh on 23 September by members of a crowd. The crowd, which was assembling for Mass, pursued him and another witness with the loud cry, "There goes Hoskins's spies." Some of "the most respectable farmers in the parish" looked on complacently as Buckley was being murdered.[89]

Murder without Remorse

Two other murders gave extensive popular satisfaction and furnished a significant stimulus to the growth of organized discontent. One was the fatal shooting of the Palatine Christopher Sparling at Ashgrove near Newcastle in mid-October 1821.[90] Sparling's was a celebrated case because of its special circumstances. Sparling had earlier been persuaded by Hoskins to take the lands of Rooskagh on the Courtenay estate in Ardagh parish. He had invested a large sum in building a new farmhouse and out-offices and in making other improvements, but he fell into a dispute with Hoskins about his rent, as a result of which the agent had Sparling's substance seized and auctioned at Newcastle. Normally, this characteristically harsh treatment might have elicited sympathy for the victim, but not in this instance because the Palatine Sparling had come to the Courtenay estate in response to Hoskins's "fair promises to establish a colony of loyal Protestants" on the property. This kind of activity was always likely to provoke a violent response from the Catholic side. In addition, Sparling had also earned hatred in another way. It was said that in an earlier attack on his house at Patrickswell, he had "defended himself and [his] family most nobly, shooting some and wounding others severely." The consequence was that Sparling and his family were marked down for destruction: "the lower orders seemed to think their assassination [a] meritorious and laudable act."[91]

The second murder that promoted the expansion of the Rockite movement was that of Major Richard Going, the former head of the police establishment in County Limerick. As he was riding from Limerick city to Rathkeale on the evening of 14 October 1821, Going was waylaid by a small party of men who fractured his skull and shot him five times.[92] In certain quarters Going's death was sincerely lamented. The author of *Old Bailey solicitor*, who always vilified Alexander Hoskins, insisted that Going was "a truly honest, harmless man" whose "great humanity" prompted him to show unusual solicitude for the welfare of prisoners in his charge. In addition, he was said to have been "kind and courteous to the men under his command, and his loss was deeply regretted by them."[93] But the reaction at the popular level could scarcely have been more different. Going's death was proclaimed by "a joyous shout through the country which re-echoed from place to place; lighted heaps of straw were also at night exhibited on the different hills in triumph" over the accomplishment of the deed.[94]

Why was there such popular exultation over Going's death? His killing illustrates two themes, further elaborated below, which ran prominently through the Rockite movement: bitter hostility to the police and sectarian animus against Protestants and especially Orangemen. Going had permitted an Orange lodge to be formed within his police establishment, and when this fact became widely known, it helped to make the corps "much disliked," to put it mildly.[95] But what really instilled hatred for Going was his close association with a bloody tithe affray near Askeaton on 15 August 1821 as well as with the events and rumors that flowed from it. In this affray a large force of over two hundred Whiteboys who had attacked the house of a tithe proctor named John Ivis were surprised by a much smaller party of police numbering fewer than twenty men. In the ensuing exchange of gunfire the police shot two Whiteboys dead on the spot, mortally wounded a third, and took several prisoners. (A subconstable of police was also killed in this engagement and two others were wounded.)[96] When the bodies of the insurgents were brought to Rathkeale, Going arranged, as a terror to Whiteboys in general, that they should be buried in quicklime, with their captured associates digging the graves, instead of allowing the families and friends of the deceased to claim the corpses and to wake and bury them in the traditional way. What made this affront to popular custom far worse were two alleged aspects of

The lighting of signal fires on mountaintops or hilltops (with heaps of turf carried up in baskets) was a long-established practice among the Whiteboys. It served to communicate the news of some notable event over a wide geographical area and was usually done to celebrate a popular triumph. While political successes might also be trumpeted in this way, the Rockites were among those agrarian rebels who sometimes used signal fires to celebrate and broadcast the news of the murder of a hated enemy. They did so in the case of Major Richard Going, killed in mid-October 1821 after he had served as head of the police establishment in County Limerick, a post in which his repressive and sectarian behavior made him bitterly detested at the popular level. (*Illustrated London News*, 20 Jan. 1844)

the police action: first, the rumor—unfounded but "universally believed among the lower orders"—that one of the Whiteboys shot by
Going's police had been buried alive; and second, the apparently erroneous charge that at Going's instigation the bodies "were committed in the good old style of ninety-eight to a croppy-hole, which the
prisoners were compelled to dig."[97] Whatever Going's exact role in the
aftermath of the bloody engagement, the Whiteboy casualties and the
disputed circumstances of their burial could not "be obliterated from
the minds of their brethren," who brooded and then plotted revenge.[98]
Given all the circumstances, the bonfires on the hills that greeted
Going's death were not so surprising. The tithe proctor John Ivis was
also murdered shortly after Going.[99]

By the end of October 1821 most parts of west Limerick—not just
the Newcastle and Rathkeale districts—were manifesting the classic
signs of a serious supralocal agrarian movement. The robbery of arms
and ammunition, the exaction of financial "contributions" to the
cause, the swearing of oaths, and the posting of threatening notices
were all much in evidence, along with a striking amount of collective
violence against the enemies of "Captain Rock."[100] (The first mention
of a notice signed "Captain Rock" appeared in late October. Posted on
the door of John Ambrose near Newcastle, it threatened him "with destruction if he did not surrender the farm which he had the presumption to lease after a tenant who was ejected had abandoned it.")[101]
Other signs of organization and discipline also appeared among the
insurgents, including special emblems and courts. "Such active partisans as were deputed officers and serjeants under Captain Rock" took
to "decorating their necks with a scarlet worsted band, in the nature
of Caravats," the general name for the agrarian rebels of 1813–16 and
a celebrated faction name as well.[102] To maintain discipline in the
ranks, certain Rockites also decided "to erect a high court of justice
with power to try, by a select committee chosen by their leaders, such
felons as may be guilty of robbery or plundering of houses or any other
criminal act derogatory to the pure system of Whiteboyism."[103] It was
obvious to the authorities that the disturbances which had so far occurred were not, in the main, a series of unconnected protests with
aims "of a local or insulated nature." On the contrary, insisted the police magistrates Richard Willcocks and George Warburton, "We must
repeat our opinion that there is a systematic proceeding in all these

disturbances, and that the people are unfortunately worked on by invisible agency, brought into a degree of organized and mysterious affiliation, all tending to an ultimate object, and that object must be the total upset of the established order of things."[104] With Pastorini's prophecies and notions of a grand division and distribution of "gentlemen's estates" both circulating widely,[105] this analysis could not be dismissed as far-fetched.

The Hardest of Hard Times

Although the conflict on the Courtenay estates should properly be seen as the seedbed of the Rockite movement of 1821–24, the unrest there would almost certainly have remained a local affair had it not been for the punishing economic crisis that began in 1819. This crisis was one of the worst in modern Irish history. Its most obvious manifestation was the ruinous collapse of grain prices. At Dublin market, the biggest emporium for corn in the country, the average "middle price" of oats slumped from over 18s. a barrel in 1818 to less than 11s. by 1821, a decline of 42 percent. The value of barley and wheat plummeted even more steeply. When the price of barley reached its nadir at Dublin market in 1821, it was as much as 58 percent less than it had been three years earlier, and when wheat hit bottom in 1822, its value had plunged by 56 percent since 1818.[106] This round of price reductions was the second since the Napoleonic wars, so fondly remembered by Irish farmers, had come to a close. Between 1812 and 1816 the average prices of wheat, oats, and barley at Dublin market sank by 39, 51, and 53 percent respectively.[107] They rose modestly in 1817 and 1818, only to plunge once more, throwing Irish grain growers into a slough of despondency.[108]

Livestock producers from graziers to cottiers also had cause for despair at price trends in the aftermath of war. Like the battered tillage farmers, they faced two rounds of debilitating reductions separated by an intervening period of advance. First came an enormous fall between 1813 and 1816, reflecting the drastic contraction in military demand for Irish salted beef and pork. In 1813, a year that livestock producers would long remember for its prosperity, the Victualling Board in London awarded to Irish provision merchants contracts totaling 47,000 tierces of beef at an average price of nearly £11 per tierce, and over 56,000 tierces of pork at an average price of £12 4s. By 1816, however, the Victualling Board was able to secure all the salted beef and pork that it needed from Ireland—a mere 13,000 tierces—for as little as £5 5s. per tierce.[109] The cattle enterprise was considered so unremunerative in 1816 that new calves were killed as soon as they had been dropped, or were sold as veal, a situation akin to "the universal slaughter of calves and young cattle" in 1807.[110] A partial recovery in livestock prices took place from 1817 to 1819 through a fortuitous combination of diminished supplies and increased naval demand. Stocks of cattle, sheep, and pigs all declined sharply during 1816 and 1817— calamitous years of crop failure leading to famine, an epidemic of typhus fever, and extreme scarcity of fodder, and forcing many thousands of tenants to sell or slaughter their farm animals, often without having the means to replace them.[111] In conjunction with enlarged purchases in Ireland by the Victualling Board, these shortages on the side of supply pushed up prices under the naval-provision contracts to £7 12s. 6d. by 1819—admittedly, a far cry from the prices of 1813, but still as much as 46 percent above the level of 1816.[112]

By the summer of 1820, however, another downturn was in full swing. At the great October fair of Ballinasloe in 1820 the prices of both cattle and sheep had already fallen enough to rob many graziers of any profit on the summer's grass. The decline, according to one report, amounted to about 30 percent since October 1819. Those graziers who "could not draw on their banker were nearly ruined," and as many as a third of the cattle shown in the fair were driven home unsold by their disappointed owners.[113] This dismal pattern of many would-be sellers and few actual buyers firmly established itself at livestock fairs throughout the country during 1821.[114] The announcement of an extremely limited naval-provision contract—less than 6,000 tierces of beef and pork combined, in contrast to 28,000 in 1819—did nothing to lift sagging prices or to fill empty purses.[115] The October fair of Ballinasloe in 1821 was tersely described as "extremely bad," with even less demand for sheep than in 1820.[116]

As daunting as conditions already were, they worsened considerably for all livestock producers in 1822. To read the reports of transactions, or to note the lack of them, at fairs around the country during these months is to gain a vivid impression of the unrelieved gloom that prevailed in all quarters. "Most ruinous prices and trifling demand" characterized the fair of Bennettsbridge, Co. Kilkenny, in February.[117] The fair of Castledermot, Co. Kildare, in the following September "presented such a picture of [the] misery of our country as never before met our eyes; stock of every description in great abundance, yet no demand for anything!"[118] What faint hopes remained of improvement were completely dashed by the October fair of Ballinasloe, which gave "a finishing blow to the farmers."[119] The choicest fat wethers, "of a description which, four years ago [i.e., 1818], produced £3 12s. 6d. per sheep," now realized a mere £1 11s.; prime cattle could only be sold at £3 to £5 per head, and inferior cattle at £2 per head, "under last year's price."[120] Graziers were clearly operating at a substantial loss: one who had purchased a lot of forty bullocks for £5 5s. per beast in October 1820 sold them for £5 15s. apiece at the fair of Newport, Co. Tipperary, in October 1822, "leaving him only 1s. 3d. [per head] for two years' feed."[121] Similarly painful experiences persuaded numerous graziers to abandon or curtail business by surrendering land. "We know that thousands of acres have been thrown up," declared an editorial writer for the *Dublin Evening Post* in December 1822; "we know that whole tracts of country are at this moment untenanted."[122]

Grazier ranks were thinned even further by yet another losing season in 1823. The driving of cattle and sheep to one market after another in hopes of a stir in prices led only to endless frustration.[123] The fair at Rathkeale, Co. Limerick, in April "was not, these forty years, known to exhibit such a feature of depression and distress"; scarcely any business was done.[124] The story was the same at the Kilkenny fairs of Kells and Callan in the following July. They were "the dullest ever remembered in this county"; less than £200 worth of property . . . was transferred at both fairs."[125] Flockmasters, at least, saw their long nightmare begin to recede at Ballinasloe in October, when sheep prices rose by almost 20 percent above their value a year earlier, but cattle prices apparently reached a new low, falling by an additional 10 or 15 percent.[126] And the important fair of Newport later that same month furnished yet another "miserable example of the state of the country," with bullocks of a grade worth 12 to 14 guineas a few years previously now realizing £6 10s.[127] Only in the last two months of 1823 did a recovery in livestock prices finally begin.[128]

Alone among the major sectors of Irish agriculture, dairy farming weathered the depression of 1819–23 without serious loss. In fact, for two main reasons butter production remained at least moderately profitable during these years of general economic crisis. First, the price decline was not nearly as steep in the case of butter as it was for corn and livestock. Dairy farmers had suffered grievously enough in the immediate transition from war to peace. Between 1814 and 1816 the price of Irish butter in London dropped from slightly over 130s. per hundredweight—a record long unsurpassed—to less than 90s., or by about one-third.[129] Following a season of stability at this much lower level in 1817, there occurred a severe drought, European-wide in scope, throughout the last three quarters of 1818; "scarcely any rain fell for the space of nine months" in Ireland.[130] The sharp reduction in butter supplies during the summer and autumn of 1818 drove up prices almost to the fabled peak of 1814. But if this unusual postwar aberration is ignored, it will be found that the average price of Irish butter sold in London between 1819 and 1823 was less than 10 percent below that of 1816–17.[131] Second, output was growing as farmers shifted their resources to dairying. Indeed, Irish butter exports attained record levels during the early 1820s, exceeding 500,000 hundredweight in three of those years (1820, 1821, and 1824). Average exports during the five seasons 1819–23 amounted to nearly 481,000 hundredweight per annum, or about 15 percent greater than those in the preceding quinquennium.[132] Of course, dairy farmers who reared calves and bred pigs, as most of them did, shared in the general depression in livestock prices, but at least their principal line of business continued to be somewhat remunerative.

While exports of butter rose significantly in the early 1820s, those of certain other major farm commodities fell substantially in one or more years of the depression, thus exacerbating the adverse effects of low prices on agricultural incomes. Corn growers were heavy sufferers in this respect in 1822 and 1824 because of crop shortages. As the deficiencies affected potatoes as well as corn, however, much more was involved than the level of grain exports. Of the five harvests between 1819 and 1823, one (1819) was "uncommonly abundant" and two (1820 and 1822) were probably well above average.[133] But the remaining two (1821 and 1823) were deficient in varying degrees. The harvest of 1821 was almost but not quite a disaster. The early summer of that year was remarkably hot and dry, seemingly promising a fourth successive season of plentiful food. Then in late August came heavy rains that continued in torrents throughout the rest of the year, causing repeated floods and enormous destruction among the corn crops, potatoes, and hay. Much of the corn either could not be saved or was not considered worth saving, and what was harvested, especially the wheat, was of very inferior quality. In large parts of the south and the west, where dependence on the potato for food was heavy, that crop was seriously deficient to begin with, and because of the fearful inundations in the last quarter of the year, great quantities of potatoes soured in the ground undug, rotted in their pits, or in low-lying lands were simply swept away by floods.[134]

Famine Conditions

This calamity of nature led to the horrendous subsistence crisis of 1822. Though much mitigated by private charity on an unprecedented scale and by public relief (extended to an estimated 260,000 persons in County Mayo), the famine of 1822 hastened the death of many people in the more remote areas of Munster and Connacht; even in accessible regions it produced appalling misery and semistarvation among small farmers, cottiers, and laborers for five or six months.[135] (It was no accident that during this same period Rockite activity in the south practically ceased altogether, to resume immediately after the plentiful harvest of 1822.) In the aftermath of the harvest failure of 1821, grain normally exported was diverted to home consumption. The smaller farmers were forced to buy meal to feed their families and livestock, thus joining cottiers and laborers in the retail food market, and even the larger tillage farmers retained some of their marketable corn for use at their tables and in their barns. As a result, Irish exports of oats and oatmeal to Britain in 1822 slumped to only 570,000 quarters—a reduction of as much as 50 percent below the shipments of the previous year.[136]

The same atrocious weather that so badly injured the grain and potato crops in the autumn of 1821 also led to a punishing "fuel famine" in the following winter and early spring. Since turf had been almost impossible to save, many communities were compelled to find their firing wherever they could. Often enough, this meant both collective and individual theft from the woodlands of neighboring landowners. What happened on a wider scale was demonstrated in miniature by the events of 18 and 25 February 1822 on the demesne of the Copleys at Ballyclogh near Askeaton in Limerick. About noon on 18 February, at least two hundred people assembled on the wooded part of the demesne with the intention of cutting and carrying off the timber. The local magistrate Edward Hunt arrived on the scene with a body of troops and ordered the crowd to disperse. When his order was ignored, Hunt seized one of the ringleaders; at this point "the whole body cried out, rescue him, cut the rascals to pieces, *and stand to your cause.*" Upon Hunt's command five of the soldiers responded to stone throwing by the closing crowd with a volley of shots that killed one person and severely wounded two others. Only then did the crowd disperse, but its members did so with bitter threats that "they would return with an armed force . . . and cut the rascals to pieces."[137] They were as good as their word, barring the violence. The distressed community defiantly returned one week later. Mustering five or six hundred people this time on Copley demesne, they proceeded to cut down the timber and carry it away on carts "in all directions." In fact, they made a special point of carting a considerable quantity of the timber through the town of Askeaton in full view of "two country magistrates, the police, and the army." The magistrates simply declined to stop the offending crowd, giving the limp excuse that they had received no call for legal assistance from Mrs. Copley, whose husband was away in Dublin.[138] The truth was that neither the Copleys nor the authorities wanted the public ignominy that would have resulted from the rigorous enforcement of the law in the face of this acute fuel famine.

Compared with the mammoth deficiencies of the 1821 harvest, the shortages that resulted from an exceptionally wet summer in 1823 were a much less serious affair. In fact, oats were pronounced "considerably beyond an average crop" across the country, an assessment amply confirmed by the record exports of the following year.[139] Barley, also generally undamaged by the rains, was regarded as "*perhaps* an average crop." But it was otherwise with both wheat and potatoes. The former turned out badly all over the country; the quantitative deficiency was estimated at about one-third less than normal, and the quality was deemed generally poor. In the case of potatoes the losses differed greatly from one region or county to another.[140] Though reports of shortages came from parts of all four provinces, the worst-affected areas were apparently Mayo ("seriously deficient, particularly in the mountain districts"), west Galway ("almost a total failure"), and to a lesser degree east Galway as well, much of Clare, and four counties in south Leinster.[141] Clearly, the most serious harvest deficiencies in 1823 occurred outside the region of the Rockite movement or on its periphery, but in other respects that movement was very much the product of the agricultural crisis analyzed in this chapter.

Two

Expansion and Retreat

Like their predecessors in the long tradition of Whiteboyism, the original Rockites soon set out to enlarge the territorial scope of their fledgling movement. In Ireland as elsewhere, the effectiveness of collective popular protest heavily depended on strength of numbers. It appears that the earliest Rockites instinctively recognized that in several critical respects they could advance their cause by swelling the number of their adherents and by widening their geographical base. In gaining broad and active support for their aims and methods, they would obviously be taking a major step in affirming their moral and political legitimacy in the eyes of ordinary country people. With great numerical strength they would also be able to shape and, when necessary, coerce popular and elite behavior in the desired directions. It seemed to these agrarian rebels that the kinds of changes in economic behavior that they wanted could not be achieved without violence or at least the realistic threat of it. Intimidation and violence could be powerful forces in persuasion or dissuasion at the local level, and they might even prompt a reluctant central government to initiate meaningful reform. For decades, in spite of repeated batterings by a succession of Whiteboy movements, the Irish tithe system had stood as an impregnable fortress. It was the Rockites who would make the first serious breach, who would extract the first significant concession. But before this could be achieved, it was essential to create sufficient space for the movement to grow. Repression must be made as difficult as possible. Obviously, this aim would be greatly advanced if the rebels could disarm their adversaries, actual and potential, and arm themselves. Insofar as local magistrates, other members of the gentry, and perhaps large farmers could be deprived of their weapons, growth in numbers and territorial expansion would be considerably enhanced.

Penetration into Kerry

It was in a quest for weapons that followers of Captain Rock first entered County Kerry. Early in August 1821, Thomas Sandes of Sallow Glen reported a series of arms raids "by [I believe] some people from about Newcastle." The strangers also enlisted fresh recruits in the standard way—by means of an oath. "I have discovered," said Sandes, "that the oath is to pay no tithes or country charges [i.e., county charges] or rent except to the owner of the estate."[1] Two months later, in October, other Rockites descended on horseback from the mountains around Abbeyfeale, Co. Limerick, into the lowlands of the Castleisland neighborhood, where they forced the inhabitants "to supply arms and powder under the penalty of having their houses burnt over their heads."[2] Toward the end of October the posting of notices and the tendering of oaths began in the vicinity of Tralee,[3] and a committee that sat regularly in this town helped to mobilize the population of the surrounding countryside.[4] By mid-November the efforts of emissaries from west Limerick to gather recruits in the two north Kerry baronies of Clanmaurice and Iraghticonnor were said to be enjoying such success that some intimidated members of the local gentry were taking refuge with their families in Listowel, while others were barricading their houses and preparing for the worst.[5]

Over the course of the next three months the area of Rockite agitation widened considerably and the disturbances assumed a much more serious aspect. Unrest spread southward to the district of Killarney, where in a dramatic act of violence reminiscent of 1798, the Protestant church of Knockane was burned to the ground in late November in protest against the collection of church rates.[6] Turmoil also spread to the vicinity of Milltown and Killorglin, where in January 1822 a process server was beaten and forced to swallow his tithe notes and processes, and a gentleman's haggard was consumed by fire.[7] Substantial portions of south Kerry also rallied to the Rockite standard as the new year opened. There were arms raids at Ardtully near Kenmare and the attempted burning of the Protestant parish church of Templenoe in the same district.[8]

Even more extensive were the operations of Captain Rock's adherents in the southernmost barony of Iveragh, especially in the Cahersiveen district, where Daniel O'Connell's property was located. The valuable correspondence of his brother James, who helped to manage the lands there, shows that the troubles originated in a conflict between the local parson and magistrate, the Reverend Michael Dowling, and his parishioners over his tithes, which were said to be grossly excessive. Shortly before Christmas in 1821 a large body of Whiteboys, estimated at two hundred, attacked the house of Dowling's tithe proctor, thrashed him, and carried off his tithe notes and processes.[9] This was a prelude to far greater violence. Less than two weeks later, one of Dowling's drivers, a hot-tempered man named Sugrue, who was apparently drunk, murdered an unarmed peasant youth by shooting him in the head after some women and boys had rescued a few cows seized earlier for nonpayment of tithes. A crowd of people who had witnessed the murder avenged it by killing Sugrue and beating his two fellow drivers severely.[10]

This tragic incident was followed by an aggressive and extensive popular mobilization. Rockite activity greatly increased in Dowling's parish (Caher) as well as the neighboring ones of Dromid, Filemore, and Glanbehy. Parties of men marched through the countryside administering oaths, posting threatening notices, and in one instance leveling a pound on Lord Headley's estate—a practice that was then rampant in north Kerry.[11] James O'Connell warned his famous brother to expect little money from the Cahersiveen property, since "*every peasant in the barony of Iveragh is a Whiteboy* and, as such, are determined neither to pay *rent, tithes, or taxes*. This information I had from a priest who lives there and whose name I cannot even tell you"—the priest evidently feared a reprisal if suspected of the least disclosure.[12] At considerable personal risk James O'Connell took "energetic measures" to repress the disorders; in mid-February he succeeded in apprehending and lodging in Tralee gaol seven alleged offenders.[13] The novelty of this event "created a great panic," and his district seemed to return to peaceful ways.[14] Whether it would remain so must have appeared doubtful; the movement now possessed strong local roots. In fact, most of Daniel O'Connell's own tenants in the parish of Caher, as his brother told him, "were actually leaders and principals among the infatuated wretches who composed *Captain Rock's corps.*"[15]

Rockite Missionaries in Clare

Like Kerry, Clare also received evangelizing missionaries from west Limerick. The police official George Warburton reported in mid-February 1822 that there had recently been "several very considerable assemblies" of Clare peasants, "attended by persons from Limerick, who are making every effort to procure a diversion in this county in favour of Limerick."[16] But these efforts often proved unavailing. West Clare long remained almost completely unaffected. In large part this was because a military cordon had been put in place by November 1821 along the north side of the river Shannon from Kilrush around to Ennis.[17] In addition, guard boats patrolled the Shannon below Limerick city and prevented Rockite emissaries from crossing into Clare.[18] In the whole area west of Ennis there was only one recorded public protest concerned with either rents or tithes during the first phase of the Rockite movement. It occurred in November 1821 near Milltown Malbay, where a crowd of peasants relieved several proctors of their records and told them never to return for tithes, because the people there "were determined . . . not to pay any."[19] With this exception, however, Rockite activity was practically nonexistent in west Clare until 1823 and 1824.

In portions of east Clare, on the other hand, significant efforts at mobilization were made in the last two months of 1821 and the first three of 1822. These took the usual forms: raiding for arms, tendering oaths, and posting hortatory or threatening notices.[20] One notice posted in mid-November on the door of Killaloe chapel called on the parishioners to meet with Captain Starlight of Limerick and Captain Moonlight of Cork at the end of the month in order to "form the same plan and methods as ours," or else "we will rank all of them in the number of Protestants and will pay them off the same day."[21] Another notice attached to the gate of Tomgraney church urged the people of that barony (Upper Tulla) in the name of Captain Rock "to be ready when called upon, as the time is now at hand." "My friends," it proclaimed, "you shall not want for arms, ball cartridge, and money. . . . The county of Clare will not fail as they did when wanting before."[22]

On the whole, however, the response to such appeals was unenthusiastic. In sharp contrast to other counties there was little serious violence against the person in east Clare. The murder of two suspected informers—one named Stack near Cratloe in December 1821, the other named Guerin near Sixmilebridge in June 1822—were exceptional occurrences, as were the beatings in January 1822 of some keepers guarding distrained goods near Ennis and of an apparent land grabber near Sixmilebridge.[23] Equally unusual in Clare at this stage was violence against property. It was a rare case of incendiarism when early in April 1822 six houses on the Jackson estate near Limerick city were burned down to punish the new tenants who had just taken possession.[24] And although a report from the Scarriff district at the same time spoke of "the slaughtering of sheep as well as the plundering of every description of property" as general complaints in that locality,[25] there is good reason to believe that this violence was largely restricted to the property of the Reverend Sir William Read and his family, against whom there had been long-standing animosity.[26]

It is not altogether clear why the country people of Clare were so slow to answer Captain Rock's siren. No doubt, as already mentioned, the military cordon along the Clare side of the lower Shannon and the water patrols served as an effective deterrent to would-be emissaries from Kerry and Limerick. In addition, magistrates, police, and soldiery in east Clare scored several early successes that probably helped to dampen the Rockites' courage there.[27]

The North Cork Heartland

While appeals for mobilization went largely unheeded in Clare, the goals and methods of Captain Rock's legions found ready acceptance in north Cork. Indeed, by January 1822 the clandestine, nocturnal, and often small-group activities typical of traditional Whiteboyism were to give way to something very like mass insurrection in the open day. The central role played at the outset in Cork by activists from adjacent portions of Limerick impressed itself on contemporaries. The first substantial incursions across the county border took place early in October 1821, when large parties of undisguised insurgents staged attacks on Newmarket and Kanturk, carried off practically all the firearms possessed by the inhabitants of these towns, and raided the houses of gentlemen and farmers in the vicinity for the same purpose.

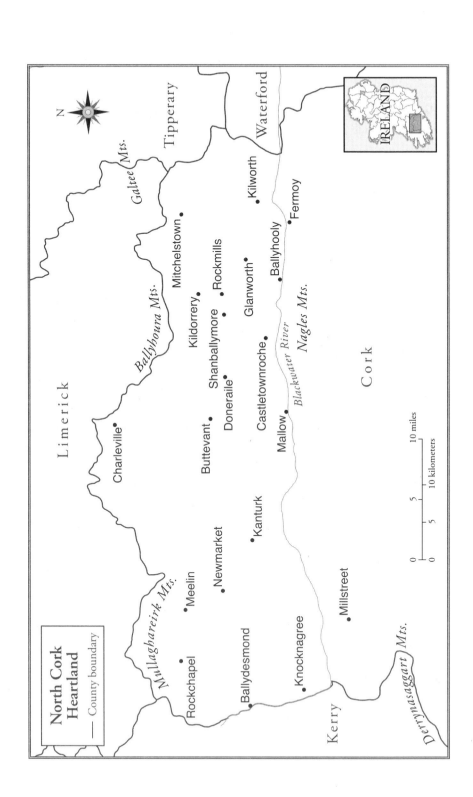

North Cork Heartland

— County boundary

Limerick

Tipperary

Waterford

IRELAND

Galtee Mts.

Ballyhoura Mts.

Mullaghareirk Mts.

Nagles Mts.

Blackwater River

Derrynasaggart Mts.

Cork

Kerry

Mitchelstown

Kilworth

Fermoy

Rockmills

Ballyhooly

Glanworth

Ballyhooly

Kildorrery

Shanballymore

Castletownroche

Doneraile

Charleville

Buttevant

Mallow

Kanturk

Newmarket

Meelin

Millstreet

Rockchapel

Ballydesmond

Knocknagree

10 miles

10 kilometers

5

5

0

0

The Catholic priests of Kanturk and Newmarket were both said to be certain that the raiders had all come from County Limerick.[28] Similar incursions, accompanied no doubt by the posting of notices and the tendering of oaths, occurred often in subsequent months.[29] In one instance a gentleman reporting a series of arms raids in the Mallow district in January 1822 identified the participants as Limerick natives from the fact that they wore the distinctive gray frieze jackets usual in that county.[30]

The torch of protest had been well passed, however, and independent operations by Cork activists had become so numerous by the second week in November 1821 that the viceroy Lord Wellesley proclaimed the two northwestern baronies of Duhallow and Orrery and Kilmore.[31] In this rugged upland region the magistrates and gentry scarcely had time to concert countermeasures before they were deprived of their weapons. Parties of Rockites aggressively seeking firearms attacked every gentleman's house between Kanturk and Millstreet, two excepted, well before November was out.[32] To resist could be dangerous: one gentleman who tried to do so, a half-pay officer named Watters near Newmarket, was seriously wounded.[33] To offer information to the magistrates could be fatal: a laborer named Scully suspected of this treachery was murdered near Kanturk.[34] To send arms to a place of presumed safety might only simplify the task of the insurgents: a large force of perhaps seven hundred men attacked the house of a clergyman in December in the village of Donoughmore, where most of the arms of the parish had been collected, and carried off as many as sixty-six stand.[35]

The Path to Insurrection

The same passion for collecting weapons seemed to overshadow all other concerns as the movement spread not only to the northeastern baronies (Fermoy plus Condons and Clangibbon)[36] but also to most of the southwestern ones.[37] To some extent the appearance was misleading: acts of violence against persons or property did occur occasionally in disputes over tithes or rents,[38] but they were vastly outnumbered by rapidly escalating arms robberies. One observer counted thirty-five houses that had been raided for firearms in three parishes near Bandon in less than two weeks in late December—a rate of plunder sufficient to cause almost all the gentry for several miles around that town to quit their country residences.[39] Such a loss of courage, so openly displayed, naturally emboldened the insurgents. In the districts of Bantry and Skull parties consisting of as many as five hundred to seven hundred men were reportedly scouring the countryside for weapons in January 1822.[40] Indeed, disaffection was said to have spread all along the southwest coast from Rosscarbery and Glandore down to Ballydehob and Skull, as well as into the interior, with meetings held nightly for the purpose of enlisting recruits.[41] The story was much the same at the opposite end of the county. The Reverend Matthew Purcell informed Dublin Castle on 22 January that about sixty stand of arms had recently been seized in each of the districts of Mallow, Doneraile, and Castletownroche. In Purcell's own district of Charleville "marauding parties are out every night firing shots" and "posting notices for nonpaym[en]t of tithes and taxes and reduct[io]n of rents." Every preparation, Purcell solemnly warned, "is making for open-day work."[42]

The insurrection that was erupting even as Purcell was writing had long been expected by the alarmist section of the southern gentry. Yet even the coolest heads had become convinced by the beginning of 1822 that a rising at no distant day was in contemplation. The gentry all over County Cork, declared a military officer early in January, "are preparing for defence, some even blocking up the whole of the lower windows with brick & mortar; no one will be out of an evening; there is an end to all convivial intercourse for some time to come."[43] The siege mentality of the gentry was heightened by the deliberate withdrawal of army units in the southern counties from their dispersed rural outposts into the chief towns and regular depots. This shift in strategy to a policy of concentration, first urged upon the government by the military authorities early in the previous month, was adopted precisely to guard against "any extensive insurrectionary movement or general rising of the people." It was recognized that a system of concentration would expose loyalists in the countryside to danger, and the new policy certainly facilitated Rockite mobilization, but army leaders professed to believe that the problem of endangered loyalists could be solved by a judicious distribution of police and yeomanry.[44] What made an insurrection seem so likely during the winter of 1821–22, in the eyes of those to whom its suppression would fall, was that the certain prospect of famine in the coming spring and summer had

destroyed all sense of caution among the lower orders. "*Short nights &
starvation* may do much bye & bye" to restrain the current movement,
thought Lord Ennismore,[45] but meanwhile all the circumstances
pointed toward an explosion.

A number of fatal clashes just prior to the insurrection—clashes in
which the authorities appeared to use excessive force—added considerably to popular desperation and contributed to a widespread desire for vengeance. With little provocation a detachment of soldiers
passing by Ballyagran chapel between Charleville and Croom, a highly
disturbed district, fired on a group of country people who had assembled for Mass on Christmas night in 1821. One peasant was killed on
the spot, another was not expected to recover from his wounds, and
two more were injured.[46] Two deadly affrays west of Bandon also inflamed feelings in that troubled neighborhood. In the first, on 7 January, a party of the East Carbery yeomanry under Captain Arthur
Bernard encountered near Newcestown chapel what was described as
an advance guard of over five hundred Whiteboys and shot four of
them, two fatally. In a second encounter with a smaller body of Rockites two days later, Bernard's brother killed a man named Harrington,
who lived near Ballydehob, over thirty miles away.[47] Overshadowing
these two affrays, however, was another confrontation that took place
at Ballinard near Clonakilty in the third week of January. This caused
what was termed, with some understatement, "a strong sensation in
the country."[48] A crowd of several hundred people tried to rescue two
prisoners and an illicit still that had just been seized by a revenue
officer and a unit of the Ibane and Barryroe yeomanry. When the
yeomen were stoned, they responded by firing into the crowd and by
shooting one of the prisoners dead, even though his hands were tied.
Altogether, the yeomen discharged over sixty rounds, but accounts of
the incident differ as to whether any of the country people were killed.
A physician who attended at the scene concluded that the incident
demonstrated "the melancholy results of letting the half disciplined
loose without an experienced head, for the young gentleman who led
them here made his military debut."[49] By their overreaction the
yeomen became unwitting recruiters for Captain Rock: "The whole
country [thereabouts] was thrown into a state of insurrection. Men
ascended the hills, sounded horns, and raised the peasantry."[50]

"The Battle of Keimaneigh"

The most lethal of these preinsurrectionary clashes, and the one that most directly sparked the uprising, took place on 21 January in the Shehy mountains, about a dozen miles northeast of Bantry on the road to Macroom. A party of thirty-five or forty gentlemen and tenants under the Earl of Bantry, accompanied by a small detachment of soldiers, had just taken some suspected Rockites into custody when they were assailed by a much larger force of "about 500" insurgents, many of them well armed and showing great tactical agility. The attackers, who had been speedily collected by shouts, horns, and bugles, endeavored to rescue the prisoners by trapping Lord Bantry's party within a mountain glen. The troops foiled this effort, however, and made a safe retreat with the loss of only one man. The severed head of this victim, according to one report, the insurgents placed in triumph on a pole, carrying it along with them.[51] Under the circumstances there could be no precise accounting of the casualties suffered by the rebels, but the military officer in charge at the scene reported that he and his troops "must have killed about twelve" and wounded many others.[52]

This encounter quickly became locally renowned as "the Battle of Keimaneigh" and has remained so to this day in that area of northwest Cork, since it was apostrophized in *Cath Céim an Fhia*, the best-known poem of the illustrious poet Máire Bhuí Ní Laoire, several of whose sons participated in the episode (one may later have been hanged and two others long eluded the authorities). In her ballad poem the insurgents achieved a significant victory ("The heroes joined the Clanna Gael at a mountain recess, / And they drove the fat rabble away down the slope."). Ignoring the retribution that followed, the poet also proclaimed: "A hundred great praises to Jesus that we didn't pay the penalty for the rout, / But lived to make a joke of it, and tell the story at our ease." But the most dominant notes of the poem are its bitter sectarian consciousness ("May they come to no better end [than the dead soldier with the severed head], those foreign cubs of Calvin's"), its desire for bloody retribution, and its millenarian expectation that Protestant rule was on the verge of extinction:

> In this present year of ours, every boor will be put to rout,
> They will be knocked into the dikes, gutter be their shroud.
> We won't hold court or inquest, the gallows is a-building,
> And the rope with vengeance twisting for their ugly throats.

They have the power, 'tis ill they rule, they are well appointed
    in coaches too.
All sorts of food have this bear's brood for partying with
    pleasure.
An authority has informed me that before the harvest ends,
The prophet Pastorini is declaring their measure.

Dear beloved sons of Erin, do not stop or retreat,
For the task undertaken will soon be complete.
Keep up the courage, those runts must be routed,
In hell-fires to flounder and roasted apiece.
Have your long pikes cleaned and polished,
Go into battle, don't stay from it.
Help is at hand, that is God's promise.
Pulverise these porks [the "English" and Protestants].
Regain possession of your ancestral abodes,
There to be seated and remain for evermore.[53]

Combating the Insurgency

In spite of the large number of participants, Major General Sir John
Lambert refused to view this skirmish as an extraordinary occurrence
presaging insurrection. Three days later, he asserted that "there was
nothing more in the character of the business than might be expected
amongst the inhabitants of this mountainous country, who from habit
are easily assembled."[54] Easily indeed! Writing in considerable alarm
to Lambert on 25 January, Captain Arthur Bernard declared that the
whole country between Palace Anne near Bandon and Greenhill, south
of Mallow, stood "entirely deserted by the male population." Collected by the sounding of horns and the discharging of firearms, the
men had marched on that morning and throughout the previous night
to Inchigeelagh, not far from the scene of the recent attack on Lord
Bantry's party.[55] In the wide district between Bandon and Dunmanway as well, a substantial portion of the male inhabitants had gone
"to join the insurgents on the mountains."[56] There were other risings
at the same time in the northwestern part of the county. The Rockites
reportedly dispatched "expresses" in all directions around Kanturk,
calling upon the people to gather at a place about three miles from
that town and threatening "vengeance against all who should neglect
to obey the summons."[57] Many laborers in the Doneraile district suspended work and, avoiding the main roads for fear of interruption by

the military, made their way across the fields to join forces with the rebels in the neighborhood of Kanturk and Newmarket.[58] For the landed elite, some of whom now anticipated a massacre of the propertied, the situation was all too reminiscent of an earlier time: "petty quarrels, drinking, and rioting at fairs has totally subsided, as previous to the rebellion of 1798."[59]

Just as that earlier rebellion had begun with the stopping and destruction of mail coaches near Santry and Naas,[60] so too this insurrection started with similar incidents. On the afternoon of 24 January near Millstreet, where over five thousand insurgents had collected on the surrounding hills and formed two encampments, complete with beds and "every appearance of taking the field," a large peasant force halted the Cork-to-Tralee mail coach; they wounded the coachman, the guard, and one of the passengers, cut the mailbags to pieces, battered the coach, and disabled one of the horses.[61] On the following day the Tralee mail-coach agent William Brereton, having learned of the incident, decided to accompany the mail on its route between Killarney and Millstreet, a mountainous tract to which peasants were flocking on the orders of Captain Rock. As Brereton reached Shinnagh on the Kerry side of the border, he was cut down by a throng of country people who murdered him and mangled his body, despite the frenzied efforts of a local parish priest, Father Sylvester O'Sullivan of Knockacoppul, to save his life.[62]

These two days (24 and 25 January) also saw a whole series of encounters between large bodies of Rockites and units of the army. The Rockites, in general poorly armed (muskets were scarce, pikes and scythes plentiful), were invariably worsted by much smaller but better equipped and disciplined parties of troops. Two major clashes took place on successive days near Macroom, where again the insurgents crowned all the heights. In the first, at Carriganimmy, the soldiers scattered the rebels, killing (according to one report) at least seven, wounding about thirty, and taking over twenty prisoners.[63] (While its owner was away at Carriganimmy, a force of no fewer than three thousand Rockites stormed Kilbarry, the residence of James Barry, and completely destroyed almost every article in the house.)[64] In the second clash, at Deshure, south of Macroom, about half of a rebel force of "at least 2,000" people rushed down from the hills, "urged on by a man holding his hat upon a pike," but they soon beat a hasty retreat under the deadly fire of the troops, who killed at least six insurgents and captured thirty.[65]

The Rockites also staged attacks on the towns of Millstreet and Newmarket. At the first, a body of over one thousand rebels marched into the town and attempted to break open the bridewell with the intent of rescuing some prisoners, but they departed after sustaining slight casualties at the hands of the fewer than fifteen soldiers stationed there.[66] At Newmarket, also defended by only a small military force of about forty men, the one thousand or so insurgents suffered much heavier losses; the troops there reported that they had killed thirteen, wounded many others, and taken three prisoners.[67] A similar onslaught was apparently contemplated against Kanturk, where just twenty soldiers were stationed. The Rockites "posted themselves in great numbers" on Dromcommer mountain outside that town, but when the military went forth to engage them, they fled from their perch.[68]

The chastening lessons of repeated defeat led to the complete collapse of the insurrection by the end of January 1822. The rising had lasted barely a week. So disastrous an outbreak would not again be repeated. Had the insurgents of Cork been as well armed as those of Limerick, asserted one observer, they might have given a better account of themselves, but only "a small proportion of them were at all prepared for battle, & by far the greater part were totally unarmed, driven like sheep to a slaughterhouse."[69] In fact, casualties in the field had been relatively small, but the county jail at Cork was crowded with prisoners, who would shortly be tried before a special commission. By mid-February there were slightly more than three hundred prisoners on the calendar of the commission at Cork, nearly two hundred of them charged with Whiteboy or insurrectionary crimes.[70] Their spirit thoroughly broken, most of the thousands of country people who had earlier decamped to the mountains of West Muskerry and Duhallow were now "returning to the duties of husbandry."[71] This process was aided by the limited amnesty that the government extended to those who had been neither guilty of murder nor seen in arms against the king's troops.[72]

Rockite Militancy in Tipperary

Despite the collapse of the insurrection in northwest Cork, the Rockite movement elsewhere remained quite strong until it was overtaken by famine during the late spring and summer of 1822. Large parts of Tipperary declared their allegiance to Captain Rock's cause. A county with an ancient Whiteboy tradition, Tipperary began to be stirred up in the latter part of November 1821. National attention focused on the county toward the end of that month because of an especially gruesome atrocity that instantly gained notoriety. In the early morning hours of 19 November, at Gurtnapisha between Fethard and Mullinahone, local Whiteboys seeking weapons burned the thatched house and out-office of an opulent farmer named Edmund Shea, incinerating seven members of the family, five of whom were children, along with six male laborers and three female servants—a total of sixteen persons![73] Various motives associated with Shea's treatment of undertenants on his farm and with his methods of acquiring land (he was stigmatized as a "land jobber" in one report)[74] were initially assigned for the burning.[75] But there are good grounds for believing that what was conceived as a simple arms raid went horribly awry when Shea and other occupants of his house stoutly resisted the Whiteboys' demands, and that the attackers, though they set fire to the farmstead, did not intend to incinerate all the inmates.[76] There is also reason to treat this incident as a local one, not directly connected at the time of its occurrence with the Rockite movement. Subsequently, however, "the fate of the Sheas" became a common intimidating watchword as Captain Rock and his laws gained adherents in the county.

The location of the earliest systematic disturbances in Tipperary strongly suggests that, once again, emissaries from County Limerick were at work. Riding horses taken at night from farmers' fields, Rockites traveled up along the river Shannon toward Nenagh in the barony of Owney and Arra.[77] At the same time some of the mounted parties that were seen almost nightly to "parade the roads" from Croom to Hospital apparently crossed the border and began to administer oaths near Tipperary town in the barony of Clanwilliam.[78] Other Limerick Rockites, based in the southeastern barony of Costlea, seem to have slipped over or around the Galtee mountains into the district of Ardfinnan.[79] Within two weeks or less the usual signs of Rockite presence appeared in the districts of Cashel, Caher, and Clonmel, where notices were posted demanding large reductions in tithe rates, forbidding the payment of tithe arrears, and setting forth an approved scale of rents.[80] The speed with which the movement spread is largely attributable to the hallowed Whiteboy practice instinctively adopted by the Tipperary Rockites: "They take the farmers' horses in almost every instance to mount their parties."[81]

Nevertheless, strangers from the county in which the troubles originated continued for months to assist in the mobilization of Tipperary. The Protestant minister of Ballingarry parish in northwest Tipperary reported early in February 1822 that three men from County Limerick were then "organizing and perhaps hiding from justice" in his immediate neighborhood, and that in the adjacent parish of Terryglass "there are several [men] from the same place and similarly employed."[82] To the west of Tipperary town, Limerick Rockites were still active at the beginning of March. The police at Emly actually expected them to attack the barracks there. Although no raid occurred, "the people came from the county of Limerick in numbers," giving understandable grounds for police concern.[83] Individuals as well as bands promoted this kind of mobilization. Before he was taken into custody in March 1822, William Ryan "was doing a great deal of harm" in Tipperary by swearing people and threatening them "under the title of Captain Rock from County Limerick."[84]

Not all of Tipperary, however, showed the same enthusiasm for Captain Rock's aims and methods. Rockite activity in the northern portion of the county above Thurles was quite light during early 1822. In particular, there was very little unrest in the three northwestern baronies (Upper Ormond, Lower Ormond, and Owney and Arra), which were separated from the rest of Tipperary by a chain of mountains. The same area had also been little affected by the Caravat movement of 1813–16. On the other hand, mid and south Tipperary, which had been highly agitated in the days of the Caravats, once again became cockpits of agrarian activism in the years of the Rockites. This broad region included not only the poorer, less fertile land south of a line from Caher to Clonmel but also the rich limestone tracts around Cashel, Fethard, and Thurles.[85]

Activism in Kilkenny

The Rockite movement in Tipperary had not progressed much beyond its infancy when, toward the end of 1821, it was exported to neighboring Kilkenny. For at least one local landlord, Francis Despard of Killaghy Castle, the burning of the Sheas in November of that year appeared to have been the catalyst. Looking back in 1824, Despard remarked that very soon after this grisly incident, there "commenced the Rock system, which so notoriously displayed itself . . . along the range of hills from Clonmel to Knocktopher. . . ."[86] The coming of Captain Rock to Kilkenny, as to other counties, was marked by an avalanche of threatening notices and a wave of arms raids.[87] Especially at the outset, much of the responsibility for the notices and raids lay with Tipperary Rockites, who made frequent incursions into Kilkenny during the early months of 1822. A magistrate residing in the southwestern barony of Iverk spoke in January of stopping "the ingress of the insurgents from the Co. Tipperary into this vicinity."[88] Early in February a detachment of the Rifle Brigade patrolling from its base at Callan traced arms raiders in that area "to the bounds of the County Tipperary, when we thought it useless to continue our pursuit any further."[89] This pattern was duplicated around Knocktopher, according to the Earl of Carrick and two other local landlords in late March. Insurgents from Tipperary, they declared, had recently "made repeated inroads" into the Knocktopher district, where they had broken open houses, seized arms, administered illegal oaths, and "put the well-affected inhabitants of the whole neighbourhood in terror for their properties and lives."[90]

The authorities were no match for the invading Rockites. Instead of taking the offensive, many magistrates simply remained within the safety of their demesne walls and shuttered houses. As one such magistrate in the Carrick-on-Suir district acknowledged early in January 1822, "in no one instance has any resident gentleman been able to associate with his neighbours or leave his residence after dusk these last 6 weeks, but *without any exception* [they] have been compelled to keep regular guards in their houses."[91] Directed to Dublin Castle, this forthright expression of timidity also contained the usual plea for military assistance. Of that, however, there was very little beyond the main force barracked at Kilkenny. In March 1822 the military detachments at Callan, Mullinahone, and Kilmaganny each included about twenty men—hardly enough to overawe local Rockites. Small in numbers, the troops often could not count on surprise. Their barracks at Callan and no doubt elsewhere were closely watched, and their movements were promptly signaled to the agrarian rebels.[92] Even if the authorities had not labored under these serious handicaps, the nature of the terrain of south Kilkenny and adjacent parts of Tipperary greatly favored the insurgents. As the landlord John Flood remarked in April 1822, the incursions from Tipperary were almost impossible to check in "such a wild and extensive range of country" as that from the uplands beyond Callan south to the Waterford border. In any case the movement in Kilkenny was no longer mainly a matter of outside agitators; the Rockites were "now well organized in parts of this county."[93] Indeed, an army officer conceded in the same month that much of south Kilkenny was "now considered as almost under the Whiteboy dominion."[94]

In north Kilkenny, by contrast, incursions by Tipperary Rockites must have been rare, if indeed they occurred at all, for the sources do not mention them. Nevertheless, the Rockite movement was quick to establish itself in much of the northern region on the basis of acute local grievances and sometimes preexisting organization. In the northwestern barony of Galmoy, which was soon to become almost a byword for murder,[95] agrarian and otherwise, the troubles actually started before the name of Captain Rock was first invoked in County Limerick. In late June or early July 1821 the Peace Preservation Force was sent into the three Galmoy parishes of Erke, Fertagh, and Glashare to deal with what local magistrates described as "many atrocious outrages."[96] The main, if not the only, target was a large farmer named John Marum, who by 1824 was to accumulate nearly 1,200 acres in separate farms. A brother of the Catholic bishop of Ossory, Marum was reportedly "a bigoted hater of Protestants."[97] But he himself was despised by local Catholics as an avaricious land grabber and an oppressor of his undertenants. His taking of certain lands in Erke parish led to threats of assassination against him, and though he was long able to avoid this promised vengeance, other violence was repeatedly directed at him and his property.[98] Indeed, until he was finally murdered in March 1824, his proceedings and behavior helped to keep Galmoy in a greatly unsettled state.[99] Probably not all the violence that swirled around Marum came from his enemies, many though these were. For he reportedly surrounded himself with hired guns for his protection: he "had taken the Ryans, Campions, and some other of the most notorious murderers into his employ and pay, and was generally attended by some of them armed."[100] Since tithes were also a bruising grievance in parts of Galmoy barony,[101] it is scarcely surprising that so many of its inhabitants embraced Captain Rock's cause at an early stage.

Also notable for its early Rockite allegiance was the district around Kilkenny city. The city itself was both a source of recruits and a center for Rockite planning and coordination. To a considerable extent the lower-class Catholics of the town were predisposed in favor of the methods and ideology of the Rockites. Militant urban trade unionists used violence in ways that bore a strong resemblance to the tactics of rural agrarian rebels. Thus the directing committee of the Kilkenny Union of Trades was accused early in 1822 of having ordered two outrages in the environs of the city: an attempt to burn a slater's house and the beating of a man who had returned to work during a strike by woolen workers.[102] If militant trade unionism was a potent force, so too apparently was militant anti-Protestantism in the particular shape of Ribbonism. The well-known Protestant evangelical preacher and moral reformer, the Reverend Peter Roe, furnished William Gregory with evidence in March 1823 as to why he was "fully convinced that the Ribbon system prevails to a great extent in this city and neighbourhood."[103] And in an obvious reference to Ribbonism in April 1824, a well-informed military officer remarked that he was "quite satisfied . . . that the city of Kilkenny is the point where disaffected people from the counties of Waterford, Wexford, Tipperary, and Kildare meet to discuss subjects of greater moment than common outrages."[104]

Rockites as well as Ribbonmen commonly found meeting places in the town, especially in its many public houses. It was said in 1824 to be "the general opinion that most of the mischief done throughout the county is planned in the city."[105] Even though this statement involved considerable exaggeration, it contained a substantial element of truth. For example, David Quin, one of the principal leaders of the so-called Templemartin or Lyrath "gang," which was based on the estate of Sir Jonah Wheeler Cuffe just outside the city, drew its followers from both town and country.[106] Quin reportedly had "a very strong party in as well as out of Kilkenny [city]."[107] It is not known exactly when the Templemartin band was formed, but it seems to have been operative at least since the early months of 1822.[108] Its members were blamed for a long list of violent crimes in the Kilkenny district over the next two years.[109] As an army officer aptly remarked, "the gang which infests Sir Wheeler Cuffe's estate . . . is a little county of Limerick in itself."[110]

The West Limerick Heartland

In west Limerick, the original fountainhead of insurgency, the scale of
Rockite activity and the level of violence accompanying it reached new
peaks in February and March 1822. The insurrection in County Cork
may have failed dismally, but for many weeks afterward fears and rumors of a general rising in County Limerick were widespread.[111] Some
of the Limerick gentry felt certain that an insurrection would shortly
occur there too. Most, however, were of a contrary opinion, either because they regarded as a sufficient deterrent the military precautions
already taken or because they believed that the Rockites lacked ammunition, money, and capable, respectable leaders. Nevertheless, it
was agreed on all hands that "the people" looked forward to such an
event and were "almost universally determined to join in it" if a rising
got off the ground; that the Limerick Rockites were far better armed
and organized than those of Cork; and that they were displaying
greatly increased boldness, partly because of the withdrawal of troops
from dispersed country outposts.[112] The daring of the Limerick rebels
was demonstrated by the size of some of their musters (bodies of five
hundred to a thousand men were appearing in places) and by the frequency with which they now conducted their operations in the daytime.[113] But their worst sin in the eyes of the authorities was their
readiness to engage in both murder and incendiarism.

Limerick's record for murder in this period was remarkable by any
standard. The appalling frequency of this crime prompted the *Limerick News* to comment on 7 March: "No murder has been done in this
county since our last [issue]. A cessation from this horror, for even half
a week, is consoling."[114] The figures spoke for themselves: the county
coroner held no fewer than twenty-five inquests for murder between the
close of the summer assizes in September 1821 and the first week in
March 1822.[115] Not all of these killings were connected with the Rockite movement or even with agrarian disputes, though prevailing conditions probably helped to raise the toll. Reporting a homicide near
Rathkeale that he was inclined to attribute to private vengeance, the
police official Richard Willcocks declared, "In fact, they are now so
trained and inured to murder in this quarter that I am certain many
private disputes and hatreds will end in assassination. . . ."[116] Still, at
least ten murders in February and March alone were apparently the
work of Rockites. The most alarming case was that of Major Thomas

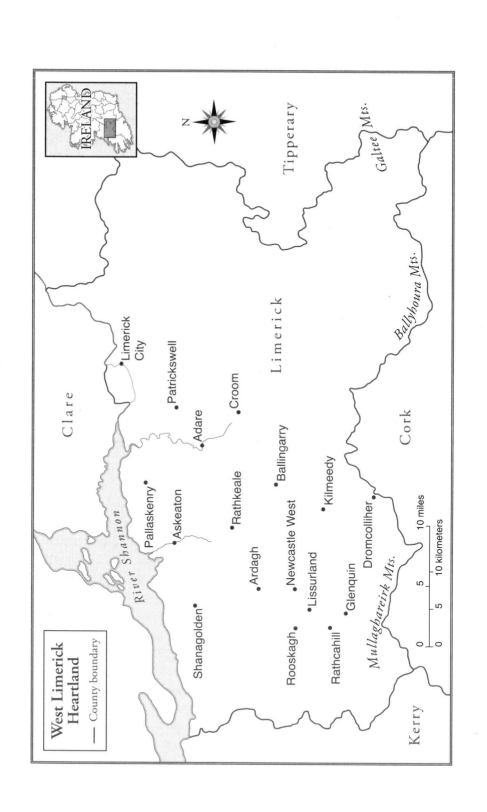

West Limerick Heartland
— County boundary

IRELAND

N

Clare

Limerick City

Patrickswell

Adare

Croom

River Shannon

Pallaskenry

Askeaton

Rathkeale

Ballingarry

Shanagolden

Ardagh

Newcastle West

Kilmeedy

Limerick

Tipperary

Galtee Mts.

Ballyboura Mts.

Cork

Rooskagh

Lissurland

Glenquin

Dromcolliher

Rathcahill

Mullaghareirk Mts.

Kerry

10 miles

10 kilometers

0   5

0   5

Hare, who was shot dead at Mount Henry, his residence near Rathkeale, when he resisted a raid for arms by a small party of Rockites.[117] Apart from Hare, no gentleman was killed anywhere in the county during these months, though it was not for want of effort, as three or four had close encounters with assassination.[118] Hare's was not the only murder in the Rathkeale district. A gentleman's servant, a farmer, a postboy and a dragoon escorting him, and an informer also met death there at Rockite hands.[119] Elsewhere in west Limerick, a former underagent employed by the infamous Alexander Hoskins was murdered near Newcastle, and two stewards of farms were also deprived of life, one near Ballingarry, the other near Limerick city.[120]

If some Limerick Rockites showed no great horror of murder, many demonstrated positive enthusiasm for acts of incendiarism. In rapid succession in February, Rockite bands burned three Protestant parish churches. The first to be consumed was that of Killeedy near Newcastle on the night of 8 February.[121] Twenty-four hours later, the Rockites returned to the scene, broke into the glebe house of the absent rector, cooked themselves a meal of potatoes and bacon, drank their fill of wines from the cellar, and left only after setting fire to the house and the adjacent out-offices. Neighboring peasants, instead of dowsing the flames, busied themselves by carrying away the moveable property.[122] (The night of the burning of Killeedy church also saw the parish church of Abbeyfeale stripped of its lead—for ammunition—and otherwise damaged.)[123] Less than two weeks later, near Limerick city, Rockites set on fire and partly destroyed Ballybrood church along with the attached rectory. The curate's house nearby, guarded by two soldiers, was left alone, but the incendiaries made off with the communion table.[124] And toward the end of the month, in another corner of the county, a party of about fifty men put the parish church of Athlacca near Kilmallock to the torch and ended the night's work by burning to the ground the deserted barracks at Ballynagranagh.[125] It would probably be wrong to ascribe these deeds simply to anti-Protestant animus. The destruction of Ballybrood church stemmed primarily from the depth of local resentment against tithes, and the burnings at Killeedy were apparently a reprisal for the rector's cooperation with the authorities, which had led to the capture and punishment of several Rockites.[126] Still, the strong Catholic sectarianism of the time almost certainly helped to stimulate these attacks on Protestant churches.[127]

Church burnings of course made prominent news, but they constituted only a very small part of the wave of incendiarism that swept Limerick early in 1822. Some thirty to forty houses were reportedly burned in this county—more than the combined total for Kerry, Cork, and Tipperary—during the months of February, March, and April.[128] In some instances dwellings were destroyed because they had been employed, or were capable of being used, to quarter soldiers or policemen, especially the latter. In other cases thatched houses were set on fire when their occupants resisted demands for weapons. But since a few lighted coals placed in the roof could produce a blaze in minutes, opposition by farmers was rare.[129] In most instances the burned dwellings had been occupied by stewards, herds, and especially those whom the Rockites designated as "land canters," a group that included but also extended well beyond persons who had recently supplanted evicted tenants. Together with threatening visits and beatings designed to achieve the same end, burnings aimed at forcing tenants—regular or temporary—to surrender their holdings became the main preoccupation of Limerick Rockites in these months.[130] At the end of March there were said to be twenty-five families residing in the town of Newcastle West who had been driven from lands in that district alone.[131]

Retreat of the Movement

By April 1822, however, the Rockite movement was in retreat almost everywhere, and in some areas its decline had become evident even earlier. In Kerry tranquility was said in mid-March to be returning rapidly to the northern barony of Clanmaurice, "except [that] it may be disturbed by small bands of robbers, who are beginning to plunder the farmers."[132] A month later, the southern barony of Iveragh was described as "perfectly quiet," apart from the Glenbeigh district, "which has at all times been infested by desperate characters."[133] Practically the whole of Cork had seen an end to organized disturbances by early April, the only significant exception being the region north of a line from Newmarket to Charleville.[134] Even in County Limerick peace had been substantially restored. With some minor qualifications military officers and police officials there were reporting during April that tranquility now reigned in their respective districts.[135] The best proof of the transformation, declared one observer, was that "the *payment of rents had begun*," even on Lord Courtenay's estate around Newcastle, where little money had been expected from the mountain tenants.[136] Agrarian conflict did not cease altogether in the southern counties, and activists in certain districts were still capable of seizing the offensive occasionally.[137] But the five months from April to August were marked by a drastic decline in Rockite activity throughout almost all the regions that previously had been so highly disturbed.[138]

To account for the abrupt eclipse of the Rockite movement, some elite commentators pointed to what they considered the effectiveness of judicial repression. It was easy to be impressed by all the arrests of alleged Rockites (including numerous leaders), by the large number of capital convictions at the special commission and the assizes, and by the wholesale transportation beyond the seas of persons convicted under the summary procedures of the Insurrection Act of February 1822.[139] But wiser, more experienced heads knew better that after a time hunger and starvation became the allies of the forces of order. Intense distress, by turning human energies to the immediate and urgent task of finding sufficient food to preserve the lives and health of one's family and kin, dissolved the broader bonds of collective action that had earlier made the Rockite movement so formidable and extensive. A subsistence crisis like that of 1822 also sharpened the antagonism between farmers of means and poor cottiers and laborers, two groups whose cooperation in agrarian warfare had previously been evident, at least on a limited scale.[140] The food crisis also weakened that part of the Rockite *mentalité* that perceived landlords, agents, and parsons as unmixed oppressors of the poor, since in times of dearth relief flowed from their hands (and those of Catholic priests) when it flowed at all.[141] Banditry may flourish well enough in periods of famine or near-famine, as it did in some southern districts during the spring and summer of 1822,[142] but for a variety of reasons supralocal or regional movements of the Whiteboy type could not and did not thrive. A good harvest, on the other hand, could do more than put food on the tables of the poor; it could resuscitate the legions of Captain Rock. The police official Major Samson Carter admitted as much in June 1822 when he remarked that while the prevailing distress had suppressed Whiteboyism temporarily, there was no "decided prospect of permanent tranquility."[143]

Yet the approach of famine and the impact of government repression together produced some startling consequences in one well-defined region where the name of Captain Rock had been held in high esteem since the outset of the movement. This was the triangular area whose points were the three towns of Newcastle West, Abbeyfeale, and Newmarket, within the confines of which lay most of the Mullaghareirk mountains. As the authorities came to realize, the vast extent of mountainous terrain in this region afforded "an excellent place of concealment to outlaws and banditti of all kinds."[144] The roads there were few and bad, in some parts being impassable for horses even in good weather. Broken bridges, as, for example, on the road between Abbeyfeale and Newcastle, interrupted travel. More than one elite observer urged the need to drive new roads from the outlying towns through the district so that "it shall no longer be to the rebels one extensive fastness."[145] Such a haven, located near the point where Cork, Kerry, and Limerick meet, was calculated to attract those fleeing from the authorities of all three counties.

As their own localities became unsafe for well-known Rockites, many of these wanted men went into hiding in the Mullaghareirk mountains, where they were joined by others who, though not yet accused, still feared exposure, capture, and punishment.[146] Just how many took to these hills (and others of lesser note elsewhere) is impossible to discover. Rumor had it that as many as a thousand armed men had gathered in the mountains west of Newcastle, but this figure is almost certainly inflated.[147] Knowing that a famine was near, the resourceful fugitives adopted the simple expedient of driving whole herds of cattle, invariably the property of gentlemen graziers and rich farmers, to their remote mountain retreats. Activity of this kind was first reported from the vicinity of Newcastle in late February 1822, and similar accounts were soon arriving from the districts of Charleville, Newmarket, and Kanturk.[148] Six gentlemen in the Newmarket neighborhood were said early in March to have lost about 130 head of cattle during the previous week alone.[149]

The raiders, who in some cases ranged as far as twenty miles from base, did not limit themselves to seizing livestock.[150] Also carried off to the mountains, besides money, was "a vast quantity of bacon, bedding, and other articles," including timber. The timber may have been used, as some claimed, to make pike handles, but it was certainly employed for firing as well as to fashion tubs in which the freshly slaughtered and salted beef could be stored.[151] Tales were soon circulating among country people down in the lowlands that the fugitives up in the mountains were living well, feeding on the choice stolen beef boiled for them daily by admiring local women.[152] These stories no doubt incorporate an element of envy among the deprived, but the poor sometimes had reason to be grateful for the outlaws' thoughtfulness. In one instance near Ballingarry, Co. Limerick, while the raiders carried off part of the stock on a farm belonging to two gentlemen, they also killed twelve cows and a bull, and instructed the herdsman on pain of death to distribute the carcasses "among the neighbouring peasantry."[153] Throughout the late spring and summer of 1822 parties of troops and police made regular forays into the mountains of west Limerick and northwest Cork in search of the outlaws. Though the authorities succeeded in apprehending some noted Rockites, including four men charged with the murder of Thomas Hoskins,[154] the great majority of the fugitives eluded capture by means of their own wits and the protective loyalty of the local inhabitants.

Three

Ideology and Organization

Although a great deal can be learned about the ideas and motives of the Rockites from examining their behavior, at least as much can be discovered about their ideology by analyzing the notices, proclamations, and threatening letters in which they directly revealed the character of their mental world. Hundreds of threatening notices and letters are extant for the Rockite movement of the early 1820s. Dozens of them were printed in contemporary newspapers, and a much larger number survive, either in their original form or as copies, in the State of the Country Papers at the National Archives in Dublin. One thing that the Rockite movement did not lack was scribes. Many of these scribes were only half literate, but many others, especially country schoolmasters formed in the hedge-school tradition, had no trouble expressing themselves clearly, forcefully, sometimes eloquently, and often at considerable length. Virtually all of this material is in English, the language of public discourse for the lower-class Rockites as well as for the landed elite, although the bombast, exaggeration, and stilted style of much of the English used betrays a society in linguistic transition. For many of its members English was still a somewhat strange tongue or a secondary language with which they were not yet comfortable.

Rockite Conceptions of Justice

Any discussion of the Rockites' ideology must begin by stressing that they articulated their own standards of justice. These took concrete form in the laws, constitutions, and regulations that they promulgated. Earlier agrarian rebels had also been legislators, but the Rockites carried this old tradition to new heights of seriousness and comprehensiveness. Providing a striking illustration is a proclamation by "his excellency John Rock, captain general and supreme director of the Irish liberators," which was posted on the church gate at Kinneigh in west Cork in January 1822. It began by roundly denouncing a magistrate who had torn down a previous notice, and then proceeded to list other decrees:

> We also enact that rents in general be reduced so as to render the tenant solvent, and that the tithe system be utterly abolished. We also enact that all rackrents and backrents be forgiven, and that any person or persons driving or distraining for any of the aforesaid charges, or processing or executing decrees for tithe or any illegal charge whatsoever within our jurisdiction, shall suffer capital punishment. We also enact that any person or persons invited to the standard of liberty and refusing to comply therewith shall suffer such punishment as is attached to that offence. We also enact that all bidders at auctions for rent or tithe shall suffer accordingly. We also enact that any person or persons proposing or agreeing for any farm or farms of land until the same be 3 years unoccupied shall suffer death. It is further enacted that any individual attempting to take down or in any wise disfigure this advertisement or any other [of] the like tendency shall, without regard to age, sex, or function, suffer capitally.[1]

Other notices were equally insistent in demanding strict adherence to Rockite ordinances. In one of February 1822 setting forth a variety of regulations concerning rents, tithes, and the letting of land, the writer spoke of "enforcing and making the following tempo[r]ary laws perpetual."[2] A second notice sent early in 1822 to the Cork landowner and agent William Stawell declared that "in consequence of a new code of laws recently given out by General John Rock, legislator general of Ireland, it is unlawful for any gentleman to hold any more lands than that which immediately adjoins his dwelling residence."[3] The writer of a third notice in March 1823 warned Charles Haynes, a land grabber in the Mallow district who also employed strange laborers, that "your infidelity and your transgressions against the new established laws of

85

this land, which were founded on the laws of God, must be represented before the senators at the illustrious senate house."[4] And a fourth notice posted in at least three different places in north Cork in September 1824 cautioned "all land canters" about "a law enacted [in] the third year of his majesty's reign and second year of General John Rock, Esqr., that no man is allowed to take tithes or overhold any farms or dispose [i.e., dispossess] poor families until the year 1825 is over."[5] Even though the Rockites promulgated laws that were clearly intended to supersede parliamentary statutes (they often spoke of their own ordinances as statutes), they sometimes disclaimed any subversive designs or any disloyalty to the legitimate monarch and government. Thus one Rockite legislator assured his audience in a notice of February 1822 that he did not seek "any division of any part [of the kingdom of Ireland] whatsoever against his present magesties crown or dignity, and that he and his followers are as loyal subjects as any other person or persons now existing under his magesty."[6] One of the many notices placarded in County Limerick early in 1822 proclaimed, "We war not against our king nor his government. Had he occasion for our services, he should find us true to the last drop of our blood."[7] And as the writer of still another notice declared, after demanding better treatment from farmers for "the very poor and distressed labou[r]ers," "It is not against the king or his constution [sic] that I take arms, but against cruel tyrants."[8]

Prevalence of Nationalist Emotions

But alongside these rather occasional disclaimers of subversive or revolutionary designs must be set the more frequent expressions of nationalistic and anti-English sentiment also found in Rockite notices and proclamations. Providing the Rockites with justification or rationalization for their readiness to use extreme violence were deeply held convictions about oppression, and these were commonly rooted in a keen sense of the historic political wrongs inflicted on Ireland. The need to throw off the yoke of English tyranny was sometimes cited explicitly, as in the following notice of December 1821 posted in County Tipperary:

> We are pressed by the tyrannical laws of the English government which we cannot endure no longer, to which they crowd us up with rent, tythes, and taxes which we daily sustain the burden of, without the

least abatements of rents we expected at the coming of George the fourth to Ireland, that he would grant a small privilege to the farmers and husbandmen of Ireland and then to the poor peasantry of this nation. English laws must be curbed in, for we will never be satisfied until we have the Irish parliament and king crowned in Ireland as there formerly was; the sword is drawn and the hand of God is with us and tyranny will be soon swept away. . . .[9]

It is quite wrong to suggest, as some historians have done, that prefamine agrarian rebellion can be explained more or less exclusively in terms of economic rationality. Though the political ideas of the Rockites may not have been very sophisticated (surely, the expectation of sophistication is misplaced), the political content of their thinking deserves greater attention than it has so far received. The millenarian and sectarian strands in Rockite ideology, both of which were highly charged politically, are of such importance as to require a separate chapter in this book. Besides the many notices containing distinctly millenarian ideas, numerous others expressed political sentiments hostile to English rule and heavily tinged by revolutionary nationalism. A notice affixed to Cratloe chapel in Clare early in February 1822 urged the people "to assert their rights, as in 1798, and to reward the man who shot Stack, the informer," soon after he was murdered.[10] A second notice posted outside Adare chapel in Limerick in the same month ordered the mobilization "in the name of the Irish" of all men aged sixteen to forty, "in order to march in mass against the common enemy, the tyrant[s] of Ireland, the English whose destruction alone can insure the independence and the welfare of the ancient Hibernia."[11]

Admittedly, such sentiments were most pronounced during the short period of insurrection and millenarian frenzy in the winter of 1821–22, but they were certainly not confined to that season of extraordinary turbulence. If the official laws were not obeyed, argued the writer of a notice about tithes in northwest Cork in July 1822, that was because disobedience had been "naturally planted in the people of Ireland; the oppression they suffered by the abominable tyranny of the English made them desperate and urged them on to attempts which they would never have thought of, had they been well treated with a tenderness and humanity which the circumstances of their case so justly deserved."[12] At least some Rockites drew encouragement from foreign revolutions in the early 1820s. "Irish Patriots!" declared the author of a notice placarded on the roadside near Cork city in May 1823, "Rouse from your lethargy, embrace the standard of your patriotic general [i.e., Rock] . . . ; take notice of your patriotic brethren in South America and Greece, shaking off (with success) the yoke of tyranny. . . ."[13] In similar fashion the writer of an earlier notice posted in the Fermoy district had remarked, "You see that the Greeks are overturning their infidel masters, and is not it a shame for us, when we are so powerful, that [we] don't chrush [sic] the serpents of this nation."[14] It may fairly be objected that the contemporaneous revolutionary events in Greece and South America were far from furnishing a serious inspiration for the agrarian violence of the Rockites. But the same cannot be said of the perceived oppressions stemming from English rule or from the actions of the agents of that rule in Ireland—gentry magistrates, Anglican parsons, the police, and the army. The behavior of the Protestant landed and clerical elites as well as the conduct of the forces of repression elicited responses from the Rockites that were heavily colored by nationalist and sectarian feelings.

Nevertheless, it would clearly be wrong to exaggerate the role of nationalist or sectarian ideology in providing motives and justifications for Rockite violence. What is needed is a balanced treatment that also takes into account the main kinds of rural social conflict at the time: between the landed or clerical elites on the one hand and farmers or laborers on the other; between the larger farmers and smaller tenants, cottiers, and laborers; and between different members of the same nonelite social groups. Here our primary concern must be with the values and standards by which Rockite farmers and laborers assessed the social behavior of members of both the elite and the nonelite.

The Ethics of Localism

One central value repeatedly asserted in Rockite notices, as in those of earlier Whiteboys, was localism. This was the notion that the inhabitants of a particular locality, and especially the poor, had a superior or even exclusive claim over that of outsiders to the land, employment, and food available within that locality. The rules against strangers in local labor markets were invoked most often against spalpeens, or seasonal migrant workers, from other counties. "I do tell any strangers, or Kerrymen or Limerick men, to be off immediately," declared the writer of a County Cork notice in March 1823.[15] "Remember," an east Waterford farmer was warned in April, "if you do not part with that straing man, that we will leave him in the condition that the priest will not overtake him, and we will use the man of the house [i.e., his master] worse. This is the following oath that all the men have been sworn by— I, A.B., do solemnly swear by our Lord Jesus Christ, who suffered for us on the cross, and by the Blessed Virgin Mary, that I will burn, destroy, and murder all straingers and their masters, and [wade] till up to my knees in blood. So help me God."[16] An east Cork landholder was ordered in the following September to "cashire [sic] and discharge the strange and deluded foreigners that you're collecting and keeping, as it were in opposition to the law of our country; now be certain that we are fully bent and determined to allow no such proceedings in the neighbourhood, like all other places."[17]

The strong sentiment in favor of local workers also operated outside agriculture. A smith living in Clare village near Ennis, but a native of King's County, was the object of three notices in January 1822 threatening him "with a watery grave under the bridge if he did not quit the country immediately, and cautioning the country people from dealing" with him.[18] The proprietor of the Castle Matrix Flour Mills near Rathkeale was denounced in November 1823 for dismissing a miller "pleasing to all the neighbours"; the owner's intention to hire an outsider was rebuffed with the curt remark, "We want no vagabond strangers amongst us."[19] Even persons long settled in a given district were sometimes directed to leave. Thus a notice posted on the door of Killorglin chapel in Kerry in February 1822 commanded all strangers of fewer than fourteen years' presence to return to their "respective countrys."[20]

This localist ethic embraced not only employment but also food. After a disastrous harvest in the previous autumn the farmers of Clanwilliam barony in Tipperary were warned in December 1821 "against selling or sending" livestock, dairy produce, or grain of any kind to markets outside the barony.[21] Farmers with marketable potatoes in a north Kerry parish were instructed in the following month "to sell them to their neighbours at home . . . , as our design is to establish some just regulation in the country for the poor."[22] Dairy farmers near Tipperary town were cautioned in April 1822 to send no milk to market "until such of the neighbours as are in necessity of buying it are supplied at a reasonable price," namely, seven quarts for a penny. And

While this sketch shows a group of Irish migrant workers in the early 1880s traveling to England for harvest labor, a depiction of the internal migrants from south and west Munster who worked during the prefamine decades for the larger tillage farmers in digging up potatoes or reaping the grain crops would not have been very different. Such migratory laborers (spailpíní fánacha in Irish) from outside the local district had aroused much hostility and sparked recurrent violence in the northern and eastern regions of the province ever since the 1770s, if not earlier. The Rockites targeted these spalpeens again in the early 1820s. (*Illustrated London News*, 28 May 1881)

for every dozen cows in their dairies, these farmers were ordered to rear no more than two calves, the aim being to increase milk supplies and to reduce prices.[23] Rockites imposed very similar localist regulations on dairy farmers in the Listowel and Tarbert districts of Kerry in the following month.[24]

Access to land also came within the ambit of the localist ethic. The Rockite rules that gentlemen could not hold and stock land unless it immediately adjoined their own residence, and that farmers with multiple holdings could not employ dairymen or herdsmen to care for livestock on outlying farms, were meant to increase the amount of land available to other local inhabitants.[25] In the Kanturk district "gentlemen and rich churls," insisted the author of a notice in February 1822, were "to keep no milkwomen, dairymen, or herdsmen but to let the lands to honest, industrious tenants for the value."[26] This regulation was announced in notices sent to two gentlemen in the Doneraile district about a month later. William Stawell was told that he must dismiss all herdsmen, dairymen, and milkwomen from certain specified lands and "afterwards set the same to poor, industrious people at a reasonable rent."[27] And Arthur Creagh was taxed with not caring about "how the industrious poor of your neighbourhood live" when it appeared that he intended to appropriate to his own use land recently surrendered by a tenant; Creagh was ordered to relet the holding to "some industrious man who is not a stranger in your parish."[28] Farmers and gentlemen were also expected to satisfy the needs of local laborers and cottiers for potato ground at rates below the market value, even if this meant breaking up valuable grassland in their own hands and allowing it to be tilled. Many Rockite notices directed that pasture be converted for conacre, and specified the rates that would be acceptable. The following notice from the Doneraile district in March 1822 was typical of those sent to individuals: "Mrs. Arkinse [sic], you are ordered by Captain Rock to give two of the best fields you have at 4 pound[s] an acre, or else if you don't, you will rue the day."[29]

The Gravest Sin: Eviction

No Rockite regulations were more sacred than those designed to frustrate evictions and "voluntary" surrenders and to maintain existing tenants in secure possession of their holdings. Whereas the localist ethic was essentially a weapon wielded by the landless and the land-poor against the larger farmers and the landed elite, most farmers could support Rockite efforts to thwart evictions. With the Rockites it was almost axiomatic that a vacant holding was one from which the previous tenant had been removed against his will, and anyone who took it or even offered to take it was automatically branded as a land canter or land jobber—hateful terms indicative of what would later be called land grabbing. In order to curb this social crime and to exact a penalty for it, the Rockites enjoined that any holding from which a tenant had been evicted must be left vacant for a period of years. One notice mentioned two years, another as many as seven,[30] but three years was the interval usually specified. A notice posted on the door of Inchigeelagh chapel in west Cork in March 1823 required strict obedience to what was presented as a recent national law: "Canters are hereby cautioned to take particular notice that they are not to take the lands of a reduced farmer for the space of three years, as it's a regulation lately taken place all over the kingdom."[31] Notices with a similar stipulation had been placarded elsewhere over a year earlier.[32] Indeed, the injunction was not a Rockite invention at all. As long ago as 1786 the Rightboys had promulgated the same regulation providing for a vacancy of three years after an eviction.[33]

But the Rockites were not content merely to require such a period of avoidance when evictions occurred in the early 1820s. In this regard their laws were often retrospective as well as prospective. Land canters already in possession were to be made to atone for their sins by surrendering their ill-gotten gains. An obvious question was what length of possession might relieve them of this requirement. On this point the answer varied from one locality to another. A notice posted at Lehenagh in Cork in March 1823 specified that those guilty of canting within the past two years must surrender their farms.[34] But elsewhere older violations were also declared punishable. A County Tipperary notice of December 1821 demanded surrenders if the canting had taken place within the past five years,[35] and several notices in Kerry early in 1822 instructed violators within the last seven years to give up their farms.[36]

The intensity of popular hatred for land canters is plain from the strong language in which their crimes were discussed. "Any purse-proud rascal that is so ambitious . . . as to unhinge any poor man of his ground will suffer in the severest manner possible," promised the author of a notice posted at Kildimo in Limerick early in 1823. For throwing a family out on the roadside, this writer declared, "by the King of Kings, you will suffer sooner or later."[37] The author of another notice in March of that year observed: "Humanity shudders at the thoughts of the severities intended [for] a few individuals in Donoughmore [parish, Co. Cork]; this uncharitable system of land canting is practised there without the least idea of that divine precept, 'to love thy neighbour as thyself. . . .'" Violators of "my constitution," he warned, were in danger of becoming "a lasting monument of destruction."[38] The messages sent in April to two land canters near Millstreet were more succinct. One was told "to be clear of that farm . . . , or if you don't, you will be burnt to ashes, for we are not allowed to have any man in another man's farm."[39] "Go by this notice," the other was directed, "for we are sworn not to alow any man to do it while you have a place of your own, and by this oath you will not be spared a minute . . . , for we have [it] in for you. . . ."[40] The author of a much longer notice excoriated land canters in general:

> These worldlings live as if created for the mere enjoyment of this life. . . . All threats and annihilation of property are not able to put down the accursed practice; no, their spirits are buoyed up by the specious promises of landlords, who promise them a remuneration for any losses sustained, and that this Rock system is only a flame that will immediately cease. But I ask these [canters], will these landlords restore them to life, or will they ensure their salvation if they meet with an untimely end? Now, my dear friends, I hope you will not be so hardened as not to believe me when I tell you that . . . I am only an instrument in the hands of the Almighty, and that I cannot cease and will not cease [in punishing land canters].[41]

Mimicry of the Established Order

The promulgation of their own laws and regulations was part of a more general and elaborate mimicry by the Rockites of the legal and judicial apparatus of the established government. Just as the government did with lawbreakers, the Rockites proclaimed their right to punish transgressors. Their notices informed violators that unless they took corrective action within a matter of days or (at most) weeks, the sentence of Rockite law—usually a grisly form of death—would be carried out against them. Incorrigible offenders might just "as well prepare their coffins," could be "sure of their doom," would "die the most unmerciful death," or would be "shot, quartered, and beheaded."[42] A kind of gallows humor was sometimes evident. "In compassion to your human weakness," one intended victim was told in March 1822, "and in consideration of the enormous weight of your corpulent fraim, I mean to rid you of these inconveniencies by a decapitation. Do not, I pray you, be alarmed, for your life must be a load to you, and I think that you ought to thank me for my good intention to deprive you of it."[43] Notices often mentioned multiple punishments. One farmer was warned, "Your house will be consumed, your cattle will be houghed, and yourself gibited [sic]. This sentence was passed by the last committee. . . ."[44] "Mark the consequence," a gentry offender was cautioned, "for houghing of cattle, burning of houses, rooting [up] of trees, flogging of herdsmen, and losing of your life will be the consequence of slighting this notice."[45]

The Rockites gave special emphasis in their notices to the sentence of property destruction and death by burning. "My devouring fir[e]y laws will swallow up all cursed transgressions," one writer proclaimed with a flourish.[46] Posted on the town pump of Mullingar early in 1822 and falsely heralding "Good News for Butchers" was a notice that declared, "Any person taking the contract for soldiers' meat, or any process-server, shall be met with fire and destruction. . . ."[47] A tithe proctor in Rathlynin parish in southwest Tipperary was advised in January 1822 to

> receive the ceremonies and rites of the church and occupy your time diurnally in prayer and fasting, and also have your burial place settled; but as for the latter, we believe you need not trouble yourself, as you may be positively assured we will burn and serve you as the unhappy victims near Fethard were served [a reference to the unforgettable "burning of the Sheas" in the previous month].[48]

Certain canters of land in north Cork were warned by a classicist writer in March 1823:

> If this caution is not immediately complied with, Vesuvius or Etna never sent forth such crackling flames as some parts of Donoughmore will shortly emit, so that to a distant spectator the whole parish will seem a solid mass of fire. Oneen has his matches and combustibles prepared at all times. [The notice was signed, "The Fireman, General John Rock, K.C.B."][49]

In the same style "the Debonair Captain Rock" told a land canter near Mallow, "I will dispatch the Blessed Salamander with his unextinguishable flames which he has deposited in his shrine to punish the wicked; he will command his flames to feed annual[1]y on your houses, hay, and corn."[50] Countless other offenders were informed without fanfare that they would be "consumed to ashes."[51]

For the Rockites as for the government, penal sanctions were necessary not only to satisfy the demands of justice in particular cases but also to serve as examples of what lawbreakers in general might expect. In other words, like the authorities, the rebels firmly believed in making their harsh punishments exemplary. "I would make a public example of him to the county of Kerry," insisted the author of a notice in December 1821, speaking of anyone who violated the interdict against paying tithes in Ballymacelligott parish.[52] A farmer near Doneraile who ignored Rockite tithe regulations was ordered to obey them, "or b[y] God you will be punished so far that the smallist youth in the parrish will remember it, you vagabone."[53] Notorious victims of the Rockites were repeatedly depicted as terrible warnings to others who might be foolhardy enough to persist in breaking Rockite laws. In December 1821 a County Limerick offender received the terse message, "John Reilly, I desire you to quit this place, or else you may expect the same usage Sparling got. . . ."[54] (The Palatine gentleman Christopher Sparling had been shot and fatally wounded in the previous October.)[55] For having surrendered their arms to magistrates (instead of the rebels), and for having rackrented their cottier tenants, certain Tipperary farmers were threatened in the same month with "a worse death than Going."[56] The parish priest of Kildimo, Co. Limerick, who had disobeyed Rockite orders, was informed in January 1822 that unless he followed them, "you will absolutely meet the same fate as Father Mulqueeny [sic],"[57] who was murdered by reputed Whiteboys in November 1819. A County Tipperary land grabber named Michael Dwyer, denounced in September 1823 as "one of the greatest tyrants in the province of Munster," was solemnly warned, "If you and your family together do not quit immediately, ye shall suffer the same punishment that the Sheas did a few years ago. . . ."[58] And in a notice posted in September 1824 in various parts of the Cork barony of Fermoy, all violators of Rockite regulations concerning tithes and land canting were told, "We will not be going to the trouble of houghing or burning but [will] finish you off, the same as Thomas, Margaret, and Henry Mandfield [*sic*] Franks, if you put me to the disagreeable necessity; it is no historical metaphor but the downright truth."[59]

The Rockites also let transgressors know that they would enforce their laws even more rigorously than the established government enforced its own statutes. The idea was inculcated that if any injustice were to be committed, the Rockites would be sure to hear of it. "Captain Rock," declared the author of a notice posted in the Croom district in November 1821, "has formed a firm resolution to punish . . . with the greatest severity" all violators of his regulations about rent reductions, "and as the aforesaid Captain Rock's corps are so very extensive, and his army distributed all over Ireland, nothing can be done unknown to him."[60] The psychology of fear was cleverly manipulated in other ways as well. A hated tithe proctor in Tipperary was warned in January 1822, "If you were to be guarded by all the Peelers and soldiers in the county, we will [still] go to you if you do not give up the aforesaid claims [to tithes] . . . as soon as you receive this letter. . . ." Retribution was certain if he failed to comply, for as the postscript boasted, "there are eight thousand men of us sworn by the above commanders."[61] Another Rockite writer struck this same note of irresistibility a year later. As he told the violator, "Do not figure to yourself that military detachments will protect you, for they are no more to the invincible forces that I intend to dispatch to execute the aforesaid decree than dust is in the face of the wind."[62] No one could know when, where, or how the stealthy Rockites would strike, a point that was cunningly made in January 1823 to some would-be land canters in the Kildimo district: "You may say to yourself, 'the army is near me, John Rock is gone away,' and so on, but take care, life is sweet, don't meddle with any man's ground . . . ; Rock is still alive, don't disturb him, leave him as he is in a state of tranquillity; if you don't, the consequence will be murder."[63] Even heavy government repression, insisted the author of a notice in County Cork in the following March, would not save transgressors from punishment:

> Let no person consider that the rigours of English laws will suffer to protect an offender of my regulations. No, for a gallows filled with radicals or a [convict] ship laden with Whiteboys cannot intimidate the heart of a true United Irishman, inflamed by the spirit of liberty. 'Tis

true we have been deprived of some of our associates by hanging and transportation, but this has only increased our strength.[64]

Inherited Techniques

Like threatening notices, initiation oaths throw much light on Rockite ideology, but the oaths are also important for what they reveal about matters of organization. Initiation oaths had been used by Irish agrarian rebels ever since the first Whiteboy movement of the early 1760s. By the early 1820s, after many decades of intermittent agrarian upheaval, the accumulated stock of such oaths was very large indeed, and it is hardly surprising that the Rockites availed themselves of formulas borrowed from the detritus of previous movements of protest and revolt, including those of the Defenders and United Irishmen in the 1790s. A prime illustration of this imitative tendency was the manuscript book that the police discovered in the possession of Denis Egan, a wealthy farmer's son, when he was arrested near Roscrea, Co. Tipperary, in April 1822. The book bore the legend "Signed by the Delegate to Captain Rock," and it comprised, according to a police report, "the oaths or tests of 'United Men, Ribbonmen, and Caravats,' commonly so called," together with "several questions and answers and devices relative to the said illegal combinations."[65]

Another instance of borrowing was Rockite use of a document entitled "The Royal United Irishman's Test," copies of which circulated in some parts of Cork during the early 1820s. This set of fourteen initiation oaths comprised the long-customary pledges to preserve strict secrecy, to obey summonses to action, to proceed only with proper orders, to report to constituted leaders any unauthorized or forbidden activity, to avoid fighting and plunder, to assist brothers in distress and never to do them injury, especially by informing against them or by giving hostile evidence in any kind of judicial proceeding. But also included were vows of a sectarian and political nature. The initiate was sworn to assist in destroying "all tyrants, kings, suppressors, and hereticks not of the true Roman Catholick and apostolick religion," to "plant the tree of liberty in as many hearts as I can," and to "fight knee deep in Orange or any other heretick's blood," without ever being daunted by "the crying of children, the moaning of women, or the groaning of men."[66]

The sectarian features of initiation oaths stirred controversy both at that time and later among skeptical historians. Alarmed Protestants cited them as indisputable evidence of the Catholic supremacist and revolutionary intentions of the agrarian rebels and warned that another Saint Bartholomew's massacre or a reenactment of the horrific atrocities of 1641 and 1798 was in contemplation.[67] Respectable Catholics and especially clerics, on the other hand, were inclined either to dismiss such oaths as fraudulent or to insist that their lower-class coreligionists, though Rockite partisans, were innocent of these vengeful sectarian vows, which they identified with Ribbonism—an influence held to be largely alien to the rural south.[68] Historians as well have sometimes been too ready to draw a sharp distinction between the so-called primitive nationalism and sectarianism of the Ribbon underground on the one hand and, on the other, the allegedly apolitical or prepolitical mentality and the purposively economic motives of Whiteboy movements.[69] The kinds of sectarian oaths attributed to the Rockites and other Whiteboys have at times been treated as ritualistic carryovers from the 1790s and as lacking in serious practical meaning for later agrarian rebels.

This approach has serious flaws. It seeks to impose a tidy compartmentalization on very untidy and complex realities, and it is at variance with a large body of surviving evidence. Admittedly, most of this evidence comes from feigned Rockite supporters or outright defectors from the movement, and thus a healthy degree of skepticism is perfectly in order. But the general consistency of reports from different areas strongly suggests that sectarian initiation oaths were neither unusual nor merely ritualistic exercises, devoid of serious practical import. A "very decent, intelligent farmer" informed the police official Thomas Vokes in November 1821 that in parts of County Limerick the leading features of the oath administered to new recruits were the "destruction of all Protestants" and "to have a Roman Catholic king and a French government."[70] Charles Bastable, Jr., whose access to blasting powder as the superintendent of some road construction in northwest Cork brought him to the notice of local Rockites, was persuaded to swear "to be faithfull to the Whiteboys' cause and to do everything in . . . [his] power against all Protestants and tythe proctors."[71] A Protestant parson near Ballingarry, Co. Tipperary, asserted that the oath proposed by the Catholic rebels to one of his informants "enjoined his assisting to overthrow the government, to kill all the Protestant ministers, to burn all the churches, to put down all heresy, and to have Ireland to themselves."[72] Perhaps this parson or his informant exaggerated, but the Cork farmer Patrick Donovan, forced at gunpoint to become a Whiteboy, admitted in February 1823 to having sworn an oath against tithes in which he vowed to murder, if necessary, all the Protestant clergymen in Courceys barony as well as anyone who paid this clerical tax.[73]

The circumstances under which the Rockites tendered initiation oaths varied. During an intensive campaign of mobilization, as in the initial expansionary phase of the movement, the oaths seem to have been administered indiscriminately and often under some duress, throughout whole districts; those who refused to be sworn might be punished with a beating.[74] On these occasions when the immediate goal was to secure the allegiance of large numbers, the oaths used were probably short and fairly simple. But ideally, and in practice after the movement had become routinized in a given neighborhood, initiation ceremonies were supposed to follow a prescribed form. The aims were to instill a sense of the solemnity of the occasion and to guard against betrayal. A small group of men, often called a committee, was in charge of the proceedings; it determined the suitability of the candidates, and its leader or captain administered the oaths. A County Cork informer who had feigned a desire to join the rebels claimed in June 1822 that a Rockite captain from the Kanturk district had told him that he could not be sworn without the approval of a six-man committee, which had to certify his worthiness prior to the tendering of sixteen oaths.[75] Darby Connor, a young farmer in Brosna parish on the Kerry-Limerick border who did join the Rockites, confessed in October 1822 that he had been sworn in the house of a local smith by a County Limerick activist who was assisted by four other farmers, one from Kerry, two from Limerick, and one from Cork.[76] Another lapsed Rockite in the Abbeyfeale district of Limerick asserted in April of the same year that his initiation ceremony was attended by ten men, all local farmers who obviously knew and trusted him. He was sworn not to divulge their secrets, to be ready to take up arms when summoned, not to deal with Protestants, and to use every possible means to "put down" all persons connected with the Anglican church.[77]

At initiation rites new recruits were also made familiar with the passwords and signs by which to identify their "brothers" in public places on the roads, in the fields, and at fairs. Such devices had their origins in freemasonry and had been further popularized by the United Irish and Defender movements of the 1790s.[78] Valued for the air of mystery and the sense of comradery that they promoted, passwords and signs continued to flourish among the agrarian and sectarian secret societies of the early nineteenth century. Compendia of these passwords had become known as catechisms because of their traditional question-and-answer format. The sources of the early 1820s refer to them as "Whiteboy catechisms" or "Ribbon catechisms," though like Rockite initiation oaths, many of them were apparently holdovers from the 1790s or redactions of booklets printed prior to the 1820s. The sets of queries and replies were often quite long, but they were regularly committed to memory, a task easily accomplished in a society with such a strong oral tradition, and informers sometimes recited them for the edification of magistrates. Commonly, the responses displayed the same sectarian invective found in Rockite oaths. For example, in one catechism the prescribed reply to the question "Who baptised you?" was "St. John [baptised me] with a drop of St. Barnaby's blood, which is in the uppermost part of my breast, who would rather see the Devil than to see an Orangeman who never crossed his forehead."[79]

The Influence of Ribbonism

Ribbonism influenced the Rockite movement in some of its organizational and ideological features. This was particularly the case in the vicinity of towns, for it was usually in cities and towns that Ribbon lodges or committees were centered, to the limited extent that they existed in the south. Cork city was a hotbed of Ribbonism in the early 1820s, and activity there spilled over into the surrounding countryside. At the very beginning of 1822, Edward Newsom, the mayor of Cork, called the attention of Dublin Castle to the need for "the suppression of those catechisms w[hi]ch are published and circulated from Dublin, some of w[hi]ch have been seen [at Cork] by McCarthy, the informer."[80] A magistrate near Bandon reported with alarm in December 1822 that the Ribbon oath had been "generally administered in Cork [city] and its vicinity, and that he expected a general rising in less than three months."[81]

Other Cork towns and their hinterlands were also affected. In February 1823 a magistrate at Charleville forwarded to Dublin Castle

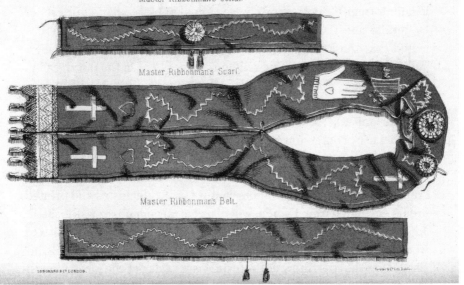

Master Ribbonman's Collar.

Master Ribbonman's Scarf.

Master Ribbonman's Belt.

LONGMANS & C° LONDON.

Scholars have generally made fairly firm distinctions between the activities of Ribbonmen and those of agrarian rebels like the Rockites in the early nineteenth century. Certainly, in other regions of Ireland and in certain urban areas of Ulster and Leinster, the Ribbonmen had an extensive system of lodges and somewhat elaborate rituals of initiation, including such sartorial elements as the master Ribbonman's collar, scarf, and belt depicted here. The religious symbols on the scarf were meant to mark the Catholic identity of the Ribbonmen. But if these features of a more highly developed Ribbonism were generally missing from the southern part of the country, the Rockites did draw on Ribbonism, to a limited but significant extent, in the matters of ideology and organization. (William Steuart Trench, *Realities of Irish life* [London, 1868], facing p. 47) copies of a Ribbon oath and catechism that he had received from a local informer.[82] Such reports were ridiculed by the parish priests of Doneraile and Castletownroche (among others), who solemnly declared that the Ribbon oath had never been taken in their parishes, and that "there was nothing religious or political in the outrages" in their districts.[83] The latter statement, though made with the utmost sincerity, was wide of the mark (north Cork had more than its share of millenarianism),[84] and even the former statement is open to question. In his lengthy confession of April 1823 the noted Rockite John Hickey indicated that committees regularly sat at Doneraile, Mallow, and Buttevant, and though he did not specifically identify them with Ribbonism, he observed in the same context that catechisms and printed orders emanated from Dublin, where a central committee was located. Just as interesting was Hickey's assertion that the local committees were linked together through the joint activities of their heads. He insisted that the overthrow of Protestantism "was not the idea amongst any of us [i.e., the leaders]," but he acknowledged that the "lower fellows" vented their sectarian bitterness.[85] It is highly doubtful that Hickey's remarks would have taken this shape unless Ribbonism had acquired at least a modest presence in north Cork.

Links between Ribbonism and the Rockite movement were also evident in other southern counties. Michael Hurley, a stonecutter arrested late in 1821 for tendering oaths in the Ballyheige district of north Kerry, confessed that he himself had been sworn in Tralee, where a committee sat and issued instructions. Included in Hurley's initiation oath was a promise "to join the French," whose reputation as liberators long outlived the Jacobin regime in popular Irish political mythology.[86] Before the Insurrection Act was extended to County Clare, the house of Thomas Jones, "a man whose influence in Cratloe and its neighbourhood was most powerful," often served as a meeting place for Ribbonmen and Whiteboys. After his trial and conviction under the Insurrection Act at Sixmilebridge in August 1823, "thousands of people flocked to the road to see him on his way" to Limerick under military escort.[87] In disturbed west Limerick earlier, an informer painted a picture of incipient rebellion with a distinct Ribbon tinge. He identified Patrick Heffernan of Garryowen as a Ribbon agent who had traveled up to Dublin for a new "test oath," which he then took down to County Kerry.[88]

This informant's claim that every town and village in four of the southern counties had a regular clerk for the transaction of rebel business was wildly exaggerated, but it was not completely without foundation. In parts of north Tipperary, where an Orange presence stimulated a militant Catholic reaction, Ribbon lodges or committees based in the towns were certainly active. Early in 1822, as Rockites using a sectarian oath were mobilizing recruits around Ballingarry, the local Protestant parson bitterly complained of the conduct of George Canny, by trade a cloth dyer and presser at Borrisokane, whom he called "a village clerk and village lawyer" and identified as a Ribbon captain.[89] Clearly, in this parson's view Canny was connected with the Rockite mobilization in the district. Ribbonism also touched the ancient Whiteboy country around the towns of Templemore and Borrisoleigh. Among the valued recruits of the Rockites in this area was a tailor named William McDonald. An army veteran and pensioner, McDonald reportedly supervised military drill in the mountains by local Whiteboys, who sometimes numbered in the hundreds on these occasions. Their leaders initiated McDonald with a Ribbon oath in which he allegedly promised "to aid and assist in the destruction of Protestants and the Protestant system." His wife was said to have incautiously told friends in Templemore that the intention "was to rise at night and destroy the Protestants, the boys of each place to destroy those [Protestants] around them."[90] In the northwest corner of Tipperary, between Toomyvara and Nenagh, old sectarian feuds between Orangemen and Ribbonmen were revived in the early 1820s,[91] and Rockite activity in that area was apparently based in part on Ribbon ideology and organization. The meeting of sixteen men (including a secret informer) that the authorities raided at Toomyvara in January 1824 was probably that of a Ribbon committee with Rockite proclivities. The authorities confiscated "prayer books" (used for swearing recruits), two formulas of oaths, and other seditious papers.[92]

Rockite Committees

The committee form of organization was also commonplace where no Ribbon presence can be detected, and the initiation of new members was only one facet of any committee's proper work. Its principal business was to decide which operations to conduct, and how and when to carry them out. Undoubtedly, much Rockite activity took place without reference to committee structures, but in mounting small-scale operations as well as larger, coordinated enterprises, committees nevertheless played an important role. Detailing a series of outrages in north Cork during the last week of December 1822, Colonel Sir Hugh Gough remarked in frustration that almost every day that week had been a holiday, "on which days their committees sit and arrange their movements so judiciously as to defeat all our attempts to come up with any of those night depredators, either collectively or even individually."[93] For as long as there had been Whiteboys in Ireland, the public house had been the most usual site for planning strategy and tactics, and the association between the public drinking place and meetings of agrarian activists was as strong as ever in the early 1820s. An informer identified three pubs in Limerick city that served as special resorts for the disaffected of that town and its environs late in 1821.[94] The core of the Rockite party that planned the murder of the Franks family in County Cork in 1823 met and discussed the deed in Patrick Power's public house at Shanballymore after Sunday Mass. At one of these meetings the knot of dedicated conspirators included seven men, almost all of whom were from Meadstown in neighboring Farahy parish.[95] These men, said to have sat as a committee, were presumably instrumental in mobilizing the much larger group that actually carried out the killings a few weeks later.[96] At Mallow the house of the publican Denis Kelleher was the regular resort of local Rockites from that town and further afield. The publican's son Patrick was, until his sudden defection in September 1823, a leading activist ("there is not a man in this country who knows more of the present system").[97] In the middle of the previous April, just two days before the scheduled execution of John Hickey, a Rockite committee consisting of Patrick Kelleher and six other men convened in his father's pub. This was not the gathering of just a local committee, for those in attendance were drawn from five different parishes in north Cork.[98]

Rockite committees also assembled frequently at private houses in the countryside, and there was a decided preference for such venues after nighttime drinking in pubs was, in effect, made illegal in proclaimed districts under the Insurrection Act of 1822. Illuminating the situation in a portion of west Limerick during the early months of that year is the detailed information of the renegade Rockite William Connell. Of the numerous meetings mentioned by Connell, all took place in farmers' houses, where the number of people in attendance varied from as few as ten to as many as forty or fifty. At least one of the larger meetings drew participants from various parishes, but the basic organizational unit apparently consisted of eight-man committees at the parochial level. The principal Rockite leader in the area was "Captain" Edward Roche, a farmer, and Roche seems to have exercised a general authority over the lower committees. In one notable instance he used this authority to obtain supralocal cooperation. This was the attack in February 1822 on the Protestant church of Abbeyfeale. Some forty men participated. For two hours before the attack twenty of them talked and drank whiskey at the house of a farmer named Maurice Leahy, and after setting off for the church, they were joined by twenty additional men. At other meetings the Rockites under Roche's command decided to "pick off" as many Protestant gentlemen as they could, for it was urged that until they had destroyed Protestant influence, "they could never succeed in their plans." No fewer than three meetings were held to discuss the intended murder of a Kerry magistrate named Robert Twiss, but in the end this project was abandoned.[99]

Large Musters

To the extent that committees directed Rockite operations, they usually did not have to coordinate the movements of large numbers of activists. In the overwhelming majority of Rockite raids and attacks, fewer than two dozen men were involved, and in many of them the number did not exceed a dozen. But certain kinds of activity stood out as exceptions to this general pattern. In the wave of arms raids that preceded the short-lived insurrection at the start of 1822, large musters were commonplace. At Shanagolden in Limerick two to three hundred Rockites forced open every house in late September 1821 and carried away all the available arms;[100] large parties of similar size did essentially the same thing at Newmarket and Kanturk in northwest Cork in the following month.[101] At the end of October the police dispersed a body of three to four hundred insurgents who had assembled near Sixmilebridge in Clare with the intent of seizing the arms of the Rosscastle yeomanry.[102] Even in their attacks on gentry houses at this stage, the Rockites brought large numbers to bear: over one hundred against a Protestant parson at Bruree in Limerick; at least two hundred against a gentleman near Kenmare in Kerry; and as many as seven hundred against another Protestant clergyman near Kanturk in Cork.[103] In January 1822 arms-raiding contingents numbering in the hundreds were reported around Bantry and Skull in southwest Cork, near Bandon, and around Killarney and Castleisland in Kerry.[104] Little is known about how these large musters were brought together, but their very size, together with reports that some of them were operating far from their home base, implies that a considerable degree of supralocal organization was involved.

In later stages of the Rockite movement the mobilization of large bodies for particular enterprises was generally restricted to the rescue of livestock or crops seized (or about to be seized) for arrears of rent or tithe. In the late summer of 1822, for example, certain hard-pressed farmers in the Rosscarbery district of Cork collected "bands of ruffians, sometimes to the number of 200 or 300, with arms and horses," and carried off the whole of the harvest as well as their livestock to a place of safety.[105] Rescues of distrained cattle, or attempts at rescue, were reported near Ballyneen in west Cork (by over two hundred men), near Galbally in east Limerick (by three to four hundred peasants), and at Ballyclogh near Mallow, where a body of four to five hundred country people broke open a livestock pound and destroyed its walls.[106] In most such incidents there is no direct evidence of organized Rockite involvement, and this sort of activity was most often a communal enterprise carried out in open day. But it was certainly consonant with general Rockite aims, and in some cases at least it must have enjoyed strong Rockite encouragement and support. The difficulty and unwisdom of drawing a sharp distinction between such communal initiatives and specifically Rockite operations may be illustrated by two cases of the turning up of pastureland in the Charleville district of Cork in 1823. In the first instance the deed was done by a "mob" of about two hundred men in the daytime; in the second case, by a body of one hundred "Whiteboys" toiling at night.[107] Apart from the time of day, there was no essential difference between the two events.

Certain other Rockite enterprises also required coordinating the movements of large numbers of activists. The audacious burning of the police barracks at Churchtown at the end of January 1822 was said to have involved as many as seven hundred men, who could not have been made to act in unison without considerable planning and coordination by the leaders of numerous smaller Rockite groups.[108] Similarly, the venomous Rockite attack on the Palatine village of Glenosheen in Limerick in April 1823 was, from all appearances, a carefully planned affair. The raiders, more than one hundred strong, came on horseback over the mountains from County Cork and returned there after the attack. They were probably invited into the area by Limerick Rockites, and they would almost certainly have needed local intelligence and perhaps local guides to carry out their mission with as much success as they did.[109] Even in much more prosaic enterprises, sizeable bodies of Rockites from different localities sometimes concerted their actions. This happened in the attack on the proctor Daniel Sullivan, who was robbed of his tithe books, promissory notes, and tithe processes and had his haggard burned near Glenville, close to Cork city, in January 1823. The Rockites who participated in this operation consisted of coalescing parties that came from three directions—Mallow, Whitechurch, and Killeagh Cross. Wearing no disguises, they had joined forces at a public house on their way to the scene and had openly avowed their objective.[110] In County Tipperary a band of Rockites said to number eighty armed men was active early in 1822 in driving away Kerry spalpeens from farms near Clonmel, but they also combined forces with another Rockite band of about the same number in the district of Carrick-on-Suir who were struggling to reduce tithes and rents and to abolish taxes in that locality.[111]

How the Rockites of different districts might come to cooperate with each other can be illustrated by the well-documented case of one of their renowned commanders, David Nagle, who was betrayed and taken into custody in July 1823.[112] From statements that he made soon after his capture, the authorities were firmly convinced that Nagle was acquainted and connected with practically all the leaders of Rockite operations around Glanmire and in the North Liberties of Cork city. In one newspaper report he was called "the Captain Rock" of the North Liberties.[113] But Nagle was also busy elsewhere. He was a native of Annakisha in Clenor parish, located between Mallow and Doneraile, and no doubt had relatives and many friends there. Most of the information that he conveyed to his captors concerned Whiteboy activities in that locality. He gave details of specific incidents and named committeemen and other leaders in his native district.[114] He was also friendly with Patrick Kelleher, the Mallow publican's activist son.[115] The crime for which Nagle was eventually convicted, however, was an arms raid in May 1823 by thirty men on a gentleman's house near Watergrasshill, far from Annakisha and a good distance from Glanmire as well.[116] Clearly, Nagle was not bound to a single neighborhood; he circulated among several districts and was intimate with the principal Rockites in each. He was, in short, the kind of Rockite captain who was able to bring disparate groups of agrarian rebels into common action.[117]

Solicitation of Outside Help

The committees that often directed Rockite operations did not always rely on local activists to carry out their instructions. At times they employed other, indirect methods to accomplish their objectives, thus making it difficult for the authorities to trace the real source of local violence. Committees and other Rockite groups were known to hire the services of poor men to carry out specific missions, including the murder of their enemies. Thomas Hoskins, the hapless son of Lord Courtenay's hated land agent, fell victim to a small band of hired killers in July 1821. At the trial of four men for this murder in August 1822, Patrick Dillane, an accomplice turned approver, testified that he had been paid 50s. to slay Hoskins. One of the four prisoners declared in court that it was Dillane who had "committed the murder and brought them into it." The accused were all Irish speakers and all apparently poor.[118] The same technique was also used against James McMahon, employed by Alexander Hoskins as subagent until Hoskins's dismissal in late 1821 and pursued by the Rockites long afterward. In a report of October 1822 noting that two houses belonging to McMahon had just been burned down, the writer remarked that hired assassins had tried to waylay the former subagent several times within the past two years. He also observed that McMahon and his brothers had been stoned almost to death by a party of "nearly forty" men at the most recent fair of Nantinan in the Askeaton district, many miles from McMahon's residence.[119] Since stone throwing in faction fights at fairs was a quotidian occurrence, and since it was virtually impossible to determine who had struck any fatal blow, this was yet another way to avoid culpability for premeditated murder.

To offer public rewards for the killing of intended victims was a standard Rockite practice, and strangers were particularly welcomed as perpetrators. The Limerick gentleman Henry Harding of Harding Grove near Bruree was one of the many targets of this device. He had become a marked man by publicly tearing down threatening notices, defending his family's arms against seizure, and mounting patrols against local insurgents. For his pains the Rockites of his district advertised a reward of 40 guineas for his assassination, and Harding informed the viceroy early in 1822 that strangers from County Cork were now stalking him. His nerve broken, he sought protection and a job outside Limerick from the government.[120]

For numerous other purposes besides the hiring of killers, Rockite groups often perceived advantages in importing parties from outside a given locality to perform deeds of violence or intimidation within it. In countless instances it was the Rockites' intent to frighten victims through nocturnal domiciliary visits into complying with their mandates. Facial features and voices could of course be disguised, but there were numerous cases in which victims recognized local assailants and, what was worse, were prepared to identify them in court. If some of the attackers were killed, wounded, or captured in a raid, and they were local men, the authorities might well be able to pierce the veil of secrecy that the Rockites always tried to throw over their organization and activities. Almost invariably, the risks of detection were much less whenever activists from outside the neighborhood could be persuaded to undertake menacing domiciliary visits or to perform other duties in aid of local Rockites.

The authorities suspected that this technique was employed frequently, and there is at least some evidence that confirms their suspicions. In a case mentioned earlier, the fact that the Rockites who attacked the tithe proctor Daniel Sullivan near Glanmire in January 1823 did not bother to disguise themselves strongly suggests that they were strangers acting at the behest of Sullivan's local enemies.[121] Also undisguised were the body of men who in open day visited relatives of the murdered Shea family in a field at Cloran in Tipperary in November 1822, untackled their horses, broke their ploughs, and ordered them never to till that ground again.[122] In other cases magistrates or gentlemen reported incursions by strangers under circumstances that suggest local complicity. The well-armed Rockites who beat some of the wealthiest farmers in the Tarbert district of Kerry in January 1822 and ordered them to reduce the rents of their undertenants apparently came from the Shanagolden area of Limerick.[123] Arms raiders in north Cork in the same month were recognized as Limerick natives from their style of dress, but "some ruffians" from Duhallow barony were said to be aiding them.[124] "The alarms of last year," declared Lord Carbery from Castlefreke in July 1823, "and the few bad deeds that were done in this country [around Clonakilty in west Cork], were all effected by strangers from the northern districts" of the county. Although Lord Carbery absolved "our [local] labouring poor" from any complicity with the northern intruders, his complete exculpation hardly seems warranted.[125]

Rockite Disguises

But since by far the greater number of Rockite raids were carried out by activists engaged in their own parishes or neighboring ones, disguises were essential. Like other Irish agrarian rebels, the Rockites adopted a variety of disguises to conceal their identity when on mission. The most common disguise was the blackening of the face with some burnt cork or perhaps the cold ash of a turf or coal fire. Many accounts of Rockite domiciliary visits indicated that the intruders had blackened their faces.[126] Occasionally, a different face coloring was used. The members of the small band who killed Thomas Hoskins were said to have disguised themselves with "yellow paint," which was probably ochre.[127] Other Rockites covered rather than colored their faces. The headdress of the traditional costume of the Strawboys could be borrowed or adapted for this purpose. The party of Rockites who beat two keepers at Cunnigar in the Dungarvan district of Waterford in March 1823 "had their hats and necks tied round with straw and hay ropes."[128] In other cases straw might be stuffed into the collar of a shirt or coat so that it hid the face. More simply, Rockites might also gain concealment by muffling themselves up in their greatcoats and pulling their caps down over their faces.[129]

Symbolic meaning as well as the usual desire for disguise has been seen in the wearing of women's dress, which many Rockites were accustomed to adopt. In one of the earliest reports of this practice in the Rockite period, a band wearing women's clothes attacked a ploughman-caretaker in the Rathkeale district in late January or early February 1822, shot his plough horse, and ordered him to quit the district. They called themselves "Lady Rock and her suite."[130] In

To disguise themselves, the Rockites resorted to a variety of devices. One common disguise was the traditional costume of Strawboys—bands of young men who visited houses where relatives and friends had gathered to celebrate a wedding; the bride would dance with the Strawboys' captain, and the hosts or guests would give them drink and sometimes money. The handcrafted headdresses made of oat straw and worn by the two men with white linen frocks in this old photograph were part of the typical Strawboy costume, but the legend below the photo identified them as "agrarian terrorists." This disguise, which was closely associated with rural folk customs and symbolized elements of communal identity or solidarity, has been traced back to the Whiteboys of the late eighteenth century. (Unattributed photograph courtesy of the late Kevin Danaher)

another incident at about the same time near Wallstown in north Cork, five of the eight men who raided a gentleman's house (apparently seeking arms) were dressed as women.[131] As such incidents multiplied, it became common to refer to agrarian rebels so attired as "Lady Rocks."[132] One case that gained wide notoriety because the "Lady Rocks" were captured in the midst of their deeds involved seven "stout young men . . . , much in liquor,"[133] and dressed in white shirts, "with women's apparel underneath."[134] Another account indicated that they wore bonnets and veils.[135] After they had driven some cattle off a land grabber's farm and were about to burn the house of another tenant, they were collared by a Catholic priest and his congregation, who sallied forth from Rathcahill chapel near Newcastle in April 1822. Those captured were not local men (all of them were from the Glanduff area), and they seem to have been marginal to society. The "founder of the plot" was described as "a noted sheep-stealer," and another participant was an illegal distiller.[136] The use of female attire continued to be noted in some later cases of Rockite violence in both Limerick and Cork, including the horrific murder of the Franks family in September 1823. Indeed, besides the Lady Rocks, Whiteboyism in the early nineteenth century threw up the Lady Clares and Terry Alt's Mother as well as the Molly Maguires. As Natalie Zemon Davis has commented, Ireland provides "the most extensive example of disturbances led by men disguised as women" in a tradition dating back to the 1760s.[137]

Any thorough explanation of the adoption of female dress by the Rockites must go beyond the obvious fact that it provided a good disguise from materials ready to hand, if only because other alternatives were also available. The most convincing interpretation is that this practice is a notable example of the common inversion of the natural or traditional order in times of "misrule," though in the Irish case the misrule was not licensed or tolerated by the elite. In a reversal of customary social and sexual roles, women—or men dressed as women— were now "on top," paralleling the social reversal or inversion that exalted the lower-class Rockites as the real lawmakers and governors of the new agrarian order. It may also be the case that in donning female attire, the Rockites were seeking to appropriate such reputedly female attributes as passion, lack of restraint, and irrationality, thus adding to the terror felt by their victims.[138] The recurrence of female dress among successive Whiteboy movements lends added weight to these views stressing its symbolic dimensions.[139]

Mainly for practical reasons Rockite captains often assumed special dress that distinguished them from their followers. "The captains of those banditti," remarked a Protestant parson about the leaders of arms raids around Kanturk in October 1821, "are described as persons of respectable appearance and good address. . . ."[140] The large party that attacked the house of a Protestant clergyman at Bruree in November of that year "was commanded by Capt[ai]n Rock in person, dressed in blue, with sash and capt[ain]'s hat."[141] The leader of a Rockite band that burned the house of Thomas Foot's steward near Shanballymore in January 1823 wore a dark-colored surtout and carried not only a blunderbuss but also a sword in his belt.[142] And helping to confirm John Hickey's status as a Rockite captain in the Doneraile district were the feather and belt recognized as having been worn by the leader of many attacks in that locality and found by the authorities in his house at the time of his apprehension.[143] Such sartorial distinction made sound practical sense. In giving orders to, and demanding immediate obedience from, the occupants of houses that he and his men visited, the Rockite captain obviously wanted to establish his authority at once, and his special dress was a reliable means of achieving this recognition.

The Significance of Factions

In some localities factions furnished the Rockite movement with the leaders and organization that in many other districts came from Rockite committees and Ribbon lodges. The complex connections between long-standing faction groups (held together by kinship, neighborhood, or class) and episodic agrarian upheavals are not easy to unravel. Partly, this is because the inner workings of factions are not well documented. But it is also because there tended to be an inverse relationship between the frequency of faction fighting and the intensity of organized agrarian unrest. Thus, as the Rockite movement acquired strength in late 1821 and early 1822, faction fights at fairs and other public gatherings diminished sharply. Conversely, as the larger agrarian struggle lost momentum beginning early in 1824, violent clashes between feuding factions increased greatly.[144] But if major agrarian upheavals like that of the Rockites signaled a redirection of human energies away from feuds between and among communities, factions and their leaders did not necessarily fade into the background at such times. On the contrary, faction groups often became agrarian rebels, with the old structure furnishing the social cement of collective action in the new setting.

The evidence for this phenomenon is sometimes only circumstantial, as in the coincidence of areas of intense agrarian warfare with districts well known for their faction feuds. The tenantry of Viscount Courtenay's estate around Newcastle West, the flash point of what subsequently became the Rockite movement, had been organized into two antagonistic factions that periodically bloodied each other; the party of the indulgent former agent Edward Carte tried to exploit this situation in waging its struggle against his demanding successor Alexander Hoskins.[145] Portions of the Cork barony of Fermoy, the scene of so much incendiarism and other Rockite activity, were also notable for factionism. Two large factions in particular, known as the Luddites and the Sheehans, were said early in 1823 to clash at all the fairs in the vicinity of Kilworth, Glanworth, Kildorrery, and Mitchelstown. Though the reporting military officer claimed that their quarrels had nothing to do with the general agrarian disturbances of his district, he did admit that the encounters promoted "a lawless spirit."[146] In October of the same year this officer drew attention to the assemblage of at least fifteen hundred people (almost all men) near Glanworth village for the funeral of one of the Sheehans, "who was considered the head of a faction in that part of the country." The incident, he remarked, showed "how prone the peasantry of this country are to support a lawless leader and the facility with which large numbers can assemble."[147] The Holycross neighborhood near Cashel provides a final example of the coincidence of an area of factionism with one of intense Rockite activity. The village of Holycross was not only the venue in September 1822 of a vicious battle between two settled factions, the Mahers and the Hickeys.[148] It was also the scene in April 1823 of a remarkable display of sympathy and respect by a great number of "decent-looking farmers" and their wives for the remains of an executed felon named Dwyer, who had been a scourge of local land grabbers.[149]

Fortunately, in other cases the evidence for an interchange of personnel between factions and agrarian bands is far less ambiguous.

Soon after the capture of certain noted Rockites named Murphy near Banemore in west Limerick during the winter of 1821–22, their uncles made overtures to the authorities for clemency in return for the surrender of arms by "their faction." The Murphy family was said to possess "strong connections" in the district where they lived, and "unless there is some more respectable person at the head of this insurrection," the Murphy uncles could be "instrumental in persuading their friends and neighbours to abjure their present illegal pursuits."[150] County Cork furnishes a rather different illustration. When a substantial farmer named James Hill was slain while returning in October 1823 from a fair in Clonakilty, his murderers were immediately identified as a party of men led by James Hayes. The killing stemmed from the taking by Hill of a farm from which Hayes had been evicted. But even if the murder was in one sense the outcome of a bitter personal conflict, it was also something more. For Hayes was "the head of a quarrelsome faction that has often disturbed the neighbourhood of Rosscarbery."[151]

The elite was so removed from ordinary rural life that for them the springs of faction feuds usually remained a closed book. Occasionally, however, the full colors of a clash were revealed. In September 1824 there occurred at Abbeyfeale what appeared on the surface to be just another ordinary encounter between two factions that had been "often before at variance." But this particular fight, in which one man was mortally wounded, was closely related to a notable Rockite enterprise and its sequel. One faction consisted of the partisans of the influential Abbeyfeale postmaster-innkeeper David Leahy and his son Daniel, a Rockite leader, both of whom had been found guilty of destroying the former military barracks in the town with an intent to defraud the government. The fatally wounded combatant, John David Roche, who belonged to the other faction, had been one of the prosecutors of the Leahys at the spring assizes of 1824. This act of betrayal had led to the continual persecution of Roche and the rest of his family.[152] As he lay dying, Roche reflected bitterly that the faction fight and his own injury would never have happened if he "had been let alone, but . . . neither he nor his brothers nor any of the family could go to Mass in Abbeyfeale without being offended and attacked by the opposite party."[153] It appears almost certain that Rockite organization around Abbeyfeale was based on preexisting faction ties.

A final example is provided by the significant and well-documented case of the celebrated Rockite leader David Nagle, whose capital conviction at Cork in August 1823 so heartened the authorities because he was the reputed commander of "all the attacks which took place in the Liberties of this city during the recent disturbances."[154] Nagle was apparently deep in factionism. His apprehension was the result of "a schism" among Rockites themselves; one party, including five young farmers or perhaps sons of farmers, "determined to betray and deliver up" Nagle and two other leaders of the opposing party.[155] Those who handed Nagle over to the authorities were "Harties," an extended-family faction or clan of varying degrees of affinity, consisting of some thirty or forty males. Exactly what Nagle had done to offend his former comrades is unclear. But their explanation that he had demanded money "for his company," that is, for the Rockite cause, was dismissed as a "fictitious plea" by a gentleman well acquainted with the Harties. For the members of this faction had a history of agrarian dissidence. Some of them, for example, were related to "a very fraudulent and disaffected tenant" of Dr. Thomas Wood; they successfully protected the tenant from the penalties of distraint and eviction. Referring to this delinquent farmer and his factionist guardians, Dr. Wood grumbled in October 1823: "I lately consulted the chief constable of Cork, whose opinion is that twenty keepers could not prevent him [i.e., the defaulting tenant] from taking the produce of the present harvest off by night."[156] It seems reasonable to conclude that in the months before his capture the Rockite leader David Nagle, who was himself "the son of a man holding forty or fifty acres,"[157] drew at least some of his followers from the young farmers belonging to the Hartie faction.

Accent on Youth

Also contributing in a variety of ways to Rockite organization was the special attractiveness of the movement to young males. Paul Roberts has stressed the general youthfulness of both leaders and followers in early nineteenth-century Whiteboyism.[158] This phenomenon is hardly surprising. The rapidity of population growth after 1780 and especially during the great wartime economic boom of 1793–1813 gave to Ireland an extremely youthful population during the early nineteenth century. In 1841 fully three-quarters of the Irish male population were under the age of thirty-six, and slightly more than three-fifths were under twenty-six. The position twenty years earlier was, if anything, even more pronounced in favor of youth. In 1821 (when the age groupings in the census were presented differently), more than 70 percent of all the people in Munster were under thirty years of age.[159] In these special circumstances the heavy predominance of young males in agrarian secret societies was fully to be expected, even if other factors had not predisposed them to participate in rural rebellions.

The ages of Rockite activists are noted frequently in the sources, and they generally confirm what would be anticipated from the skewed age distribution of the population. The rich farmer's son Denis Egan, already mentioned in connection with the manuscript book of initiation oaths found in his possession near Roscrea in 1822, was only seventeen or eighteen years old, as was the journeyman weaver who penned a millenarian notice posted in Clonakilty, Co. Cork, in 1823.[160] Denis Barrett, executed at Buttevant in the same year for his participation in an arms raid, was only nineteen.[161] James Bridgeman, hanged in 1824 for the killing of Major Richard Going and implicated in two other murders, was just twenty-two years old.[162] Michael Donovan and Bartholomew Russell, executed in the same year for their role in the long-remembered burnings at the Palatine village of Glenosheen, were twenty-two and twenty-four, respectively. In fact, of the eleven men tried for this crime up to April 1824, "the eldest of them did not exceed twenty-seven years."[163] Timothy Sheehan, long sought as a principal in the slaying of the Franks family in 1823, was twenty-five.[164] And of the three men hanged at Cork in August 1824 for agrarian murders, one was thirty-five, but the other two were both about twenty-six.[165]

As striking as this youthful pattern is, it is nevertheless important not to exaggerate it. The proportion of Rockite activists beyond the late teens and early twenties appears to have been substantial. Of the seven men condemned to death for agrarian crimes at the special commission in Limerick city in December 1821, as many as six were married, and between them they had fathered a total of thirty children.[166] It is unlikely that more than one or two of them were under twenty-five. Similarly, a group of men sentenced to transportation at Mallow in 1823 for delivering a threatening notice in connection with a wage dispute on Lord Doneraile's demesne were "all poor labourers with large families."[167] The six prisoners convicted in 1824 of murdering the rich and widely hated Kilkenny farmer John Marum were described as "powerful middle-aged men."[168] And two County Limerick Rockites executed in 1823 for their role in an attack that led to a farmer's murder were said to be "in the prime of life," in contrast to so many other agrarian rebels whose youth arrested contemporary attention when they stood in the dock or mounted the scaffold.[169] By far the oldest Rockite to be executed was apparently Patrick Ivers, who by his own admission had joined in a raid on the house of a County Limerick gentleman. Aged fifty-eight when he was hanged in August 1823, Ivis must have been at least thirty years older than the majority—perhaps the great majority—of Rockite activists.[170]

The extraordinary youthfulness of the population had important implications for Rockite mobilization and organization. First of all, the malfunctioning of the economy worked special hardships on the young. In the postwar period entry into the labor and land markets was extremely difficult. Steady employment in both agriculture and industry was contracting; wage rates were falling; land values remained higher than the postwar fall in farm prices warranted; and the cost of potato ground was rising. For young men of the laboring and small-holding classes, these adverse trends, exacerbated by the depression of 1819–23, meant that their chances of establishing or decently maintaining a family were reduced substantially. A growing proportion of the population was being condemned to grinding poverty, and of those mired in this condition, the young bore a disproportionate share of the burden. This situation must have magnified the Rockites' appeal to males ranging in age from late adolescence to early adulthood. In addition, the existing patterns of sociability among young males provided social networks and habits of association that could easily be absorbed by agrarian secret societies. Membership in factions, on hurling teams, and in other forms of athletics, public-house drinking, and special youth activities at fairs and patterns—all brought young males into ongoing social relationships that could be and were exploited by agrarian combinations.

Four

Pastorini and Captain Rock

Millenarianism and Sectarianism

Anti-Protestant Animus Disputed

Part of the reason why students of Irish history have undervalued the important role of millenarianism and sectarianism in providing an organizational and ideological basis for the Rockite movement may be that the negative testimony of certain contemporaries has been given far too much weight. A considerable number of contemporaries belonging to the middle and upper classes strongly resisted the notion that anti-Protestant animus figured prominently in the mentality of Rockites or motivated their behavior. Thus a gentleman in the Churchtown district, reporting to Dublin Castle in April 1822 that nearly the whole of County Cork had once again become tranquil, declared flatly that "there was not at any time the slightest tincture of hostility to the government nor any of the spirit of religious bigotry pervading the mass of the community in this county."[1] Although few Protestant clergymen would have credited the Catholic farmers and laborers among whom they lived with quite so much religious tolerance, even some parsons derided the idea that Protestants were considered enemies by the Rockites. Using commonsensical logic, the Reverend John Orpen pointed out in March 1822 that many of his Catholic neighbors had "suffered severely" at Rockite hands, while he—a parson, a small tithe owner, the occupant of an unguarded

119

house, and a zealous magistrate—had been left unmolested.[2] Even the receipt of "a terrible denunciation of Capt[ai]n Rock's vengeance if I persevered in having divine service for the soldiers in Kanturk barracks" failed to shake Orpen's convictions on this point. After mentioning the threat in May 1823, he also observed that "outrages appear in this part of the country to be committed with equal severity on every class, rank, & description of persons. The poorest labourer who infringes on Capt[ain] Rock's laws is treated with as much vengeance as the Protestant gentleman who has the audacity to demand a portion of his rent."[3]

Catholics of superior social status often expressed similar views even more insistently. To them, the anti-Rockite pulpit oratory of priests and their edifying speeches at public executions gave the lie to charges that the agrarian rebels aimed at the destruction of Protestantism and the ascendancy of Catholicism. As the *Dublin Evening Post*, a Catholic paper whose columns were filled with news about the Rockite movement, boasted in February 1822, "There has been scarcely a barony meeting through the counties of Cork, Kerry, Limerick, or any district menaced with disturbance in which the thanks of the resident nobility and gentry have not been voted to the Catholic clergy for their incessant and laborious attention to their duties."[4] The same newspaper was quick to notice the loyal conduct of particular priests, such as Father Prendergast of Ballingarry, Co. Tipperary, who initiated a subscription to rent a barracks for troops in that village, and Father Rochford of Monagay, Co. Limerick, who rushed from his chapel along with some parishioners and assisted in the capture of eight Lady Rocks almost red-handed.[5] Individual priests who had been victimized by Rockites for showing hostility to the cause also received publicity.[6]

The sectarian prophecies of Protestant doom in 1825 associated with the name of Pastorini presented a special problem. Many Catholics of standing denied that the country people really believed in these predictions, and scouted the notion that such addled thinking played a significant part in the agrarian strife of the early 1820s. John Dunn, a politically prominent Catholic in Queen's County and a large landholder in the Ballinakill district, who claimed to be well acquainted with the views of the lower orders, was asked about the credit given to Pastorini's prophecies before a parliamentary committee in

June 1824. Replied Dunn tersely: "The people laugh at them they meet with; nothing beyond that. The pastoral address of the Roman Catholic bishop of the diocese lately disabused them of any idea that they might have had of their truth."[7] What Dunn was really telling the committee was that no person "of the least respectability" treated the prophecies with anything but ridicule;[8] this threw into question his original assertion.

Newspaper editors sympathetic to the cause of Catholic emancipation seized upon the killing of the "opulent farmer" John Marum in the highly disturbed Kilkenny barony of Galmoy as clear evidence that the Rockites were not motivated by sectarian prejudices. Marum was a brother of the Catholic bishop of Ossory.[9] The "immediate cause" of his murder in March 1824 was his taking of a property that lay under an order of ejectment, but subject to possible redemption. Two months later, when the land was in fact redeemed and restored to the possession of a "kind and indulgent" Protestant gentleman whose family had held it for generations, the tenants reportedly greeted his return with great rejoicing. Here, declared the *Leinster Journal*, was "another decisive proof" that the "vulture press" was lying when it asserted that "the unfortunate disturbances which occasionally occur in this country spring from a belief in Pastorini's prophecies and an union for the destruction of Protestants."[10]

What emerges clearly from the historical record is that politically conscious Catholics of the middle and upper classes found Pastorini's prophecies to be deeply embarrassing at a time when they were giving their wholehearted support to a campaign aimed at gaining admission to parliament for some of their wealthy coreligionists. This embarrassment was sometimes evident in the proceedings of the Catholic Association in Dublin. At one of its meetings in June 1824 a speaker maintained that the enemies of the Catholics were seeking to poison the minds of the English people by publishing a work entitled *Pastorini in Ulster* and then actively distributing it in England. There, he said, it was represented "as having emanated from the Catholic bishops and the association," though it contained "monstrous calumnies upon the Catholic religion."[11] At another meeting in the following December a speaker claimed to know why "the enemies of peace and good order" were reporting the presence of "numerous copies of Pastorini" among the peasantry of County Clare. "It was understood," he asserted unconvincingly, "that [Protestant] missionaries were circulating them about the kingdom in cheap pamphlets for sinister purposes."[12]

Daniel O'Connell himself told a parliamentary committee in March 1825:

> I think that no effect has been produced upon the lower orders of the Irish Catholics by the book called Pastorini's prophecies. That book was written by an English bishop . . . ; and it would not have been heard of in Ireland if it had not, as we understand, been spread very much by persons inimical to the Catholic claims. There was a considerable number of copies of it printed in Dublin, and certainly not printed with the assent of any Catholic. . . . As to the book itself, it is a book very likely to excite very little attention in Ireland; it is not written with virulence so as to gratify the vulgar; it is not written with taste or talent so as to please the educated; and it has been condemned by the highest authorities in the Catholic church from the moment it [was] issued.[13]

Concluded O'Connell, "What we call the Orange party have put forward Pastorini on all occasions."[14] A more remarkable series of half-truths has not often been uttered.

"Pastorini" and His Prophecies

Contrary to the implications of O'Connell's tendentious statement, the so-called prophecies of Pastorini were certainly distributed widely, and also widely believed. There had been numerous Irish editions of Bishop Charles Walmesley's *General history of the Christian church* since the first version appeared in Dublin as early as 1790 under the author's pen name of Signor Pastorini.[15] An edition labeled the fourth was published in Dublin in 1805; another, called the sixth, was printed in Belfast in 1816; and still another, also labeled the sixth, appeared in Cork in 1820.[16] The work was essentially a commentary on the Book of Revelation, for centuries a favorite of those who believed that the second coming was an event to be expected and rapturously welcomed in their own lifetimes. Walmesley's reading of the Apocalypse led him to assert, among other things, that God's wrath would be poured out to punish heretics about fifty years after 1771, thus initiating the sixth age of Christ's church, the last before the second coming.[17] This vague assertion led others to place a definite term to the existence of Irish Protestantism. By most the limit was fixed eventually at 1825, though some put it at 1821.[18]

Copies of Walmesley's ponderous and expensive pseudoscholarly tome were generally to be found in the libraries of curious Protestant gentlemen rather than in the cabins and farmhouses of Irish country people, schoolmasters apart.[19] Roadside inns, however, often kept copies of the book, and if the story recounted by the Earl of Rosse in April 1822 is at all representative, the work was much in demand: "Mr. Daly of Castle Daly . . . told me that, travelling some time ago, he was at the inn at Loughrea, and having asked for a book to read, the waiter brought him Pastorini, supposing from his name that he was a Catholic; and that the 8th chapter was so much dirtied by reading that it was scarcely legible."[20]

But it was in the form of small tracts and handbills that Pastorini's prophecies penetrated countryside and town over much of Ireland. Exactly when and where this began to happen is something of a mystery. No reference to an acquaintance with Pastorini at the popular level has been discovered for any year before 1817. The agrarian rebels who staged the Caravat movement from 1813 to 1816 were apparently altogether innocent of Pastorini, though not necessarily of millenarian predictions in general. One piece of evidence points to the calamitous famine and typhus epidemic of 1817 as the time when short, printed condensations of Pastorini's predictions reached certain rural districts of Munster. In his book *Captain Rock detected*, published in 1824, the Reverend Mortimer O'Sullivan, sometime Protestant curate and schoolmaster in Tipperary town, recalled:

> Anyone who has been a resident in the country parts of Ireland may have observed that about seven years since, a considerable change began to take place in the nature of the little penny tracts and ballads with which the itinerant pedlars were supplied. . . . The fact is certain that love songs and stories were no longer the principal wares of the book venders; and that stories of martyrs' deaths, and judgments and executions of obstinate heretics, and miracles performed in the true church were now in very general circulation. By one class of these productions the animosity of the faithful was whetted against the b—y Protestants; in another, they learned how heretics ought to be treated; and the miracles . . . sustained them by a hope that at last God would fight for them and exterminate their oppressors. At the same time prophecy, the constant resource of a depressed people, afforded them its consolations. Pastorini, circulated in various forms . . . , became a favourite study.[21]

Origin and Diffusion of Pastorini's Prophecies

Although north Munster was affected as early as 1817, the new stream of millenarian ideas apparently had its source not there but rather in the west midlands. In August of that year an informer told a magistrate that while traveling from Birr to Mohill, he had stopped at the house of Andrew Murray of Highstreet in northeast Galway. Murray reportedly produced a "prophecy book" from which he sought to demonstrate that "all Protestants were to be murdered within the year 1817"; he also contended that nine counties of Ireland (meaning apparently the Protestants in these counties) were to be consumed in a single night, and that this massacre would be followed by a six-day war in England.[22] A few other scraps of evidence, somewhat later in date, also point to the west midlands as the original fount of the Pastorini cult. In June 1819 a Catholic inspector of yarn at Mullingar, Co. Westmeath, speaking of "the different counties around this," called attention to the growing popular conviction that the Protestant church and state were to be overthrown "in the year of twenty-five."[23] Though this informant did not mention Pastorini by name, a retrospective comment by the Earl of Rosse in the spring of 1822 was quite specific in its attribution of the millenarian notions that had taken root by 1819 among the country people of King's County and east Galway. On a visit in April 1822 to a relative living twelve miles from Birr, Lord Rosse discovered that "Pastorini had been in circulation among the lowest orders in that neighbourhood for three years, and that [since] last year a small book, being an extract from it, has been in circulation among them."[24]

There was a striking congruity between many of the districts in which the cult of Pastorini first flourished and the areas in which agrarian rebels calling themselves Ribbonmen staged a brief revolt in late 1819 and early 1820. This movement began and was strongest in Roscommon and east Galway, but it also spread to adjacent parts of Mayo, Westmeath, King's County, and Clare.[25] It was specifically concerned with the grievances of conacre rents, laborers' wages, tithes, taxes, and priests' dues, especially the first three; the participants were drawn almost exclusively from the ranks of the poor having little or no land.[26] Of the oaths used by the Ribbonmen to bind people (often "whole villages") to their cause, some called for a general and "undefined obedience,"[27] but others were sectarian, at least to the point of demanding loyalty to the Roman Catholic church, that is, invoking solidarity among Catholics.[28] Moreover, the movement was stimulated by, and no doubt helped to spread, Pastorini's millennial vision. Writing of the Oranmore district near Galway town in February 1820, at the height of the disturbances, one observer declared, "It is prophesied all the world are to be of *one religion* in the year 1825 or 1826."[29] It was from the midlands and the midwest, then, that the Pastorini cult penetrated into the province of Munster.

As Pastorini had become practically a household name throughout much of the south by 1822 or 1823, the diffusion of his prophecies there seems to have occurred suddenly and rapidly in the few years before and after 1820.[30] The extraordinary spread of Pastorini's millennial speculations over the area affected by the Rockite movement is revealed in the explicit language of threatening notices and in the often-alarmed remarks of those in authority. The active magistrate Andrew Batwell of Bowen's Court in northeast Cork received a letter from "John Rock" in June 1822 telling him plainly that if he should manage to escape death now, "the year 1825 will surely come, when the fate of all your sort will be decided, so [be] advised once for all by your friend."[31] In the following October a "respectable" Catholic involved in the collection of tithes near Kanturk received a notice that said: "You are not to join these bloody Protestants, for you know there [*sic*] time is expired. I will slatar [i.e., slaughter] them like dogs." Subjoined to this notice when it was submitted to Dublin Castle was a brief and almost nonchalant explanation of its first sentence: "The writer here alludes to a notion now universal in the County Cork among the lower classes that a certain prophecy in the Book of Rev[elation] commenced being fulfilled in the year 1821 and is to end in the universal dissolution of all Protestant establishments in 1825."[32] And early in 1823 a military officer stationed in north Cork attributed the recent increase in outrages there to a prophecy "greatly believed in by the credulous and uninformed peasant," namely, that in that year "there will be general war by land and sea."[33]

In north Kerry and in Clare as well, Protestants lived in dread because of the credence given by Catholic peasants to Pastorini's hopeful vision. A Tarbert gentleman reported in January 1822 that Pastorini's book (probably an extract) was "in private circulation amongst the lower orders of the Roman Catholics, who, according to its prophetical doctrine, expect to have the Prodestants [*sic*] exterminated out of this kingdom before the year 1825."[34] Intoxication with Pastorini may well explain two sectarian incidents that took place in the following month. In one case a band of Rockites "visited every Protestant family of the lower class in this neighbourhood [i.e., Kilflyn parish near Tralee] and, after treating them in the most unmerciful manner, ordered that they should immediately leave the parish," or else they would forfeit their lives.[35] In the other incident Rockites in the Ardfert district apparently threatened to destroy the ancient diocesan cathedral, for in a memorial to the government entreating the speedy dispatch of troops to the town, the Protestant inhabitants of Ardfert protested, "The venerable cathedral built above 800 years [ago] must, it seems, be demolished to satiate their fury. . . ."[36]

Unlike certain parts of the south, such as north Kerry, where Orangeism in Protestant circles helps to account for the currency of Pastorini,[37] Clare was free at this time from overbearing Protestantism. And yet its Catholic inhabitants had also become conversant with the new brand of millenarianism. Few elite observers were better qualified than Major George Warburton to say how much credence was accorded to the "emissaries" who circulated Pastorini's wisdom there. A high-ranking police officer, Warburton's duties brought him into frequent contact with all classes of the peasantry, and he was permanently resident in Clare from mid-1816 to early 1824. His view was categorical: "Upon my word, generally speaking, I do not think there is an individual of the lower orders that is not aware of this prophecy, and that is not very strongly impressed with the belief of them."[38] In fact, the country people of Clare spoke openly of Pastorini "in their common conversation, at their work and on other occasions when they are assembled together." One gardener was said to have told the gentleman who employed him that "there would be bloody work" in 1825, and that "he had the word of God for it."[39]

Though millenarian beliefs were also reportedly widespread in Tipperary, Kilkenny, and King's County,[40] it was in Limerick that the prophecies attributed to Pastorini enjoyed perhaps their greatest vogue during the early 1820s. This was the view of the able and exceptionally well-informed police officer Major Richard Willcocks; it was shared by the king's counsel Francis Blackburne, who was appointed in April 1823 to administer the Insurrection Act in parts of

Clare and the whole of Limerick. After spending a year there, Black-burne became convinced that millenarian literature could hardly be more pervasive. As he informed a parliamentary committee in May 1824, "When I went to Limerick, I made it my business . . . of inspecting every notice and every publication dispersed through the country and connected with seditious subjects, from which distinct evidence of what was operating on the minds of the people might be collected; and I do not think that in a single instance has one of those papers been produced to me that there was not a distinct allusion to the prophecies of Pastorini and the year 1825."[41]

Blackburne might have spared himself all the trouble of inspection there; for over a year before his arrival, Dublin Castle had been receiving accounts of Pastorini's renown, especially in west Limerick. In February 1822, a few days after the burning of the Protestant church of Killeedy near Newcastle, a local Protestant exclaimed, "This Pastorini is doing a great deal of mischief in the country."[42] Even more emphatic was the anonymous report submitted by a resident of Adare. Pastorini, he declared grandly, "has done more towards the subversion of the British empire than Bonaparte with all his legions." This writer had conversed with some of his Catholic neighbors about Pastorini's millennial certitude and was struck by their replies. One frieze-coated countryman told him: "You know, sir . . . , it must be so; the prophecys must be fulfilled; Pastorini was no liar." Another asserted, "I w[oul]d believe Pastorini sooner than the bible; the date of the Protes[tan]ts is out." A third man, said to be sympathetic to the Whiteboys, when taxed with the tragic fate of the rebels in 1798, retorted: "The devil mend the scoundrels; they began 25 years before it was the will of God they should."[43] Significantly, Limerick was the only Catholic diocese, apart from Kildare and Leighlin, in which the bishop issued a pastoral letter controverting the doctrines associated with the name of Pastorini. In this address, read from the altars of all the chapels in his diocese on Saint Patrick's Day, 1822, Bishop Charles Touhy declared, "I have reason to know that even under the pretext of religion the poor credulous people are led astray by these wicked advisers, telling them prophecies of wonderful events to happen in the years 22, 23, and 24."[44]

Other Prophecies in Vogue

Other predictions besides those attributed to Pastorini were also current in parts of Munster during the early 1820s. Among them were certain ancient prophecies generally ascribed to Colum Cille, versions of which had been widespread among Catholics, especially in Ulster and Connacht, during the revolutionary upheavals of the 1790s.[45] A variant of the old story of the "black militia," usually a reference to bloodthirsty Orangemen, was circulated in the Borrisokane district of north Tipperary early in 1824. Its propagator "declar'd that the time had arriv'd when they [i.e., the Catholics] would have possession of this island, as an ancient prophecy foretold that they should conquer when the black people came to Ireland." Asked who these black people were, he replied that "the police were the people so designated, as the dark colour of their clothing sufficiently proved."[46] This particular adaptation of a hoary anti-Protestant revelation was in fact a matter of common notoriety. The prominent Catholic barrister and politician Richard Lalor Sheil was asked whether the low proportion of his coreligionists serving in the police, about which he had complained to a parliamentary committee in March 1825, could not be explained by the fact that "they looked upon the police as the realization of some old prophecy about a black militia which was to arise at this period and to kill all the Roman Catholics."[47]

Another prediction, reminiscent of the millennial hopes of foreign invasion current among United Irishmen and Defenders in the late 1790s, was heard again in the Adare district of Limerick early in 1822:

> The children of God are to be defeated with great loss in the 2 first battles; before the 3[r]d, on their march, they are to meet with a white horse [a symbol of the Jacobite cause]. They are then to come victorious and to chase the locusts to the north. A man with 4 thumbs is to hold the horses of 4 kings or great generals at the battle, which is to be fought at Singland near Lim[eric]k. . . . 2 brothers, McDonnells by name, of Scotch extraction, are to come to the relief of Ireland with their fleets; one is to land at the north, the other at the west. When all is over, the Spaniards are to settle a frame of government. Ireland was to be in after times in the possession of the Spaniards.[48]

Similar prophecies circulated in County Cork. Whiteboys in the Liscarroll district were said to believe that "though they may be beaten back twice or thrice by the army, yet when having turned, they will meet a man in the way who comes from heaven, and [he] having sprinkled them with holy water, they will turn round again and the army will fly before them, though they [the Whiteboys] should hold up but straws."[49] The landowner and magistrate Justin McCarthy related to a parliamentary committee in June 1824 a striking instance of this type of millennial wisdom in his district, south of Mallow. From one of his demesne workers McCarthy had obtained a copy of a prediction attributed to Colum Cille, according to which "the Irish power" under General O'Donnell was to annihilate the English under the Marquis of Abercorn. Then, he said: "Two days before I received the order of the House of Commons, I was passing through one of my own fields and heard some persons talking. I found my own mason reading to my gardener. They appeared exceedingly confused; the man thrust the manuscript into his pocket and told me he was reading some accounts to the gardener. The gardener is a very faithful, confidential fellow; I ascertained from him that it was the very same prophecy that I had a copy of before."[50]

Of course, not every report of the propagation of millenarian ideas is to be accepted unreservedly. There were hoaxes, such as the enterprise of the man armed with prophecies who went into the parish of Kilmoe near Skull in southwest Cork, "pretending that he was an agent of the Whiteboys, to stimulate the people to rebellion." The gentlemen who apprehended him must have been embarrassed to discover that he was a government spy.[51] There were also millenarian notices issued in the name of Captain Rock that expressed nothing more than the personal feelings of the individuals who composed them. For example, an announcement echoing Pastorini and posted in the town of Clonakilty in May 1823 declared menacingly: "This is to let the Protestants know there is a scourge over them from the almighty God. I am the man . . . that will give it. Jack Rock."[52] The authorities were mildly relieved to find that the writer of this notice was only an adolescent journeyman weaver of Clonakilty who had "not long left school." Still, there was the worrisome implication that boys learned at "those seminaries something else besides their grammar." Moreover, even if this particular notice was simply the product of one misguided youth's infatuation with the writings of Pastorini, it was also true, as Lord Carbery lamented, that "many, very many" Catholics of the lower classes were equally taken with them, if only by indirect contact. "That diabolical book," he railed, "has poisoned the minds of thousands and tens of thousands of those wretched beings."[53] Clearly, the government spy in the parish of Kilmoe knew what he was doing when he equipped himself with prophecies.

Circulating Millenarian Ideas

Millenarian ideas in printed form circulated in a variety of ways. The most common vehicle was the penny handbill, or broadsheet. In the counties of Limerick and Clare handbills containing brief extracts from the speculations in Pastorini's tome, along with "observations trying to inculcate the probability of their being fulfilled," were distributed by the thousands.[54] One of these broadsheets, submitted to parliamentary committees in May 1824 by the police inspector Major George Warburton, bore the name of Thomas Conolly, printer and stationer of Camden Street, Dublin. It purported to explain why the persecution of Catholics, begun by the Protestant reformers in 1525, could last only until 1825, when the three-hundred-year "reign of the locusts" would be succeeded by the restoration to unchallenged supremacy of "the great, the only organized body of Christians, the Catholic church." "What a happiness," it declared, "if during this short remaining interval some part of them [i.e., the Protestants] would submit to see their errors and the great mischief that has been done to the church by their revolt against it! It is full time to lay down all animosities against their ancient mother, think of a reconciliation, and ask to be received again into her bosom."[55] The Dubliner Thomas Conolly had apparently acquired a considerable reputation in the south for such penny productions, for Warburton acknowledged that the expression "a Conolly" was used familiarly by peasants to denote his millenarian broadsheets.[56]

Other printers at Limerick and Cork furnished similar wares to chapmen who attended fairs and markets in the country.[57] It was hardly unknown for the chapmen to become lecturers. Peddlers "with light parcels or packs" were said early in 1824 to have recently dispersed "a large supply" of millenarian literature in the Borrisokane district of Tipperary. According to the Protestant clergyman who reported their doings to Dublin Castle, these peddlers also took it upon themselves to "dilate much on the pretended prophecys of Pastorini . . . and others."[58]

For those with more expensive tastes and an appreciation for visual demonstration, chapmen also offered what Warburton described as "a sort of scale or map and something in the way of the stream of time; it purported to be a history of the progress of the church of Christ, and the various schisms which have taken place were marked in different columns. I recollect particularly the Protestant was supposed to end [in] 1825."[59]

Protestant proselytizers also helped unwittingly to spread or deepen millenarian consciousness among the deluded papists whom they hoped to win to reformed evangelical Christianity. In the belief that Irish Catholics were deliberately kept ignorant of the Holy Scriptures by their priests, the proselytizing organizations flooded the country with Bibles and testaments in both the English and Irish languages. During the year ending in February 1823 the London Hibernian Society distributed gratis 13,000 testaments and Bibles, "making a total of 92,600 since the institution of the society" in 1806.[60] A second body, the Hibernian Bible Society, established in Dublin in the same year, was even more enterprising in this direction. By 1823 some 218,000 testaments and 104,000 Bibles had been issued from its depository.[61] Other organizations added to the inundation, though in smaller quantities. This outpouring was not always as unwelcome in the Catholic countryside as one might imagine, for it helped to make possible animated fireside discussions of Saint John's Apocalypse. Recalling his days as a Protestant curate and schoolmaster in Tipperary town, the Reverend Mortimer O'Sullivan remarked that "those who could not procure the book [i.e., Pastorini's *General history*], but who were instructed in the principles of it, often gave the members of the Bible society hope of making converts from the readiness with which they received the testament, of which they scarcely read any part but the Revelations. I remember when a house at my gate was a place of rendezvous for numbers to meet together and read."[62] The Catholic clergy were right after all. As they repeatedly insisted, it was indeed positively dangerous to allow the laity to construe the divine word without theological guidance.

To the circulation of millenarian ideas in oral form schoolmasters made a major contribution. The important role played by schoolmasters in the Rockite movement reprised their part in earlier popular political and agrarian agitations.[63] The salience of schoolmasters in the

Rockite movement will be discussed at some length in chapter 5.[64] Here we are concerned with their role in the dissemination of millenarian prophecies. Schoolmasters were invariably well versed in the prophecies attributed to Colum Cille and Pastorini.[65] Witness the case of Charles McCarthy Considine, a Clare native who had come to take charge of a school at Glenville near Cork city. Toward the end of 1824 he found himself standing before a bench of magistrates convened at Fermoy under the Insurrection Act. Besides being faced with the usual charges (absence from his dwelling during the hours of curfew, idle and disorderly conduct), he was accused of delivering threatening notices concerning the purchase of gunpowder. But what must have been most damning in the eyes of the magistrates was the testimony of one laborer against him: "Prisoner talked of Pastorini and said that next year would be a year of war. He talked of many other things and said that the price of labour was too low." Considine was sentenced to be transported for seven years.[66]

Some schoolmasters traversed the countryside, stopping for short periods of teaching in different places. When Edward Connors, "an aged itinerant schoolmaster and land surveyor," died suddenly in a cabin near Kanturk in June 1824, certain millenarian documents were found on his person.[67] One of these papers suggests that for Connors the mundane tasks of surveying land or teaching the three Rs were often subsidiary to the exercise of a higher calling:

> I am commissioned to unfold unto thee in part the secrets of futurity; pay attention to my relation and profit by my discourse. Know then, O Catholics, that the present captivity which will shortly determinate is but a just punishment for the cruelty of many of thy actions and the insatiable ambition of thy restless disposition. Yet it is likewise a preparatory measure towards the future exaltation and agrandizement [sic], for those shall be recalled from banishment [apparently, the Irish soldiers, or "wild geese," serving in foreign armies] and mayest again reign, and thy posterity in future inherit greatness. Yet before that comes to pass, many strange occurrences shall take place and wonderful events be accomplished.[68]

Connors was thought to have come to the Kanturk district from the village of Banemore near Newcastle West in Limerick—a trail that immediately raised the suspicion that he had been in contact with some of the earliest Rockite leaders.[69] Indeed, what eventually became the

Rockite movement had originated on Viscount Courtenay's estate around Newcastle.[70]

Connors's circuit was narrower than that of the traditional "prophecy man," the bounds of whose travels, according to William Carleton, "were those of the kingdom itself." A "rare character" in the Irish countryside by the early 1840s, when Carleton sketched him censoriously, the prophecy man had been a common figure forty or even twenty years before. Unlike those itinerant schoolmasters and peddlers for whom the circulation and explanation of millennial wisdom constituted only part of their work, the prophecy man was a person "who solely devotes himself to an anxious observation of those political occurrences which mark the signs of the times as they bear upon the future, the principal business of whose life it is to associate them with his own prophetic theories." He usually practiced his art at night, in the house of some countryman with whom he happened to be staying, when after the labor of the day the neighbors would crowd around the fireside to listen respectfully to his address, or his harangue, as it must have frequently seemed. The prophecy man enjoyed perhaps his greatest prestige during the French wars, when talk of the liberation of Ireland through foreign invasion and the revered name of Bonaparte were constantly on his lips. Bonaparte's downfall and especially his death somewhat diminished the stature of prophecy men. So did the accession of George IV to the English throne, contrary to a well-known Irish prediction. But with the sudden rage around 1820 for Pastorini's speculations, prophecy men obtained a further extension of credit until the failure of Protestantism to vanish in 1825 "nearly overturned the system and routed the whole prophetic host," at least as Carleton viewed the matter sixteen years later.[71]

Another important vehicle for the transmission of millennial hopes, and an outlet for sectarian feeling as well, was the broadside ballad or the somewhat more elaborate and expensive garland (a small booklet of eight pages, containing from two to six songs). Both of these were sold for a halfpenny or a penny by the professional ballad singers and the peddlers who thronged to fairs, race meetings, and other large public gatherings.[72] Many broadside ballads of the early nineteenth century proclaimed political prophecies.[73] Some songs during the years of the Rockite movement endorsed in vulgar terms the abstruse speculations ascribed to Pastorini, as illustrated by the following Limerick ballad of 1821:

> Now the year 21 is drawing by degrees, In the year 22 the locusts will weep, But in the year 23 we'll begin to reap. Good people, take courage, don't perish in fright, For notes will be of nothing in the year 25; As I am O'Healy, we'll daily drink beer.[74]

Other ballads sung in the early 1820s, though they belonged to the older millenarian and sectarian tradition of the 1790s, accurately reflected the current popular mood. Two men were tried under the Insurrection Act at Fermoy in January 1824 for disorderly conduct and violation of curfew at a public house in nearby Kildorrery on the day after Christmas. One of them had sung on that occasion an old Irish ballad learned twenty years before, whose words were more or less these: "Ye sons, arise and take up arms, and join if only three remain; they shall have from Dingle to Carrick-on-Suir as a property to be happy with; tear Orangemen and Protestants to pieces; dethrone King George, and his own soldiers shall rebel and join the rebels." Since the prisoners had been drunk, the court was lenient and simply admonished them, but it would perhaps be too charitable to conclude that what came from the throat did not also come from the heart.[75] The same might be said of the five men whom a police constable overheard singing mightily in a public house at Rathkeale, Co. Limerick, in November 1824. Among their "seditious songs" was one whose verses concluded with the well-known words, "We will wade knee-deep in Orange blood and fight for liberty."[76]

Multiple Explanations

Why was there such a great upsurge of millenarianism and anti-Protestant feeling at this particular time—the early 1820s? Though numerous factors were at work, the social consequences of the economic crisis, and especially the dread of famine after the harvest failure of 1821, played a crucial role. Some elite contemporaries took this view. When reporting the circulation in north Kerry early in 1822 of Pastorini's supposed prediction of the demise of Protestantism by 1825, William Lindsay of Tarbert remarked, "Numbers and myself suppose it is encouraged by distress, the principal cause of the disturbances."[77] But it is unnecessary to rely on such evidence alone. The connection between millenarian hopes of salvation and fear of imminent famine was clearly expressed in the exhortation and lament of a manifesto posted in County Limerick in January 1822:

> Hearken unto me, ye men of Ireland, and hear my voice! Arise, O! Melesians [sic]; the day of our deliverance is coming, when the trumpet beats to arms. . . . Your eyes shall have no pity on the breed of Luther, for he had no pity on us. Behold, the day of the Lord cometh, cruel both with wrath and fierce anger, to lay the land desolate. . . . Their children also shall ye dash to pieces. Before their eyes their houses shall be spoiled, and their wives ravished. . . . You see misery upon misery is come upon us. Seldom a day passes that our cattle is not canted by the roguery and oppression of our landlords. We have nothing left but to die valiantly or starve. We are the most miserable people on the face of the earth, while the sons of perdition are satisfying their appetite, luxury, and gluttony abroad. . . . Oh! when their belly is full and warm, what a feeling they have for the poor. . . . Lament and mourn, ye hereticks, for the day of your destruction is come.[78]

A widespread readiness to die if necessary in order to bring the millennium closer was reportedly manifest among the thousands who engaged in the abortive insurrection of late January 1822 in northwest Cork. Many of the "prophetic gentry," said a sarcastic Protestant clergyman at the beginning of February, "who marched to the attack on Newmarket on the 25th ult[im]o previously went to the priests to confess & get absolution, & what they called to prepare for death."[79] To the depth of the Rockites' millenarian convictions—"a spirit both political & religious"—another Protestant attributed the failure of government repression to destroy the movement. "The cause the wretched people have espoused appears in their eyes so praiseworthy," he declared in May 1823, "that neither the prospect of utter death [n]or banishment seems to have the desired effect."[80] The most unshakable millenarians were apparently the laborers, whose economic and social condition was least amenable to improvement. A rise in agricultural prices was thought to have brought about by early 1824 a much more law-abiding disposition among farmers in the Pallaskenry district of Limerick. But since prophecies of Protestant doom were still rife, laborers could "only be kept down by force" until the haven of 1826 was reached.[81]

Also contributing to the upsurge in millenarianism and sectarian hatred during the early 1820s were the extraordinary intensity of the tithe grievance and the provocative activities of Protestant proselytizers. Conflict over tithes in the past had not always led to a heavy accumulation of sectarian spleen. Indeed, the issue had come to sharply divide upper-class Protestants. On the one side were the nobility and gentry, who generally embraced the solution of a commutation of tithes; on the other side were the bishops and many of the clergy, who rejected such a solution.[82] But now the popular demand, or at least the ultimate popular aim, was often the complete abolition of tithes, and not simply their reduction, as heretofore.[83] With this shift the mixing of economic and religious objections to the system became much more prevalent. Thus the millenarian manifesto from County Limerick quoted at length above, which heralded the impending massacre of Protestant "hereticks," also declared: "Now, brethren and fathers, it is your duty not to pay tythes or taxes, for before full six months is over, I will be at the head of an army of two hundred and forty thousand men."[84] A placard appearing in March 1822 in the Tramore district of Waterford called upon all Catholics aged sixteen to fifty to muster in support of Captain Rock; they were to pay no tithes and "to fight for their religion."[85] A similar blending of motives was evident in the menacing notice sent to the Reverend Patrick Kennedy, the resident parson of Templetouhy parish near Thurles. This notice, which threatened Kennedy with murder and his daughters with rape, proclaimed, "You are doubly hateful to us for having been a tithe proctor" as well as for being a clergyman of the wrong church.[86] It is also difficult to believe that those Rockites who burned or set fire to a half dozen Protestant churches in Limerick, Kerry, and Cork were giving expression only to their detestation of tithes or church rates.[87] Tithes had been an important issue in every major agrarian rebellion in the south since 1760, but (apart from the events surrounding 1798) the burning of churches was a novelty introduced by the Rockites. As the king's counsel George Bennett put it in May 1824, the Catholic lower orders had come to believe both that "they are oppressed by those who profess the religion of the established church" and that "there is likely to be soon a great change on that subject."[88]

Compared with tithes, Protestant proselytism was less significant in stimulating popular Catholic sectarianism, but not much less. Indeed, the Reverend Mortimer O'Sullivan seemed inclined to believe that the striking change he had observed around 1817 or 1818 in the subject matter of the penny tracts and ballads sold in the countryside was a direct response to "the attempts at proselytism" by Protestant "missionaries."[89] The proselytizers certainly became more aggressive after 1815 in circulating the Scriptures, in distributing anti-Catholic literature, and in establishing schools aimed at the children of the Catholic poor. The Religious Book and Tract Society for Ireland claimed in 1823 to have issued over 1,160,000 tracts and 86,000 books since 1819 alone.[90] Some of its productions were simply "dropped by people travelling in gigs and picked up on the road by countrymen."[91] Formal schooling, however, was a far more serious and contentious affair. The controversies that raged after 1819 at the national level about schools under Protestant auspices, their management and funding, and the use of the Scriptures within them were in part a reflection and in part a cause of strife at the local level.[92]

In Munster and Connacht Catholic clerical opposition was focused primarily on the schools founded by the Baptist Society, which proselytized openly, and by the London Hibernian Society, whose inspectors required that children in its schools recite the Scriptures from memory. Repeated denunciations of these missionary academies from the altars of Catholic chapels, accompanied by orders that Catholic children be withdrawn from them, served to sharpen sectarian consciousness.[93] In February 1822 a cottage school sponsored by the London Hibernian Society on the outskirts of Tipperary town was burned to the ground.[94] Though this incident remained a singular one in the south until later in the decade, there were other manifestations of the grassroots Catholic reaction raised by Protestant evangelism, such as the stoning of a Methodist preacher in the streets of Kilrush, Co. Clare,[95] and the circulation of ballads scorning the efforts of the Protestant missionary organizations. One songster was jailed in Limerick city in November 1824 "for singing and vending in the streets an impious halfpenny ballad." His broadside, declared a deeply offended observer, "purported to turn into ridicule not only the Bible but all Bible societies"; it also sought "to convey the idea that aid from foreign powers would ultimately crush the system in these countries."[96] With the beginning of the long round of so-called Bible discussions in late 1824, religious controversialism and sectarian animosity entered upon a new phase.[97] As an editorial writer in the *Dublin Evening Post* colorfully if extravagantly expressed it in November of that year, "The country is converting into an immense theological arena, and polemics are likely for some time to usurp the place of politics, if indeed they will not be mixed together and form that delightful compound, the fermentation of which produced such remarkable consequences in the reign of Charles I."[98]

Protestantism and the Forces of Repression

Well before religious controversialists and the crusade for Catholic emancipation raised the heat, sectarian fires were already burning brightly for another reason: the popular identification of Protestantism with the various forces—yeomanry, police, and regular troops—used to suppress the Rockite movement. The association was strongest in the case of the yeomanry. This was an overwhelmingly Protestant armed force that had originated in the counterrevolutionary frenzy of the mid-1790s. Two-thirds of the 31,000 yeomen in Ireland in 1821 were concentrated in Ulster; in all of Munster they numbered fewer than 2,400.[99] But their unpopularity in the south was far out of proportion to their strength. Just how unpopular they were is suggested by the frantic plea of one north Kerry gentleman to the head of his brigade in October 1821: "For God's sake, my dear major, get arms and ammunition for the corp[s] and don't let a body of loyal men be murdered who have brought down on themselves the vengeance of the insurgents by their exertions to save their country."[100] The past history of yeomen as superloyalists and their local knowledge of the workings of agrarian combinations made their Catholic neighbors especially hostile toward them. Anticipating government reluctance to comply with his request for the raising of a new yeomanry corps in the Borrisokane district of Tipperary, the Reverend Ralph Stoney declared in March 1822: "I know that the yeomanry are unpopular, but so is everything that is constitutional and loyal. They saved the nation in the year 1798, and the disaffected dread them more than they do the regulars because they are well acquainted with their characters and know their haunts."[101] Dread there may have been, but there was also loathing. For the yeomanry were used in ways that were bound to inflame popular feeling, as, for example, in their seizure of illicit stills and their distraining of tenants for nonpayment of rent or tithes, not to mention their searches for illegal arms and for suspected agrarian rebels. When opposed in the performance of their duties, yeomen were more likely than better disciplined soldiers to fire upon menacing or stone-throwing crowds, as illustrated by the lethal clashes with country people near Ballybunnion, Co. Kerry, in late 1821 and near Clonakilty, Co. Cork, in early 1822.[102] Rarely were yeomen punished for employing excessive force or for engaging in unprovoked violence, a circumstance that naturally intensified the hatred of the local population. An outrage by members of the Nenagh corps in January 1822 put the Catholic inhabitants of that Tipperary town into a frenzy. "If you were to hear the language of them," exclaimed one gentleman, "as they now think they can get no law [i.e., no justice]; what a town to live in!"[103]

The Protestant triumphalism of many yeomanry corps was yet another source of Catholic irritation. There was no exact parallel in the south to the annual yeomanry and Orange festival on 21 June at Enniscorthy, Co. Wexford, where a symbolic tree of liberty was burned to celebrate the retaking of that town from the rebels in 1798.[104] But yeomen of course joined lustily in the Orange celebrations of the anniversary of the Boyne that occurred in some southern towns and that occasionally led to serious confrontations. In July 1821 the Bandon Orangemen, marching in procession, fired a fieldpiece loaded with stones at an opposing party of Catholics, killing one woman. A few days later at the fair of nearby Timoleague, a crowd of Catholics in reprisal murdered a Protestant identified as "one of the Bandon Orangemen."[105] Apart altogether from annual processions, almost any occasion was good enough for the playing of loyalist tunes by the yeomanry bands of Munster and Leinster: "The Boyne water" by one corps at Freshford in Kilkenny, "The Protestant boys" and "Croppies lie down" by others at Dunmanway and Bandon in Cork.[106] It was perhaps significant that John Walsh, an Adare yeoman murdered by a party of Rockites in October 1821, had served as the fifer of that corps.[107]

The most dramatic example of a Rockite onslaught against yeomen largely motivated by bitter sectarian feeling was the burning of Glenosheen near Kilmallock by a band of over one hundred men in April 1823. Like its sister village of Ballyorgan, Glenosheen was inhabited almost exclusively by Palatines, whose ancestors from the German Palatinate on the Rhine had settled on advantageous terms as tenants in various parts of County Limerick at the beginning of the eighteenth century.[108] Because of their Protestant heritage and the favor shown them by local landowners, the Palatines identified themselves wholeheartedly with the Protestant Ascendancy, and adult males in their communities usually belonged to units of yeomanry or volunteers.[109] In the attack on Glenosheen the Rockites succeeded in destroying three houses and set fire to four others. A local proprietor noted that "the object of the rebels (as they frequently declar'd) was to get possession of the arms & murder the 'Protestant Palatine devils.'"[110] The Palatine schoolmaster, whose dwelling the Rockites set alight, heard them shout that "he was a Protestant devil and that not one should be spared."[111] Palatines in other districts of the county also suffered at Rockite hands. Two who belonged to the Adare yeomanry corps were assaulted and robbed of their arms early in 1824; the house of a third was burned near Shanagolden; and a fourth was murdered after taking a farm on Lord Courtenay's estate around Newcastle West from which tenants had been dispossessed.[112] Along with other Protestants, Palatines emigrated from County Limerick in considerable numbers during the 1820s, and among the reasons commonly assigned for their departure was the sectarian animosity of the surrounding Catholic population.[113]

Though the police establishment was not as overwhelmingly Protestant in its composition as the yeomanry, it was by no means free in Catholic eyes from the stench of Orangeism. Even the new constabulary force established in 1822 leaned heavily to the side of the Ascendancy. In a total body of slightly more than 2,700, there were fewer than 850 Catholics. In the eight counties comprising the province of Leinster, Protestants outnumbered Catholics in the new force by nearly three to one. In the province of Munster the religious balance was less lopsided, with Catholic constables actually constituting a large majority in Kerry and a small majority in Waterford. On the other hand, in Limerick and Tipperary the proportions were fairly close to those prevailing in Leinster.[114]

Yet even if Catholic constables had been far more numerous, the nature of police work in many areas of the south during the early 1820s was such that the establishment could not have avoided the enormous popular hostility that it aroused. Nor can it be claimed that the opprobrium attaching to the police stemmed mostly from Catholic sectarianism. The ways in which the police earned hatred among the country people were numerous. Their anti-Rockite activities altogether apart, members of the constabulary interfered with many traditional folkways. They dispersed country dancers, scattered mourners at wakes, and drove bowlers from the roads. They seized the fishing nets of boatmen along the river Suir to prevent their taking salmon fry, killed the dogs of peasants by the roadside, and cleared public houses of their patrons, sometimes at bayonet point.[115] After sunset, dances, wakes, and public-house drinking were violations of the Insurrection Act in those counties or baronies declared subject to its provisions, and the police were routinely the enforcers of this draconian law.

To these sources of irritation must be added numerous instances of the unauthorized use of firearms or bayonets by members of the constabulary.[116] After a series of such cases the *Dublin Evening Post* was provoked to remark in February 1824, "There is scarcely a day in which we do not hear of some outrage committed by these preservers of the peace; and, we confess, we are too often disappointed in not hearing of their punishment."[117] On top of everything else, the police were the main auxiliaries of the regular army in the repression of the Rockite movement, seeking informers, setting spies, arresting suspects, coaching witnesses for the crown, and testifying for the prosecution themselves, both under the Insurrection Act and under the ordinary criminal law. It therefore comes as little surprise to find, for example, that after some constables intervened to stop a faction fight at the fair of Abbeyfeale in July 1824 and tried to take two ringleaders off to the guardroom, they were stoned by members of both factions, who "called out several times that 'Captain Rock was alive, murder the Peelers.'"[118]

Yet even if it must be acknowledged that the nonsectarian sources of popular hatred for the police were many, it should also be recognized that the Catholic country people often viewed them through sectarian eyes. A County Limerick notice of January 1822 screamed its vengeful message: "Those dog teachers, the police, have no mercy on them, for it is no sin to kill hereticks. It was never so easy to masacre [*sic*] them as now. There shall be one conflagration made of them from sea to sea."[119] This effusion is perhaps to be explained by the well-attested fact that the Orange Order had penetrated into the police establishment in County Limerick. Major Richard Willcocks admitted to a parliamentary committee in May 1824 that soon after his appointment in late 1821 as head of the constabulary in that county, he had found it necessary to suppress the Orange system among his men.[120] Blame for this state of affairs attached to Willcocks's predecessor, Major Richard Going, whose sorry fate owed much to his presumed Orange proclivities.[121]

As previously mentioned, Going's role in the formation of an Orange lodge within his police establishment had become widely known, and this connection with Orangeism was the source of considerable hostility toward the County Limerick force among Catholics of all classes.[122] Intensifying the antagonism toward Going and his constables was their involvement in the blood-soaked tithe affray of mid-August 1821 near Askeaton. In this encounter a badly outnumbered but much better armed party of policemen inflicted a maddening defeat on perhaps over two hundred Whiteboys who had attacked the house of a hated tithe proctor. The deaths of three of the attackers and the capture of several others instilled a popular desire for vengeance.[123] But the police had their own resentments. The killing of a subconstable in this tithe affray was one of the impulses for the harsh posture adopted by Going in its immediate aftermath. He arranged for the dead insurgents to be hurriedly buried in quicklime at Rathkeale and thus prevented the families and friends of the deceased from giving their bodies the customary funerals to which the Irish country people of the time attached so much cultural and religious importance.[124] A damaging rumor was also spread that one of the Whiteboys badly wounded by Going's police had been interred while still alive. Even though there was apparently no truth to this claim, attendant events gave the tale extraordinary credibility among the common people. This rumor was accompanied by a second inflammatory report—that the dead Whiteboys had been consigned to a "croppy-hole" in the same manner as the defeated rebels of 1798, with their captured comrades in this instance having been forced to dig the graves.[125] Again, this claim seems to have been unfounded, but the official refusal to surrender the bodies of the dead insurgents to their families went far to embed the widely circulated story firmly in folk belief. Within two months of the Askeaton tithe affray—an episode quickly freighted with bitter sectarian meanings in the Catholic popular mind—Going himself was killed. As noted in chapter 1, he was murdered in mid-October 1821 as he traveled from Limerick city to Rathkeale.[126] The news of this killing sparked exuberant popular celebration in west Limerick and beyond.[127] To "beware the fate of Going" became thereafter a stock reference in Rockite threatening notices in numerous parts of the south.

Limerick was not the only county affected by the Rockite movement in which members of the constabulary were associated in the
public mind with Orangeism. In May 1823, when the yeomanry and
Orangemen of Bandon buried a comrade in the local churchyard, they
paraded from the Orange lodge through the streets of that town, carrying Orange flags, playing loyalist tunes, and sporting the full insignia
of the Order. Along with them the police quartered at Bandon
marched openly.[128] At the assizes in County Cork, according to Daniel
O'Connell, police magistrates were accustomed to interfere in the selection of petty juries in criminal cases by "attending particular prosecutions, setting aside the Catholic jurors, and endeavouring to pick
out, as much as possible, a Protestant jury, some of them Orangemen."[129] And a resident of Freshford, Co. Kilkenny, reported at the
end of 1823 that "with few exceptions" the police stationed there were
Orangemen; they belonged to a lodge established in the town, whose
head was widely hated for having killed a local farmer, and one of
whose patrons was a local magistrate.[130] Senior police officials were
certainly conscious of the need to purge the constabulary of the dis-
crediting odor of Orangeism, and serious steps were taken in the early
1820s to rid the force of Orange lodges and other flagrant displays of
anti-Catholic bigotry among some of its members.[131] But as demonstrated by the widespread currency of the prophecy identifying the
police with the "black militia," it was no easy task to eradicate the
sectarian image of the constabulary implanted in the minds of many
Catholic country people in the south.

Like the police, the regular army also gave a fillip to Catholic sectarianism. The great military buildup in Munster and adjacent parts
of Leinster that began toward the end of 1821 was largely achieved by
transferring to the Irish establishment regiments from Britain and some
of its overseas possessions. After modest reinforcement from abroad
the army in Ireland reached a level of sixteen thousand men by mid-
1822; it was further increased to about twenty-one thousand during the
following year.[132] English or Scottish cavalry units like the Rifle Brigade
were especially prominent in military operations directed against the
mobile and stealthy followers of Captain Rock. Catholics in disturbed
districts were reminded of the religious allegiance of many of these new
troops when the soldiers marched on Sundays to what had once been
sparsely attended services in the local Protestant church. The filling of

Protestant churches with soldiers sometimes elicited a Catholic sectarian response, as indicated, for example, by Rockite notices ordering Protestant clergymen to discontinue the holding of divine service for troops stationed in their parishes.[133]

A common Rockite view of the army (and of the police as well) was to see it as a force designed especially for the protection of Protestants and Orangemen. In a notice posted on the door of a Catholic chapel at Killaloe, Co. Clare, in November 1821, Captain Rock offered to "promote & reward" anyone who would shoot eight named persons, apparently all Protestants. On the back of this notice were scribbled the words, "With this + we will conquer any reg[imen]t [of] cavalry; it is fixed to begin to end in the year 25."[134] Even plainer was the attitude toward soldiers and police expressed in another notice that a body of Rockites left behind for a Protestant resident of Pallaskenry, Co. Limerick, after burning his out-offices in June 1823: "This is to show all hereticks the way that we will serve them, and to show them that they are not to put their trust in the Peelers or army, for when they are asleep, we will be awake, and let all the bloody Orangemen of Pallas know that we are preparing for them against the first of July, and let them not boast of the forty thousand, for this is the death they may all expect."[135] What helped to solidify such views of the army was the fact that temporary barracks for troops, especially those stationed in dispersed rural outposts, were usually provided by Protestant gentlemen or clergymen. To those Rockites who burned buildings that served or were capable of serving as makeshift barracks, it must have seemed more than a coincidence that their owners were so often members of the established church.[136]

Rape as Punishment

One notorious incident that underscored the depth of Rockite hostility toward the army, and that seems to have had sectarian overtones, was the raping of certain women whose husbands belonged to the 1st Rifle Brigade. This episode occurred near Kildorrery, Co. Cork, in mid-February 1822. A group of women traveling with children in three cars or wagons in advance of their soldier husbands was stopped by a band of about forty men after passing through Kildorrery on their way to new quarters. According to the earliest reports of the incident, seven or nine of the dozen women in the group were taken from the

cars and violated in what amounted to a series of gang rapes.[137] From
the account of a later trial of three men accused of participating in
this brutal assault, it appears that fewer women than originally reported may have been abused, and that in each of the rapes only two
or three men were involved, though others assisted in holding down
the victims. Clearly, the assault was intended as an act of Rockite
vengeance against the absent husbands of these women. (Rape as punishment was not unknown in other contexts at this time.[138]) In fact,
one of the assailants boasted to his victim that "he would let the Ri-
flemen know that it was Captain Rock's men" who did the deed.[139]
Whether these self-proclaimed Rockites were also prompted by sectarian animus is not absolutely certain, but a number of circumstances
raise the possibility of sectarian motivation. First, the Rockites reportedly threatened to kill every English or Scottish woman in the
group. Second, a woman who was ravished testified that one of her
attackers had specifically asked if there was an officer's or a sergeant's
wife riding on her car, a question perhaps designed to isolate a Protestant victim.[140] The talk in some millenarian notices of ravishing the
wives of heretics could also be taken to lend support to this conjecture.
But whatever was in the minds of these Rockites, there can be no
doubt that the great influx of Protestant soldiers into the disturbed
southern counties during the early 1820s thickened the strands of millenarianism and sectarianism in popular Catholic culture.

Daniel O'Connell and Millenarianism

One final aspect of the intriguing subject of Catholic millenarianism—
an epilogue to its role in the Rockite agitation—should be treated
here. As the Rockite movement faded almost everywhere in the south
during the second half of 1824, the O'Connellite campaign for Catholic
emancipation began to take firm hold at the grassroots. Despite certain qualifications that should be made, the newfound position of
priests on the popular side being the most important, there is much to
be said for the view that energies previously channeled into a great
clandestine movement, mostly agrarian in its objectives and quite violent in its tactics, now came to be funneled into an open, nonviolent
crusade for political rights. What is significant for our purposes is that
millenarianism continued to be a factor of some weight in the new political agitation as it had been among the Rockites. Catholic priests

and bishops might dismiss millennial predictions as a heap of "unintelligible farrago,"[141] while middle- and upper-class supporters of the Catholic Association might attribute their persistence to the schemes of Protestant proselytizers and Orangemen.[142] Nonetheless, the transition of 1824 was considerably facilitated by this cultural continuity between the two movements.

Most, though not all, of the evidence for this interpretation comes from members of the Protestant elite or from those in positions of authority. Some reported that a general rising in the near future was being planned or at least feverishly discussed; others observed that Pastorini's prophecy about the destruction of heresy in 1825, which had subsided for a time, had once again been revived.[143] No doubt a certain discount must be made for the nervousness or paranoia of numerous Protestants as the year of their long-heralded doom finally approached.[144] But to dismiss this evidence out of hand as the product of overwrought imaginations would be a serious mistake. The villains of the piece were Daniel O'Connell and the priests, especially the latter, who were so assiduously directing the collection of the so-called Catholic rent, the already-famous penny a month.[145] The gravamen of the charges against the clergy, apart from their alleged use of spiritual terrors to enforce payment, was that their personal assistance in the collection of the rent, accompanied by "inflammatory harangues" and a decided reticence about how the money would be employed, contributed to a climate thick with talk of Pastorini and revolution.[146]

Nightmares of Massacre

As the year 1824 neared its close in December, the level of paranoia reached new heights of shrillness. The Protestant proprietor John Palliser, who owned estates in Tipperary and Waterford, was obsessed with the idea that the Catholic lower classes fully intended to fulfill the most sanguinary prophecies in circulation. He unburdened himself of his extreme anxiety in a letter written to the undersecretary William Gregory on 2 December:

> The Protestants are in imminent danger; we stand upon a smothered volcano, and there is a systematic vilification of everything Protestant, and if very decided steps are not taken for our safety, our blood will flow profusely in the approaching prophetic year of blood, and it is from the priests' altar (as I am informed) that the mandate for slaughter is

THE CATHOLIC ASSOCIATION OR PADDY_ coming it STRONG _ !!.

After its founding in 1823 under Daniel O'Connell, the Catholic Association quickly became the vehicle for mass expression of Catholic demands for political, religious, and social change, including the admission of well-to-do Catholics like the famous barrister O'Connell to parliament. Shouts of "O'Connell forever" and "O'Connell in parliament" come lustily from the crowd in this unflattering 1825 cartoon, and banners proclaim the primary goal of political emancipation for "six millions of people" who have now embraced enthusiastically the "Catholic Rent," or penny a month, paid by "associate members" of the association. On the speaker's platform the pro-emancipationist British politician George Canning offers his support. But intermixed with the political demands are others venting sectarian and agrarian grievances: "Down with the Orangemen," "Religion" (translatable as "Up with Catholicism"), and "Success to Captain Rock." More than a hint of revolutionary inclinations can be detected in the cartoonist's insertion of many pikes among the crowd, in the appearance of a man with a blunderbuss marked "Redress," and in the cry, "And the good old days of King James." Like the Rockite movement, the emancipation crusade that followed it was suffused with sectarianism. The disapproving cartoonist himself (Robert Cruikshank, elder brother of the famous satirical artist George Cruikshank) hangs the bulging bag of the "Catholic Rent" from a bishop's crozier and drapes a French cap of liberty from the top of the cross carried by a monk. (Nicholas K. Robinson Collection of Caricature, held within the Early Printed Books Department of Trinity College Library Dublin) to issue, by which the Protestants will, by this simultaneous movement, be cut off defenceless at their places of worship. . . . Nothing but a formidable body of troops distributed thro' the south and west of Ireland can save us from the intended bloody rush, where so many valuable lives must fall a sacrifice.[147]

The Protestant rector Frederick Blood of Corrofin in Clare was equally overwrought and claimed on 21 December that he expected an imminent general rising, especially if the government took any step to suppress the Catholic Association: "The people are looking forward to this [i.e., to revolution], & their minds are prepared for the event by indulgences, prophecies, [and] inflammatory speeches & publications." Blood demanded to know whether it was possible that the government could "be blind to the progress rebellion has made," and whether Protestants would "be left unprotected in the power of a sanguinary rabble whose avowed object is to root out heresy."[148] Yet another alarmed Protestant called on 28 December for the immediate dispatch of troops to a series of villages in his part of County Waterford. He was particularly concerned about the danger to his coreligionists in Dungarvan, "where in one night some thousands might rise and massacre every Protestant in it."[149]

These Protestant fears of massacre were mirrored by similar anxieties on the Catholic side. "The most absurd stories are . . . circulated among the lower classes of Catholics" in County Kilkenny, reported the military officer Lieutenant-Colonel Martin Lindsay in June 1823; he specifically mentioned tales "that the Orangemen are plotting to masacre [sic] them on the 1st or 12th [of July]."[150] The story was the same elsewhere. From another observer came the news that "reports have been industriously circulated in the county of Limerick as well as in the county of Clare that the Orangemen are to rise and massacre the Catholics. . . ."[151] As some Catholics must have noted, Protestants had their own prophecies, such as the one predicting the destruction of the "Church of Rome," recounted in a book entitled *The crisis*, dedicated to a bishop of the established church.[152] This was hardly the first time in Irish history that members of the two opposed sectarian camps came to believe that massacre was on the minds and even in the active plans of their antagonists.

These mirror images of impending massacre led to some remarkably widespread defensive or protective behavior on both sides as the clock

ticked down to the opening of the year 1825. The Reverend Henry
Cooke, moderator of the Presbyterian Synod of Ulster and a minister at
Killyleagh, Co. Down, reported to a parliamentary committee in 1825
that the circulation of Pastorini's prophecies had caused people to sit
up at night "last Christmas" and in the season immediately before it.[153]
Bishop James Magaurin of the Catholic diocese of Ardagh told the same
committee that he had heard that Protestants in other parts of Ireland
besides his own had either avoided going to church on Christmas Day
1824 or had gone only with an armed force for their protection.[154] The
landowner J.S. Rochfort, who resided on the Carlow–Queen's County
border, recalled in 1825 that "last Christmas the Protestants near me all
sat up the whole of Christmas eve; they thought they were all to be murdered that night. . . ."[155] And the Reverend William Phelan, an Anglican
clergyman resident in County Tyrone, remarked that even though he
had not found Pastorini's prophecies to be in circulation in his part of
Ireland, it was still the case that both Catholics and Protestants had sat
up at night during the week before and the week after Christmas because of their fears of massacre by the other side.[156] Similar reports were
actually much less frequent in the heartlands of the Rockite movement
at the very end of 1824, though it seems highly likely that the already
mentioned Protestant expressions of paranoia in the southern counties
led to similar kinds of exaggerated watchfulness there too.[157]

O'Connell appeared to be the chief political beneficiary of all this
ferment. With some reason Protestants believed that in spite of his
repeated condemnations of Ribbonism and Whiteboyism, the peasantry flocked to the standard of his Catholic Association believing
that he was about to lead an armed revolt.[158] That some country people saw O'Connell in this light is not really surprising. After all, no
other contemporary barrister defended half as many accused Whiteboys, or saved half as many from the gibbet. No Rockite brief was so
atrocious that O'Connell refused to take it, not even the case of those
charged with raping the women of the Rifle Brigade.[159] Far more of the
peasantry envisioned O'Connell as the deliverer who would fulfill the
prediction of the destruction of Protestantism, or who would at least
accomplish its abasement in Ireland. One of the best-known songs
composed by the blind County Galway poet Raftery, "The Catholic
rent," places O'Connell in exactly this position—the scatterer of
heretics, as revealed by Pastorini.[160]

Five

Social Composition and Leadership

In its social composition the Rockite movement of 1821–24 displayed a remarkable degree of complexity and diversity. Admittedly, like most other regional agrarian movements in Ireland during the late eighteenth and early nineteenth centuries, it attracted support from the poor of the towns as well as the poor of the countryside.[1] But unlike the so-called Caravat-Shanavest conflict of 1806–11, the Rockite upheaval cannot be portrayed simply as a confrontation between substantial farmers and other middle-class elements on the one hand and, on the other, laborers, cottiers, small farmers, and artisans.[2] This is not to say that class warfare below the level of the landed elite was absent from the Rockite movement. On the contrary, conflict between wealthy farmers and the various elements of the poor in country and town was a prominent and recurrent feature of the upheaval. But the great frequency with which middling or even large farmers, and especially their sons, appeared in the ranks of these agrarian rebels, often in association with laborers and artisans, shows conclusively that the fault lines within the society were not only horizontal but vertical too. In other words, there were certain grievances that promoted social cohesion across class lines below the elite, just as there were other grievances that gave rise to acute conflict between one nonelite social group and another.

The relative weights of social cohesion and disunity were largely a function of economic conditions. In addition, such factors as the social structure of a given district or the character of relations between landlords and tenants, or between parsons and their parishioners, also affected the balance. The Caravat-Shanavest rivalry of 1806–11 had occurred within the economic context of a war-induced inflation of prices and rents that worked special injury on the poor (both rural and urban). But the Rockite movement of 1821–24 took place within an economic environment of intense depression for all sectors of agriculture, dairying partly excepted.[3] This crisis was of such magnitude as to bring insolvency or at least heavy losses to once prosperous farmers and graziers as well as to shopkeepers, publicans, and certain groups of artisans whose fortunes were closely tied to those of their rural clientele. In short, a burning sense of economic injustice was not confined now to the great multitudes of poor folk at the base of society. Instead, the conviction of intolerable oppression had moved up the social scale to embrace that relatively affluent minority of farmers, at once landholders and landlords, who stood far above the mere peasantry. It is this circumstance, more than any other, that accounts for the diverse social composition of the Rockite movement.

Social Cohesion or Class Conflict

Elite observers and members of the Catholic clergy were often keenly aware of both the existence of social cohesion on some issues among the followers of Captain Rock and the class-based antagonisms that could disrupt solidarity when other concerns pressed their way to the forefront. On the one hand, elite observers believed that there was widespread support in 1822 and 1823 among farmers of all kinds for the Rockite aims of reducing rents and lowering or abolishing tithes. Writing from Kilkenny town in April 1822, the well-known Anglican evangelical preacher Reverend Peter Roe seems to have distilled a general current of thought among the better-off farmers: "People in various parts of this county who a few months ago appeared shocked at the accounts from Limerick now say that although they have nothing to do with the business, yet they begin to think that Captain Rock will do a great deal of good in bringing down rents and tythes."[4] A military officer had acquired similar views in relation to all of south Kilkenny down to the Tipperary border, a broad swath of territory that "is now considered as almost under Whiteboy dominion." A Catholic informant had told this officer that "not only the peasantry but the more substantial farmers . . . are, with hardly any exception, partizans [sic] of the Whiteboy cause," though they "do not dare openly to avow it, and are waiting only until events enable them to do so."[5] And speaking of the area stretching from Knocktopher and Callan to the Waterford border, a third elite commentator declared that many persons "who hold land do not regard unfavourably the proceedings of those midnight legislators. Indeed, in this barony [Knocktopher] & I believe in many parts of Kells and Iverk all dominion over landed property is in abeyance, as no person can enforce payment or dare to turn out a defaulting tenant. . . ."[6]

On the other hand, the aggrieved and discontented were by no means united on all issues. Colonel Sir Hugh Gough, a commander of troops stationed in north Cork, claimed in January 1824 that when Rockite parties visited farmers' houses in the Buttevant district to levy contributions for aiding the legal defense of prisoners, such financial demands might have a highly welcome result—"by opposing the two classes, the instigators [farmers] and the actors [laborers], and by making the former practically feel that although they may easily work on the ignorant, the bigoted, and the oppressed, they themselves eventually become the sufferers."[7] As if to make Gough's point, there was the case of Daniel McDuharty, a better-off farmer at Cloghlea in Lower Bunratty barony in Clare, who was "robbed" in 1824 of £8 by a party of eight men whom McDuharty recognized. He offered not to prosecute these Rockites if his money was returned, but his offer was spurned and his house was burned down.[8] More generally, an elite commentator in County Kilkenny remarked in June 1824: "The middling farmers who have any substance have expressed their approbation of the [Insurrection] Act. These farmers, it seems, were heavily taxed by the bands of ruffians to whom I have alluded. . . ."[9]

In the same vein Colonel Gough believed that the systematic burning of farmers' tithe corn would eventually have the "good effect" of setting at odds "the two parties both deeply implicated in the disturbances, namely, the farmer and laborer." Why? Because the burden of paying heavy compensation for damages resulting from incendiary crimes would fall mainly on the farmer, in the form of much higher county taxation. As matters then stood, Gough declared in November 1823, the farmer in general encouraged the laborer to commit every act of outrage which he thinks will tend to forward his own selfish views, namely, the abolition of tithe and the giving [of] an excuse for the non-payment of rent, while at the same time, as an occupier of ground, he [i.e., the farmer] is the most unfeeling tyrant, dispossessing and distraining for houses and potatoe ground, for which he never thinks of paying his landlord rent.[10]

The Behavior of Farmers

The distinction commonly drawn between farmers as instigators or abettors and cottiers and laborers as the real perpetrators of acts of violence or intimidation did possess some validity. It derived much of its credibility from certain irrefutable facts: these were the alacrity with which farmers met demands for the provision of arms, horses, and money to the Rockite cause; the farmers' ritualistic denials of any knowledge as to the identity of agrarian rebels; and their frequent readiness to supply alibis, or testimonials of good character, in trials of peasants accused of Whiteboy crimes. Such conduct was often attributed, and rightly so in many instances, to the "thralldom" of farmers, that is, to their lively fears of reprisal if they comported themselves differently.[11] But in ascribing motives, observers frequently refused to believe that so uniform and widespread a pattern of behavior could be the result of fear and force alone. Much of the reprehensible conduct of farmers, it was held, could be explained by their genuine sympathy for certain Rockite aims and, in numerous cases, by their outright complicity in Rockite operations. In this manner Colonel Gough explained his frustrating inability to apprehend a gang of robbers (allegedly composed of laborers) who had been stealing from farmers around Buttevant early in 1824. Although the victimized farmers knew the identity of their assailants, according to Gough, fears about their personal safety or trepidation about the disclosure of their "own former insurrectionary acts" prompted such farmers to remain silent. Not only did they decline to prosecute those who had attacked them but they frequently refused even to admit to Gough or other officials that they had been the targets of Rockite intimidation or violence.[12]

Lending support to this viewpoint is an unusual abundance of evidence, some of it general in nature and not capable of conclusive proof, but much of it specific and having the solid ring of truth. Writing of the Charleville district in north Cork months before the open insurrection of January 1822, one gentleman claimed to be "well informed that the sons and dependents of opulent farmers . . . are concerned in the horrid and now systematic plan of assassination and rebellion in this part of Ireland."[13] As if to substantiate this claim, the parish priest of Newmarket reported at the end of January that he had received applications from many respectable farmers whose sons had been "forced" to take the Rockite oath; he asked a police official if the government would be prepared to extend mercy to "these deluded peasants" if arms were surrendered.[14] Lord Ennismore pronounced a negative verdict in February 1822 about the better-off farmers of the Fermoy district, an area that "is for the most part tillage land, produces a vast deal of wheat, & may be called the granary of the county of Cork." Though the district therefore abounded with "a description of farmers who might be supposed to be well conducted," the situation was otherwise, according to Lord Ennismore: "I believe [that] most of the farmers, many of whom are in good circumstances, are not only privy to the business & well wishers to the cause, but some of them are deeply engaged in it."[15]

Similar reports of the active complicity of the "better orders of farmers" came from north Kerry early in 1822; these accounts depicted farmers as the initiators of disturbances there several months previously.[16] Furthermore, in February, when ten men were taken into custody as Rockite leaders in the Listowel district, they were described as "holders of large farms" and "in easy circumstances."[17] Of the three men arrested in the Tarbert district in the following June for the murder of the process server for tithes John Conway, one named John Madden was described as "a man in good circumstances." The Tarbert magistrate Robert Leslie, Jr., understood that Madden had been considered "of good character," but he nevertheless insisted, "I don't entertain *any doubt whatsoever* of the truth of the informations I have taken against him."[18] And a former Rockite named Duggan who swore an information in mid-1822 covering the whole prior period of the disturbances in the Tralee district pointed his accusing finger in the same direction: Duggan "is a young man of good character, and the party he states to be concerned are responsible farmers."[19] Yet by the early months of 1822 the social character of unrest in north Kerry, as well as in parts of Cork, Limerick, and Tipperary, was changing significantly. The first strong signs of farmer-laborer conflict appeared in the form

Bands of Rockites and their leaders needed to plan concerted actions by meeting in secret. Agrarian rebels had long resorted to public houses and (in good weather) to remote places in the open air (hillsides and mountain redoubts) for such purposes. But especially when they enjoyed the support or participation of local farmers, they used barns as forums for discussing their grievances and deciding how to redress them. Besides providing the venues for discussion, cooperating farmers must have frequently furnished the assembled Rockites with whiskey as well. This sketch of "Ribbonmen" drinking whiskey at a meeting in a barn on the Marquis of Bath's estate in County Monaghan in 1851 faithfully represents the surroundings in which many Rockite enterprises were planned, for Captain Rock had many supporters among farmers. (William Steuart Trench, *Realities of Irish life* [London, 1868], facing p. 190)

of demands by laborers for lower conacre rents, increased wages, and reduced food prices.[20] The attacks on farmers that accompanied these demands were secretly welcomed by the authorities.[21] The plundering of farmers by the poor, one Kerry gentleman declared delightedly in March, was "a punishment they richly deserve for the encouragement they gave them at the commencement."[22]

The great bulk of those who engaged in the abortive insurrection of late January in northwest Cork were apparently cottiers and laborers ("poor wretches without character or subsistence," in the words of one report) who temporarily abandoned their lowly, ill-paid pursuits for mountain warfare.[23] These were the social groups driven to desperation by the fear of famine, and raised to a frenzy by the intoxicating millenarian prophecies associated with the revered name of Pastorini. To judge from the accounts of the trials of scores of insurgents who had been captured during the insurrection, large farmers played only a marginal role in that furious but futile enterprise. Elite observers were quite surprised when the "comfortable farmer" Thomas Goggin of Clondrohid parish was capitally convicted at the special commission in Cork city in February 1822; Goggin had reportedly always paid his rents "to the very farthing." The local Protestant rector remarked that with only one or two exceptions "he was not aware of any person so decent as Goggin having been engaged in these outrages. . . ."[24] Yet small farmers and their sons reportedly appeared among the insurrectionary hosts, "very many from compulsion" no doubt, but also "many from inclination."[25]

Examples of Activist Farmers

If few substantial farmers or their sons participated in the uprising, they repeatedly took part in many other types of Rockite activity throughout the duration of the movement. In some instances the charges against them were relatively minor. In one case heard at Mallow in November 1822, five men were found guilty of assaulting keepers guarding property distrained for arrears of rent. One of the five openly declared in court that he would rather be transported than disgrace his family by becoming an informer. "I have detailed this fact," remarked a police officer, "to shew the attachment of these people, who are comfortable farmers, to a bad cause. . . ."[26] Similarly, two "respectable farmers" (Thomas Burke and Thomas Cullen) went in October 1823 to assist a third farmer in removing corn from his land near Cashel in Tipperary and in putting this grain out of the reach of his landlord so as to extract an abatement of rent and other concessions.[27] Other substantial farmers or their sons willingly enlisted recruits to the Rockite cause. Denis Egan and Darby Guinan were arrested near Roscrea in April 1822 while carrying a Catholic prayer book and "a manuscript book" bearing the legend "signed by the delegate to Captain Rock" and containing a variety of secret oaths and tests.[28] Only eighteen years old, Egan was the son of a highly respectable farmer "worth at least £1,000 a year,"[29] and Guinan was employed as a herdsman to Egan's older brother, obviously a farmer in his own right.[30] A south Kerry farmer who claimed that he had been forced to join in an attack on a local process server for tithe arrears identified five persons as active in the administration of Rockite oaths in his district: a shoemaker, a carpenter, and three farmers.[31] Still other farmers willingly levied contributions for the Rockites. At the Cork summer assizes in 1824, Daniel McNamara and Thomas Walsh, "two comfortable-looking farmers," were acquitted of extorting money for Whiteboy purposes, but they might well have been guilty as charged.[32] Certainly, in other, similar instances the evidence appears convincing. An ingenious ruse was used to extract the names of two "comfortable farmers" on John Glover's property near Mallow in north Cork in 1823 as collectors for the cause of Captain Rock. A local landlord-magistrate named Nagle took a party of troops to the dwelling of one of his dairymen who had already paid money to the local Rockites but had refused to give Nagle their names. Under the pretense of leading a Rockite visit, Nagle rapped at the dairyman's door, ordered "his men" not to set fire to the cabin, and demanded money in Irish. When the dairyman called out that he had already made his contribution, Nagle asked to whom, and the dairyman then named the two well-off farmers.[33]

In other instances the real or alleged offenses of well-to-do farmers or their sons were far more serious. Maurice Leahy, a "very opulent farmer," was capitally convicted, along with six other men, at the Limerick summer assizes in 1822 as an accomplice in the scandalous abduction and rape of Miss Honora Goold, an adolescent heiress.[34] Among those taking part in the murder of Major Thomas Hare at Mount Henry near Rathkeale in February 1822 were the three Frawley brothers. Patrick Frawley held about 40 acres of land in 1824 from the

Studderts of Bunratty in Clare—a shift from his earlier west Limerick farm at Ardlamon near Rathkeale; he claimed as a defense witness at the murder trial in August 1824 that his two brothers had decamped to an unknown location some thirty months earlier, and that neither he nor they had been among the ten men who had participated in the raid on Hare's residence. But the jury disbelieved him and convicted the three prisoners.[35] The undoubted murderers of land agent and driver John Hartnett, killed in January 1824, were the brothers Daniel and Murtagh (Morty) Flynn and Morty Flynn's servant boy Charles Dawley. The Flynns were the holders of a dilapidated farm on Colonel Fitzgibbon's estate near Abbeyfeale; they feared eviction after Hartnett had distrained their cattle and threatened to dispossess them. Daniel Flynn gave the full story of the murder—not under the seal of confession—to his parish priest Father David Fitzgerald, who testified against the three men at their trial.[36] Darby Hayes was seemingly a less financially distressed but a hunted farmer. Wanted in County Tipperary for the murder of a man named Burke, he not only eluded his pursuers but could take solace in the fact that many country people in the barony of Kilnamanagh went to considerable lengths to "screen him and assemble in numbers to till his farm."[37] And among those accused of slaying three of the five Kinnealy brothers near Cashel in 1824 because they were considered land grabbers was William Daniel— "what we call here 'a snug farmer,' far beyond the lower class."[38]

It is impossible to be certain whether such examples as these indicate that substantial farmers and their sons were represented in the Rockite movement in roughly the same proportions, relative to their share of the population, as activists drawn from lower social groups. There are sound reasons, however, for thinking that they were not. Perhaps the most telling is the strong tendency of elite observers to focus special attention on Rockites whose status was in some way superior to the commonalty.[39]

Among those put on trial at the Cork spring assizes of 1824 for the grisly murder of the Franks family was Arthur O'Keefe, "a person far above that class of the peasantry who are usually found in the dock."[40] A "wealthy farmer," Keefe was able to offer "enormous security" if given bail, but the judges refused to consider it.[41] On the other hand, the sources portray certain events in such a way as to leave little doubt that the Rockite activities of strong farmers or their sons enjoyed the warm approval of all nonelite social groups in the relevant districts, including the sanction of their own class. County Kerry provides a revealing example. During an attack early in 1822 on the residence of William Twiss, a gentleman in the Dingle district, by a large body of Rockites, one of them, a man named Casey, was shot dead. Apparently the leader of the expedition, Casey was "the son of a farmer . . . above the ordinary class." Within a short time of the fatal raid nearly one thousand people reportedly came to view Casey's remains at Twiss's house, and at popular insistence his body was eventually handed over to the multitude for burial.[42] In a parallel case in the Mallow district of Cork, members of the Ring family repelled an attack on their residence in March 1823, killing one of the raiders on the spot. When his body was carried into Mallow, "it was recognised to be that of a comfortable farmer who resided within a mile of Doneraile."[43]

Two other striking illustrations of a similar kind are also worthy of notice. The first pertains to the brothers Patrick and James Minnane, long sought by the authorities for the murder of Major Thomas Hare during an arms raid early in 1822 on Hare's residence near Rathkeale, Co. Limerick. The Minnanes, who themselves belonged to "a respectable class of farmers,"[44] were finally discovered near Shanagolden, hiding under a bed in the house of William Power, "one of the richest farmers in this country."[45] As he was being taken under military escort to Limerick jail, Patrick Minnane "in the very streets of the town . . . harrangued [sic] the mob, desired them never to despair, [and said] that he was true and would be true to the last."[46] The trial had to be postponed until an accomplice turned approver could be shipped home from Quebec, where he had fled.[47] But when at last the Minnanes were put in the dock along with another substantial farmer, William Green, at the Limerick assizes in 1824, all three were found guilty, sentenced to death, and promptly executed.[48] Vast throngs appeared not only at the scaffold but, more significantly, for their funerals. Patrick Minnane was hanged, a police officer at Limerick remarked, before "the greatest assembly I ever beheld at an execution."[49] A few days later, the bodies of the three men were conveyed from Limerick city to Rathkeale in a funeral procession said to include "more than 7,000 persons."[50]

The wake and burial of another victim of the law were accompanied by an equally impressive demonstration of solidarity embracing substantial farmers. Following the execution of a man named Dwyer at Clonmel in 1823 for burning the house of a land grabber, a great number of "decent-looking farmers" attended the corpse in procession to Holycross village near Thurles, where it was waked all night. At the interment of Dwyer's body on the following day, a local police officer was quite astonished to see "such a number of comfortable men and women assembled to pay respect to the remains of a culprit who acknowledged his guilt when on trial and also on the scaffold, not only as to the offence for which he suffered, but also two other houses on the same lands, which were burned last year."[51]

Farmers' Transgressions of Rockite Code

Of course, the typical social composition of Rockite bands that wreaked vengeance on land grabbers can hardly be judged from the Dwyer case alone. But what may safely be concluded from this episode, and from other evidence relating to particular instances of resistance to distraint or eviction, is this: because of the wide impact of the agricultural depression many substantial farmers were predisposed to accept the legitimacy of the Rockite prohibition against the "canting" or "jobbing" of land. Some of them were prepared to enforce this ban personally and did so in association with members of lower social groups. Yet it is also abundantly clear that many well-to-do farmers transgressed the code, either by evicting their own cottiers or by appropriating the nearby lands of bankrupt smallholders, and thereby laid themselves open to violent reprisals.[52] Indeed, this type of Rockite activity was usually conducted by small farmers, cottiers, and laborers. Along with certain other kinds of agrarian violence, it wore the aspect of class conflict.

As their comments show, this was no mystery to elite observers. Numerous undertenants—"all paupers"—had recently been evicted from a farm near Fermoy, Co. Cork, a military officer reported in April 1823, "and I fear will act like those similarly circumstanced, and by whom it is supposed most of the outrages are committed."[53] The assumption in official quarters was almost axiomatic that those cottiers and small farmers who had been dispossessed took revenge by destroying their former cabins. Unless this species of incendiarism was accompanied by personal violence, the authorities often adopted a rather lackadaisical attitude toward it.[54] Colonel Gough, for example,

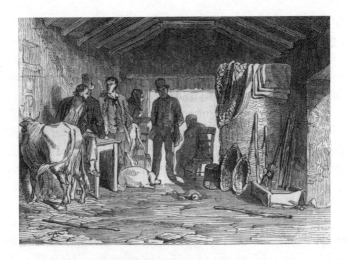

The class of smallholders known as cottiers contributed heavily to the ranks of the Rockites in the early 1820s. Their grievances—evictions, exorbitant rents, excessive tithes (especially on their potato gardens), high food prices, underemployment, and low wages—figured prominently among the complaints that the Rockites sought to redress. Cottiers endured a hardscrabble life in mean-looking cabins of a single room, often sharing space with their animals. This sketch, dating from early 1846 and showing the exterior and interior of a cottier's cabin in County Kerry, provides a fair specimen of the miserable housing conditions among the rural poor noted by so many prefamine commentators. (*Illustrated London News*, 10 Jan. 1846)

was quite indulgent about a rash of burnings in north Cork in the spring of 1823. Burnings were now to be expected, he remarked on 1 April, since hundreds of houses had been vacant locally since 25 March, "and a much greater proportion [of incendiary fires were predictable] where the tenants have been changed."[55]

The incendiarism that became conspicuous in parts of County Clare in 1823 and 1824 seems to have had a distinct social basis in the sometimes novel claims to the land made by cottiers and small farmers who were often in the position of subtenants. Rockites burned a house, three haggards, and some cattle belonging to an apparent middleman named King near Bunratty.[56] A threatening notice found in the vicinity of the burnings clearly indicated that the violence was directed against "land canters" of a distinct kind, for "if the lands were divided fair between the poor, there would be no sort of complaint."[57] The especially well-informed police magistrate George Warburton put the point more generally a month later, when commenting with alarm on the extension in east Clare of "the system of burning the habitations of outgoing tenants." Declared Warburton: "The peasantry seem determined to destroy all property in land, except when they are themselves the immediate occupiers under the head landlord, as they term the proprietor."[58] Land canters were later made to feel equally unwelcome in the baronies of Upper and Lower Tulla bordering County Galway, a mountainous district including Scarriff, Killaloe, and Tomgraney. The crown solicitor Francis Blackburne described this wide district in August 1824 as "in a perfectly lawless state";[59] it was the view of one of his correspondents that "all the mountain peasantry" of the area had been sworn into the movement.[60]

In aggravated circumstances of distraint and eviction, however, lower-class sufferers or their friends and neighbors did not stop at the destruction of property. Witness the murder and mutilation of a land grabber named Daniel Connell on Lord Stradbroke's estate in east Limerick at the end of 1824. This slaying was apparently the culmination of a series of little-noticed attacks against property sparked by the wholesale eviction of twenty-six families in the previous April.[61] Connell had taken a farm of 88 acres cleared of its former tenants; he was assailed by a small party of Rockites, who cut off his nose, dashed out his brains, and severed his head from his body. His brother, his sister, and some of his other female relatives were badly cut and beaten in the same raid. The surviving victims identified a former tenant named Ryan as a member of the attacking party, and of the eleven men arrested for the crime by mid-December, two at least were evicted tenants.[62]

Unifying Grievance: Tithes

Class conflict, however, was largely absent from disputes over tithes, the abolition of which was a principal aim of the Rockite movement. As in cases of incendiarism stemming from evictions, so too in those related to tithes, the fires were usually set by small parties, often by only one or two people. But little is known directly about the social status of the burners of tithe corn. Very few of them were ever discovered, and the accounts of judicial proceedings that have been examined identify no one (by name *and* occupation) who was convicted specifically of this offense. It was observed, however, that when tithe corn was burned, farmers living near the fires almost never helped to put them out. The police official Major Samson Carter expressed initial bewilderment in the autumn of 1822 with the way in which so many farmers in the barony of Fermoy "viewed the destruction of their [tithe] corn with seeming apathy." His inquiries demonstrated, however, that some farmers burned their own grain, with the intention of recouping themselves by presenting fraudulent claims for compensation to the county grand jury.[63] In fact, such farmers actually burned only the straw but planned to enter claims for their wheat and oats.[64] On the other hand, many farmers were thought to be hostile to the tactic of incendiarism because, as much as they detested tithes, they also resented county taxes swelled by compensation paid for malicious fires.[65] To the authorities the distinction between farmers as instigators and laborers as agents seemed particularly relevant to the burning of tithe corn.[66]

Nevertheless, the prominence of farmers in tithe disputes strongly suggests that in practice there was often no such dividing line. The small party of three men who attempted in November 1823 to shoot the Reverend Sir William Read, the rector of Tomgraney parish in County Clare, may have all been laborers, as his legal information stated,[67] but the opposition to his extravagant claims for tithes embraced most local farmers. "I am sorry to add," declared the rector's son when reporting this latest incident, "that the people all appeared glad, & though sev[era]l were in the fields & the road, none ever made any exertion to seize on the delinquents."[68] In an earlier incident in September 1822 a whole crowd of country people had attacked and "dreadfully wounded" three of the rector's bailiffs as they were distraining cattle for tithe arrears. Read's use of troops to apprehend two of the participants in this collective resistance sparked an attempt to shoot him later on the same day as he returned to Tomgraney village.[69] In fact, the hostility toward Read embraced all classes, not excepting members of the local magistracy and police. To judge from the "feeling of dislike most strongly evinced at yesterday's meeting [of magistrates and other local notables] towards Sir William by every class," declared the police official George Drought on 23 November 1823, "I do not conceive his life or his property secure."[70] The fact that virtually every window in the Protestant church of Tomgraney had been demolished in March 1822 was the least of this rector's problems.[71]

The social composition of one Rockite committee in the Mallow district, a center of tithe-related incendiarism, lends further support to this view that farmers often engaged with enthusiasm in resistance to tithes. The committee convened in the house of the Mallow publican Denis Kelleher, and among its six members were two farmers, three laborers, and a leather cutter.[72] The farmer John Mansfield of Derresbrack in the barony of Courceys served as the "captain or leader" of an anti-tithe band that included a publican and seven other farmers.[73] The defection of the Rockite leader David Nagle also sheds some light on this question. Soon after Nagle was taken into custody in 1823, he revealed the names and activities of many of his associates and followers. Most of the information that he initially provided to the authorities pertained to Rockite operations in the district between Mallow and Doneraile, his native area and one renowned for the burning of tithe corn.[74] His disclosures did not keep him from being tried at the Cork assizes and capitally convicted of leading an arms raid on a gentleman's house. But because of Nagle's "humane conduct" on that occasion, the jury strongly recommended him to mercy, a proposal that the government accepted to the point of pardoning him in return for his continued cooperation.[75] This bargain was apparently worthwhile, for in May 1824 Colonel Sir Hugh Gough declared:

> The sensation respecting David Nagle's appearance in this country continues in full force; . . . the disaffected in a large tract of country

find that all their proceedings have been disclosed, so much so that most respectable farmers—respectable from their situation—have come forward and acknowledged the part they acted or, as they say, were obliged to act, and the correctness of David Nagle's information.[76]

Of course, the enactment of tithe-composition legislation in 1823 (to be discussed in detail in chapter 6) erected a new and significant barrier to cross-class collaboration between dairy farmers and graziers on the one hand and the rural poor on the other. Tithe composition on an acreable basis, as became possible under the 1823 act, would have lightened the tithe burdens of laborers, cottiers, and small tillage farmers while raising the share of clerical taxation borne by those better off. For this reason the act was hailed by some members of the Catholic Association, such as George Lidwill, as a great popular boon because it had the effect of repealing the ill-famed tithe-of-agistment "law" of 1735, under which pastureland in Ireland had generally been exempted from tithe in the first place. As Lidwill declared at a meeting of the association in January 1824, graziers—"the peer, the landlord, and the [large] tenant"—could not complain, "for they have enjoyed what was not their own long enough."[77] Always sensitive to the fractures within Irish rural society, Colonel Sir Hugh Gough well summarized the new situation in November 1823:

> The general feeling of the poor man throughout the country is that it is not a dispute between the clergyman and his parishioners, but between the great and little landholder, or as they call it, between the rich and poor man. It does not fortunately partake of any religious dispute, for in many places the persons who have voted in the [parish] vestries against the act are the Catholic grazier and the holder of dairy farms.[78]

In fact, the 1823 act was widely expected to be abortive precisely because when exercising their property-weighted votes at the vestries, graziers, demesne farmers, and the holders of grass farms would strenuously resist efforts to lighten the burdens of conacre men, cottiers, and tillage farmers.[79]

Outright Class Conflicts

If class warfare below the level of the elite was largely absent from tithe disputes until a late stage of the Rockite movement, it was nakedly displayed in conflicts over the availability and price of conacre land as well as over the employment of migrant agricultural workers. In activity concerned with both of these issues, the enforcers of Rockite laws were recruited almost exclusively from the vast ranks of the poor. The class antagonism inherent in the regulation of conacre lettings is evident not only in the reports of elite observers but also in the explicit language of threatening notices. One notice, posted in 1822 in Tipperary and setting specific rates for different kinds of conacre, stated in part: "I have no call to [i.e., no quarrel with] the estated gentlemen, for the farmers has the land from them for a trifle. . . ."[80] The steward of a Kerry parson and landlord received a clear Rockite message on this subject in April 1823: "You have sat [i.e., set] quarter ground to them that has [spud?] ground of their own, and them that were depending on their heath to support their families would get none; look to those, or upon my oath you will get a horrid death."[81] A herdsman named Daniel Leary (no doubt acting in company with others) actually served a threatening notice on his own master, the farmer Thomas Sheehan, in the Mallow district in 1823; the notice threatened Sheehan with vengeance unless he divided his potato ground into small portions and let the little plots at a certain rent.[82] The Rockites who visited the farmer Edmond Connors near Buttevant in March 1824 obliged him to swear that "he would break up one of his best fields & give it for [potato] gardens rent free."[83] From the authorities the poor obtained sympathy, if not relief. The "lower orders," remarked Colonel Gough in the same month, "are now generally employed in preparing for the potatoe crops, and I am persuaded [that] were they to get their gardens at a reasonable rent, many of the outrages which now take place would not occur."[84] There was even a tendency to dissociate the annual springtime quest of the poor for conacre land from the Rockite movement. "The turning up of ground . . . with a view to compelling a more extended cultivation of potato[e]s," commented a military officer in Limerick, "has been for many years, and in the most peaceable times, a common practice in this and other counties, and cannot be considered as connected with the insurrectionary disturbances that have existed."[85] Yet if some popular regulation of conacre lettings always occurred in the months of March and April, its scale greatly increased during the years of the Rockite movement. Large musters of Rockites sometimes occurred even during daylight hours to turn up ground, as when a "mob" of some two hundred men put their spades to work in several fields of pasture near Charleville in north Cork one morning in April 1823.[86]

The Rockite movement exercised similar effects in relation to the regular autumnal reprisals against migrant agricultural workers. A police officer asserted in October 1823: "It has been a constant practice amongst the peasantry of this county [Cork] to oppose the employment of labourers from the adjoining counties, particularly Kerry, at this season. . . ."[87] His statement is accurate up to a point. But again, this kind of activity was pursued much more extensively during the life of a vigorous agrarian movement than it ever was in quiet times. In the autumn of 1823 and 1824 there were unusually numerous reports about the beating of Kerrymen and other "strange labourers" in Tipperary, Limerick, and Cork; those farmers who employed them were often thrashed as well.[88] A few of these attacks ended in killings. A farmer named Callaghan McCarthy was murdered in October 1822 on a mountain near Dungarvan in Waterford, in front of his wife and son, by William Fitzgerald "under circumstances of the most peculiar enormity, merely because the unfortunate deceased hired a strange servant, a Kerry man."[89]

The hostility of lower-class Rockites to "strangers" extended well beyond "Kerrymen" and other common agricultural laborers. The servants of landowners, middlemen, or wealthy farmers who were regarded as too faithful to their employers also came under systematic attack. A party of more than twenty Rockites broke into Curragh House, the seat of Sir Aubrey de Vere at Adare, Co. Limerick, in January 1822, seized hold of his steward and gardener, and told them to prepare for death because "they were too loyal to their employers in preventing the poor people in these hard times from cutting the timber and taking away turf"; the steward (a local man) was directed to be less attentive to his duties, but the Scottish gardener was ordered to leave altogether.[90] Another steward working for a farmer named Keating in the parish of Newinn in Tipperary was shot in the shoulder and hip when he resisted a Rockite attack in February 1824. The soldier reporting this attack noted in explanation that the steward was "a native of North Britain; consequently, he is deemed an invader by the Rockites."[91] Similarly severe treatment was meted out in April 1824 to a recently arrived stranger named Hanrahan, employed by Nicholas Meade, apparently a middleman tenant on the Earl of Clare's estate near Shanagolden in County Limerick. As Meade's "confidential man," Hanrahan was obnoxious to local Rockites, who broke three of his ribs, knocked out one of his eyes with the blow of a musket, and ordered him and his wife to return "to their own country the next day."[92] Rockites in County Clare went to the extreme of murdering the herd and driver Michael Ryan in July 1823 while he was minding a farm near Limerick city for the landlord, who was his employer Thomas Jackson. About ten weeks earlier, Ryan had arrived at the place of his death to succeed another herdsman discharged by Jackson. One of the murderers was overheard just before the killing to have "asked the rascal [Ryan] how he dare come to fill another man's station."[93] Even persons long settled in a particular district, or with a blood relationship to local farmers, aroused the hostility of Rockites determined to enforce their rules against "strangers" of all kinds. A party of Rockites attacked the house of the "respectable farmer" John Scully near Caher in County Tipperary in August 1823 and upbraided him for employing his nephew (who lived five miles away), contrary to Captain Rock's regulations. Not only did these Rockites give Scully's nephew a beating, but they also ordered Scully to send away his sister-in-law, who had acted as his housekeeper for some years.[94] Even a poor tailor named Goggin who had been resident near Buttevant in north Cork for a few years and had married a local woman did not escape persecution. Early in June 1824, Rockites burned down his dwelling, since it was "contrary to their system to admit strangers." The fact that he was married to the sister of a Rockite "committee man" was not enough to save him; indeed, this was the second time that his house had been laid in ruins by incendiaries.[95]

Two Murders on the Earl of Ormonde's Estate

An especially revealing set of circumstances surrounded two killings of employees that occurred on the Earl of Ormonde's estate in March and April 1822.[96] The murdered men were James Kelly, the earl's newly appointed wood-ranger, who was shot, stoned, and beaten to death with the butts of guns when he refused to surrender his firearms at Kilcash on 30 March,[97] and Martin Quane, the servant of one of the earl's tenants, who was fatally shot at nearby Killamery on 11 April because he "had been guilty of working for his master after having received a notice not to do so from Captain Rock." It was noted that

Quane might have been mistaken for his master Morrissy, a tenant who had taken a farm on Ormonde's estate "over the head of another individual." This second murder "was committed at noon, in the face of the country, who do not seem to think or care much about it."[98] Suspicion immediately fell on the Sheas, who had previously held Morrissy's farm: "Everything leads to the supposition that this barbarous murder was effected by their prosecution, as sure[ly] as it was committed under their inspection."[99] It was considered improbable, however, that the Sheas themselves had done the deed. More likely, the killers in this case, as in others, were hirelings. "I have been told by a person well acquainted with Captain Rock's system," declared Lieutenant-Colonel Lindsay in April 1822, "that men are frequently hired to burn a house, give a beating, or commit a murder, and that the price of the latter work is not above 15 to 20 shillings."[100]

The murders of Kelly and Quane were soon followed by other acts of violence in the same general area; the victims seem mainly to have been "land canters" or new tenants like Morrissy.[101] In one case occurring just across the border from County Kilkenny in an adjacent part of Tipperary, two houses were "pulled to pieces, stone by stone, by a great body of people."[102] Smaller parties wounded a farmer named Landrigan and burned down his house on the Ormonde estate and threatened to punish another of Ormonde's "respectable" tenants severely.[103] The Rockites of the district displayed dogged persistence in their pursuit of one wealthy farmer in particular—John Going, who had recently laid out £400 or £500 on a large holding at Currasilla in Slievardagh barony. Even apart from his likely status in their eyes as a "land canter," Going had incurred Rockite wrath by serving earlier as a constable under James B. Elliott of South Lodge.[104] Less than a month after the murder of Martin Quane, the Rockites mounted "the last of eleven attacks" on Going by coming in "an armed party of twenty-one" late on the morning of 1 May 1822 "through an extensive line of country, perfectly undisguised, for the avowed purpose of publickly sacrificing him for holding out against Capt[ai]n Rock's mandate."[105] A partial check was given to this outburst of violence when the authorities succeeded in apprehending five men accused of the latest raid against this hated former constable and rich farmer. "The consternation here among the farmers is indescribable," crowed the landlord-magistrate Francis Despard to William Gregory on 11 May, "in consequence of the arrest of the men for the attack on Going's [house]. . . ."[106] But the authorities were unable to gain convictions when those apprehended were tried at the Tipperary assizes in the summer of 1822. Despard was nearly beside himself with irritation over the acquittal of all the men charged with attacking the houses of Going and the Sheas. These acquittals, he declared in August of that year, "may prove most unfortunate for the tranquility of the country, as from the universal feel[ing] and known guilt of the parties, they now imagine they may engage in any atrocity, with a tolerable certainty of escape, from the situation of the criminal panell [sic], which, to repeat the words of a few of its members, is a disgrace to so respectable a county."[107]

What these events seem to show about the social composition of the Rockite movement in this area, or at least on the Earl of Ormonde's estate and in its vicinity, is that the perpetrators of the violence were rarely farmers, but that farmers screened and otherwise protected laborers and cottiers as well as some artisans from the reach of the law. In the autumn of 1823 the authorities finally managed to arrest James Kelly and William Creed for the murders on Lord Ormonde's estate in the spring of the previous year. They had eluded capture for so long largely because they had enjoyed the support of so many local farmers. Francis Despard of Killaghy Castle declared in January 1824 that "from the protection James Kelly received from the respectable farmers, it was only by the most marked ingenuity I had him arrested."[108] One of Despard's informants "solemnly swore there was not a farmer in the country, with the exception of eight or ten, but regretted his arrest [James Kelly's] more than [that of] all the men that have been taken for the last ten years; they harboured, supported, and protected him, being a willing instrument to perform any diabolical act whatsoever. . . ." If it were known that informations had been lodged against Kelly, "a large sum of money wou[l]d be instantly made up by the farmers to bribe off the prosecution, from the consternation and confusion of the people since his arrest."[109]

In the end, no farmers betrayed Kelly or his two alleged accomplices (brother William Kelly and John Callanan) in the murder of Martin Quane. Described as "near neighbours" of the accused, the four witnesses were all poor people. Two of them were the adult children of an old man named Meighan: "Old Meighan and his large family are laborers, and since it was known that two of them were to prosecute, not a person of the neighbourhood would give them employment." (In Old Meighan's house at Springmount—belonging to a tenant of Lord Ormonde—lived a sister-in-law of the prisoners James and William Kelly.) A third witness named Edmond Dillon was said to be "in similar circumstances on another part of the Ormonde estate"; and the fourth witness (Michael Skelly) shared a dwelling with his mother, an uncle, and other family members on part of Garryrickin demesne, where they had a cabin and two acres of land at a rent of £4 a year.[110] Convicted of Martin Quane's murder on the basis of this strong testimony, the Kelly brothers and John Callanan were hanged in April 1824 at Killamery, the scene of the crime.[111] The strong implication of the surviving records is that the executed men were of the same low social status as those who had testified against them.

Other Combinations among Laborers and Cottiers

Combinations of laborers and cottiers also sought to improve their economic condition in other ways. By means of threatening notices they directed farmers to supply the neighboring poor with milk at a reasonable price. One document posted near Tipperary town in April 1822 announced: "Captain Rock gives . . . notice to such as you farmers that commands cattle not to asume sending milk to town until such of the neighbours as are in necessity of buying it are supplied at a reasonable price at 7 quarts a penny."[112] Seeking to increase the availability of milk for sale, these lower-class rebels even set limits to the number of new calves that could be fed per cow. Near Tipperary town the mandated ratio was one calf for every six cows; in the districts of Listowel and Tarbert the ratio was one to seven. In these parts of Kerry it was not unknown for bands of Rockites to "kill & destroy the new dropt calves in order to encrease the quantity of milk for sale & to lower the price."[113] Besides flogging the servant boy of a farmer named Condon in the Rathkeale district of Limerick in August 1823 and ordering him to leave his employment, the attacking Rockites "swore Condon's wife to give milk to whoever asked [for] it and to be more liberal for the future."[114]

In times of food scarcity efforts were also made to regulate the sale of corn. Early in 1822 in north Kerry numerous notices expressing this aim appeared in the name of Captain Rock. The following is typical: "The farmers are to prepare their corn and sell it to their neighbours at the market price, abating every expence of carrying it to market; no carrier to take corn out of the country; no forestaller to be permitted; no corn factor to buy up any more corn. . . ."[115] The sale of potatoes might also be controlled, or their export be prevented, as after the partial failure of the 1823 crop. In the spring of 1824 a large boat lying near Rathkeale and taking on potatoes for County Clare was boarded by a party of men who cut the rigging and the mast.[116] This incident was only one example of a much more general phenomenon of lower-class protest in this region at that time. As one Clare gentleman reported to Dublin Castle, "There are constantly parties of the *poor* people of this *parish* [unspecified] searching for boats which are in the habit of coming to buy potatoes here [on the Clare side of the Shannon] and taking them elsewhere."[117] Those who challenged the determination of the Rockites to lower the prices of potatoes or to control their sale were chastised by nocturnal visitors. One farmer near Kildimo, Co. Limerick, was placed on his knees at gunpoint in the summer of 1824 and sworn to return some of the money that he had received from a recent sale of potatoes; he was also sworn "to take a small quantity of potatoes to a poor woman near Kildimo and let her have credit for them 'till harvest."[118] More generally, it should be stressed, the Rockites invoked the principle that during a subsistence crisis the people of any particular locality had the first claim on the food produced there. As in County Kerry early in 1822, so too elsewhere, they endeavored to regulate the prices at which milk, butter, and corn were to be sold locally, and they tried to interdict the shipment of produce headed to distant markets.[119] A Rockite notice posted in County Tipperary early in December 1822 spelled out this general principle with absolute clarity:

> The sword is drawn, and the hand of God is with us, and tyranny will soon be swept away. . . . I caution all the farmers of Ireland, and county of Tipperary, and barony of Clanwilliam against selling or sending either beef, pork, mutton, potatoes, milk, or any kind of grain to either city, town, borough, or village to be sold. If they do, they will die the most unmerciful death.

This notice was "signed" by "General Rock," "Captain Steel," and "Major Rib[b]on."[120]

But if the grievances of unbound laborers thus received a great deal of the Rockites' attention, the plight of the laborer employed by the year was not ignored. Having heard complaints about exorbitant rents charged by farmers in the Croom district of Limerick, "that man of the people" John Rock issued in November 1821 "my last notice," ordering all farmers to reduce cabin rents by one-third and potato-garden rents "as has been before directed."[121] A similar notice posted a few months later in County Cork commanded that bound laborers were to have a house, a garden, and the customary "freedoms" for as little as £5 per acre and £1 a year in addition, and that no rent was to be paid for laborers' houses which farmers failed to keep in good repair.[122] An armed party of Rockites attacked the dwelling of a wealthy farmer named Edmond Daly near Kildorrery in Cork in April 1823; they warned him not to employ "strange" laborers, to retain his present workmen, to reduce their potato-garden rents to £5 13s. 9d. and their cabin rents to 30s., and to pay them wages of 6½d. per day, or else "General Captain J. Rock" would burn his house to the ground.[123] Another Rockite party broke into the dwelling of a well-to-do farmer named Sullivan near Buttevant in May of the same year and obliged him to swear that he would "let his labourers have their gardens well manured at four pounds per acre"; his laborers were sworn not to work for him unless he complied with these orders.[124] And at about the same time Rockites ordered a Protestant clergyman near Doneraile to stop violating the "new laws" of Captain Rock. Instead of lowering his laborers' wages and raising the prices he charged for potatoes, he was directed to augment their earnings and to reduce his potato prices by a third.[125]

An Urban Component

Since urban workers and artisans shared many of the same concerns as agricultural laborers and cottiers (including wage rates, food prices, and conacre rents), it is hardly surprising that numerous Rockites were drawn from the cities and towns of the southern region. While toiling to convict Whiteboys at the special commission at Limerick in late December 1821, the attorney-general asserted: "Many of the dangerous characters that infest the County Limerick and the County Clare, and more distant places, come out of the streets of this city."[126] More than a year later, another observer maintained that the Liberties and neighborhood of Cork city were disturbed by "constant nightly meetings of the insurgents which are kept so secret that it is impossible to get a clue to them."[127] Kilkenny was yet another town where Captain Rock's cause found many adherents. The military officer Lieutenant-Colonel Lindsay reported it to be "the general opinion" in April 1824 that "most of the mischief done throughout the county [of Kilkenny] is planned in the city."[128] One reason for this connection may well have been that the "Union of Trades" there had methods of operation that mirrored those of many rural Rockites. Early in 1822 the directing committee of this union was accused of having ordered two acts of violence in the environs of Kilkenny: an attempt to burn the house of a disliked slater and the beating of a man who had returned to work during a strike of woolen workers.[129]

Ribbonmen as well as militant trade unionists were active in the town, and as noted in chapter 3, Ribbonmen and Rockites were likely to overlap in urban settings. The evangelical Protestant preacher Reverend Peter Roe told the undersecretary William Gregory in March 1823, "I am fully convinced that the Ribbon system prevails to a great extent in this city [Kilkenny] & neighbourhood."[130] As will be discussed below, the formidable "Templemartin gang" drew numerous adherents from the lanes of Kilkenny.[131]

Demesne Workers as Agrarian Rebels

Certain demesne workers, relatively a well-off group, took concerted action to better their economic lot. Running battles with demesne stewards resulted. During a combination in October 1823 among Lord Doneraile's demesne laborers in north Cork, his steward John Lothian was accosted in a corn field by an armed man; the stranger demanded to know why Lothian "did not obey the orders of General Rock, viz., 'to give more potatoe gardens to the labourers, employ more of them, and treat them better than heretofore.'"[132] Lothian's cool reply was that he had received no such orders, a statement that his assailant said had saved him. In fact, it was merely a prelude to the punitive wholesale dismissal of Lord Doneraile's demesne workers.[133] Some of them, charged with delivering a threatening notice to Lothian, were convicted of the offense in the following December and sentenced to be transported.[134]

A different kind of conflict involving demesne workers revealed itself in late 1823 and early 1824 on the County Kilkenny estate of Sir

Jonah Wheeler Cuffe. Mayor of Kilkenny city, Wheeler Cuffe owned a country seat at nearby Lyrath House. In late November 1823 his steward John Phelan (alternately named as John Whelan) was shot and fatally wounded. One man charged with this offense was rescued by Cuffe's own workers while being taken to jail, and Phelan was able to identify one of Cuffe's laborers as a participant in the shooting.[135] Barely six months later, Gabriel Holmes, the steward of Abraham Ball of Castle Blunden in the neighborhood of Kilkenny city, was also murdered.[136] From the outside these events seemed to suggest the existence of a feud similar to that on Lord Doneraile's property in Cork. But outward appearances were highly deceiving. Prior to his death John Phelan had in fact been the leader, or one of the principal leaders, of what became known as the Templemartin gang, a violent agrarian combination of at least fifteen or twenty persons whose serious misdeeds in County Kilkenny allegedly included three murders, one attempted murder, six beatings, and four arms raids.[137] What might be termed the flag of rebellion had been raised early in the Templemartin district. Placards posted at Templemartin fair as early as December 1821 had proclaimed: "No tithes!" "No taxes!" "Sixty per cent . . . reduction in rent!"[138] Soon after Phelan had been fatally wounded on 30 November 1823 within a mile of the military barracks in Kilkenny city, it was revealed that he himself had been at the center of this turmoil. His murder was said by a knowledgeable military officer to have resulted from "the private pique of two [members] of the gang" and did not at all appear "to have been the general wish."[139] According to this officer, "the late steward was clearly the head of the Templemartin gang, but he had latterly made some enemies among them, & they feared his betraying them, particularly, I fancy, with regard to the proposal for shooting Sir Wheeler [Cuffe] himself. . . ."[140] The main dissidents were identified as John Wall, William Wall, and John Larrissy. John Wall, an employee on Wheeler Cuffe's property, was quite literate; he "is a good scholar & was intended for a priest." He was considered "the chief instigator" of Phelan's killing because he reportedly thought that he would succeed the murdered steward.[141] William Wall, John's close relative, toiled on the estate as a herd; his coworker and coaccused John Larrissy normally slept with William Ward in the "Garden House" of the demesne. Though Larrissy was eventually apprehended, he had at first escaped "with the connivance and assistance of the laborers and domestics at Lyrath."[142] The Wall family of Templemartin in particular had a well-established reputation for agrarian militancy. Its members, insisted a military officer in March 1824, had "been at the bottom of all the outrages committed for some years back" in their neighborhood. "There is not an honest man among them," he acidly remarked, "but William & this John Wall are two notorious nightwalkers." Associated with them was Patrick Wall, Wheeler Cuffe's stable boy and "a most foul mouth[ed] young rascal."[143]

There were several remarkable features about the Templemartin gang. One was its extension from the estate itself into Kilkenny city, coupled with its lower-class leadership. Among its main leaders was David Quin, called by one reliable observer "the captain" and identified as a herd on Wheeler Cuffe's estate.[144] Quin was said to have "a very strong party in as well as out of Kilkenny." This seemed rather inexplicable to the military officer Lieutenant-Colonel Martin Lindsay, who acknowledged his puzzlement early in 1824: "I cannot well account for the influence this fellow has over so many of the lower order."[145] A second feature of this gang was its geographical breadth. It may have had its reputed headquarters at the public house close to the Catholic chapel of Pit and near Templemartin,[146] but its reach extended far beyond its base. Lieutenant-Colonel Lindsay was impressed with "the facility with which the nightwalking gentry unite when requisite." He reported that three named Rockites came from a place twelve miles from Templemartin to attack a rich farmer whose house was located within a mile of the military barracks at Kilkenny; they were then under instructions to attack two other houses at Kilmacow near Waterford city.[147]

A third distinctive feature of the Templemartin gang was the support that it enjoyed among the tenant farmers of the Wheeler Cuffe property. Lindsay described the estate at the end of 1823 as a miniature County Limerick.[148] Clearly, the Templemartin rebels had endeared themselves to those in the lower ranks of society in Kilkenny city and its rural hinterland. "You have no idea," Lindsay told William Gregory in February 1824, "of the interest which is excited among the common people in favor of Quin and [John] Wall, who are quite well known to have been the contrivers of two other murders, & of many other outrages, & these deeds have gained them the favor of the greater part of the country."[149] But the better-off farmers of the area were also ready to support the Templemartin gang, perhaps because one of its victims had been a farmer named Thomas Max on Wheeler Cuffe's estate who had been murdered in February 1822.[150] (It is likely that Max had taken a farm from which another tenant had been evicted for nonpayment of rent; he was fatally wounded by a party of Rockites in an arms raid on his dwelling in the Liberties of Kilkenny city.)[151] David Quin reportedly boasted to a government spy after his apprehension that "all the farmers in the neighbourhood would subscribe . . . to make up a purse for the attornies." Sympathy more than fear apparently accounted for this situation. "This [boast] . . . is too true," declared Lieutenant-Colonel Lindsay, "& the general friendly feeling towards the murderers is by far the worst [aspect] of this business."[152]

Wheeler Cuffe was hardly alone in having to fear betrayal by his own employees. The domestic servants of gentlemen often collaborated with the Rockites, and the laborers of gentlemen participated in the wider movement to reduce rents and to abolish tithes. A parson at Kanturk in Cork deplored in 1821 the system of "domestic espionage" by which the agrarian rebels learned all that transpired in private houses. "It has frequently occurred," he noted, "that the parties in their nightly attacks have reproached the owners of the houses with expressions uttered by them the day preceding."[153] With good reason landlords widely believed that the Rockites actually set out to suborn or intimidate their servants into cooperation. The serious wounding of Commissioner John Ormsby Vandeleur's steward during an arms raid near Newmarket-on-Fergus in Clare in March 1822 prompted another local notable to declare that the Rockites' object was "to tamper with every gentleman's servants to assist them in the plunder of arms & money, & those who will not agree to join them on oath are doomed to death."[154] Baffled by their inability to identify the agents of incendiary outrages against gentlemen in the Fermoy district, the authorities concluded early in 1823 that "the servants are themselves often the perpetrators of the deed."[155]

Certain discoveries soon buttressed this opinion. In March a band of Rockites set fire to the house and out-offices of Hugh Norcott near Doneraile. Norcott's was the first gentleman's residence to be consumed in that portion of the Blackwater River valley, and the destruction of his premises was unprecedented in its scale for a member of the Cork landed elite.[156] One of the incendiaries lost part of his hand when his blunderbuss burst during the attack. This circumstance quickly led to the apprehension of the badly wounded John Hickey and his two brothers. What appalled the neighboring gentry was that all three worked for Norcott, and that John Hickey, his gardener, was a special favorite with him.[157] Hickey was no ordinary Rockite. He had been, said Major-General Lambert on 1 April, "for these [past] four months the active leader on the north [side] of the Blackwater."[158] Another gentleman apparently victimized by his own servants or laborers was the Reverend Edward Geratty, whose glebe house in County Limerick had been plundered and burned in 1822. In a report of the capture of several men charged with this offense, it was asserted that those chiefly responsible were "individuals particularly connected with his domestic concerns and in whom he placed the greatest confidence."[159] In an incident earlier in the same year the landowner Arthur Beamish Bernard of Palace Anne near Bandon in Cork headed a yeomanry party that skirmished with a large body of Whiteboys, killing one and taking two prisoners. The dead man, generally supposed to have been a rebel leader, turned out to be a carpenter "constantly employed" by Bernard, the son of one of his tenants;[160] the carpenter "had worked & dined at his house the day before."[161] All this evidence suggests that the social intimacy among demesne workers and domestic servants, as well as their lack of strong loyalty to their employers, contributed significantly to their prominence in the Rockite movement.

Itinerant Rockites

Certain categories of persons who were itinerant by occupation helped to win adherents to Captain Rock and his laws as they traveled from one place to another. A group of carmen, or carriers, bringing butter from west Limerick to Cork city in 1821 stopped for refreshment at a public house in Mallow. There they imprudently proposed "the health of Captain Rock" in the company of some soldiers. Incensed at this toast, one of the redcoats bayoneted a carman in the thigh.[162] This episode suggests that long-distance carriers of agricultural and other goods were instruments for the diffusion of the movement. Yet they were hardly the only ones. Among those convicted at the Kerry assizes in 1822 of tendering illegal oaths in the Tarbert district was Edmond Elliott. "The conviction of Elliott is important," stressed the crown solicitor Matthew Barrington, "as he was a travelling dancing master, swearing in Whiteboys in different parts of the country."[163] Schoolmasters were also sometimes itinerant, though most of them settled indefinitely in a given parish. As pointed out in chapter 4, schoolmasters were well versed in millenarian prophecies, such as those of Colum Cille and more recently of Pastorini, the latter being a particular favorite of many Rockites.[164] What, if anything, the old, itinerant schoolmaster and land surveyor Edward Connors did to promote Captain Rock's cause is unknown. But at the time of his death near Kanturk in 1824, two documents of clear millenarian import were found on his person. And a properly suspicious police officer pointed out that Connors had come from a village near Newcastle West in Limerick, "which at the commencement of the disturbances in this county [Cork] was, from its remote situation, the resort of many of the disaffected leaders."[165] Connors may well have been the kind of "prophecy man" described by William Carleton, that is, a patriarchal figure who would stop for a short while in some house in one locality to share his millennial wisdom with neighbors gathered around the fireside, before traveling to another district where the ritual would be repeated.[166]

Schoolmasters in fact played a significant role in the Rockite movement, as they did in other popular political and agrarian agitations before and after 1800. The elite in general and the authorities in particular were only too well aware of this. Reporting a series of charges against a man named Kelly (setting houses on fire and conspiring with others to burn a mill), a parson of Ballingarry in north Tipperary remarked in 1824: Kelly "is a land surveyor and had been a schoolmaster, and people of that description have been in general very active agents of treason and outrage."[167] According to Carleton, who once belonged to the profession, many rural schoolmasters as a matter of course carried "articles," the cant term for the Whiteboy oath and regulations, and presumably those of Captain Rock, which they used to enlist recruits as the occasion arose.[168] Besides "articles," schoolmasters also carried "catechisms," which contained the questions and answers that Rockites (like earlier agrarian and political rebels) used to tell friend from foe, the initiated from the uninitiated, when meeting on the roads, in the fields, at fairs, and in public houses.[169] To the schoolmaster, as the general scribe of the village or parish, repaired all those who had petitions or letters that needed to be written, including the innumerable threatening letters and notices that advanced the movement during times of agrarian rebellion. The bombastic or pedantic tone of some of these efforts at intimidation, as well as the graceful style of others, point to the schoolmaster's practiced hand.[170] Thus it comes as no surprise to find Edward Fitzgerald, a schoolmaster in the Shanagolden district of Limerick, described as "a dangerous and mischievous character in the country." (Fitzgerald was convicted in March 1822 of joining in an arms raid and was promptly hanged.)[171] The authorities were equally relieved in December 1823 to be able to convict Thomas Heffernan of having been absent from his dwelling after nightfall—a transportable offense under the Insurrection Act. As the schoolmaster of Kilfinny parish near Rathkeale, Heffernan "was a great writer of Rockite notices and of bad character."[172]

Leadership among the Rockites

Just as the followers of Captain Rock were drawn from a broad social spectrum, so too were the leaders of the movement. They included such well-to-do farmers as the Minnanes in Limerick and Arthur Leary in Kerry. Described as "a very decent farmer of some property & quite above the lower orders," Leary had reportedly "acted as captain and commanded the party to fire" in the attack leading to the murder of the mail-coach agent William Brereton in January 1822.[173] In addition, Michael Carroll, the son of a "very wealthy farmer" in the Kildimo district, was sentenced to be transported early in 1822 because the magistrates considered him "a leader or committee man amongst the insurgents" in west Limerick.[174] This son of "an opulent farmer" was himself a considerable landholder near Kildimo (he had reportedly returned from America with money). His violation of the curfew was considered to be among his lesser transgressions. Having "a bad character" and his high social status earned him a stiff penalty for a minor offense.[175]

Less exalted in social status were the alleged principals in the murder of the three members of the Franks family near Rockmills (between Glanworth and Kildorrery) in County Cork in September 1823.[176] The surviving evidence on this infamous case is not fully conclusive as to the identity of the leaders of the murder party, but it appears that there were eight men who sat as a "committee" to decide the fate of

"The Installation of Captain Rock"—1834 [retouched 1843], oil on canvas, 172 x 244 cm [private collection]. Easily the most complicated of the works of Daniel Maclise (1806–70), this 1834 painting aroused such controversy that the artist significantly retouched it in 1843. He did so largely in response to criticisms that he had shown too much sympathy for his subject, or combination of subjects. These include stereotypical images of Irish violence (pikes, swords, pistols, shille-laghs, and a military cap), along with common features of Irish communal life (funerals or wakes mixing grieving with celebration, rituals of drinking and courtship, the seditious Catholic schoolmaster, and the professional beggar or "boccaugh" brandishing his wooden leg and crutch). At the center of his painting Maclise depicted the death of one agrarian leader and the swearing in of a new Captain Rock. As retouched, the painting here shows an oddly well-dressed woman tying a green sash around the new leader's waist, whereas in the original 1834 version the "boccaugh" had crowned the new Captain Rock with the military cap. If Maclise's final portrayal of Captain Rock and his followers involved "odd juxtapositions and bizarre details," as one acutely observant critic has noted, this was because Maclise's art closely mirrored real life. (Courtesy of the Crawford Art Gallery, Cork, and the anonymous owner)

the Frankses.[177] Among those who formed part of this committee or were closely associated with its members were three brothers named Cremins (John, Maurice, and Patrick) and two brothers named Sheehan.[178] Timothy Sheehan confessed that he had been asked on 31 August by one of the Creminses (and six others) whether his brother Cornelius Sheehan had not been transported by Henry Franks, and whether he "would not join them to kill the whole of that family, as they were bad people in the country."[179] Prior to the murders the three Creminses had been small tenants of a landlord or middleman at Rockmills for some two decades, while Timothy Sheehan and his two brothers had worked as laborers.[180]

Other Rockite leaders drawn from this same lower end of the social spectrum were the demesne gardener John Hickey, executed for his central role in burning his employer's premises near Doneraile; William Maher, a farmer's laborer, hanged for leading the party that burned the Sheas in County Tipperary in 1821—the holocaust in which sixteen occupants of a farmstead perished;[181] and the laborer Edmond Magner, who became an approver after being sentenced to be transported.[182] Magner, reported a police officer in 1824, was "a most notorious Whiteboy leader, against whom I could not proceed capitally, wanting a link in the chain of evidence, for burning Hennessy's mills at Ballywalter in this barony—Fermoy—and for which offence his brother & four others were executed at [the] last summer assizes."[183] Of an intermediate social status were George Canny, a dyer and presser in northwest Tipperary, a man said to be "worth some money," who served as "a village clerk and village lawyer," and also as a captain of Ribbonmen-Rockites in his district;[184] John Dundon, a wheelwright, who was regarded as an important Rockite leader in northeast Cork before he was captured in 1823;[185] and one Norwood, whose father was master of the Protestant charter school at Dunmanway in the same county, and who himself was "a kind of underagent" to the principal local landlord. Norwood, by his own confession, was a leading activist in the Macroom district.[186]

Not surprisingly, deserters from the army, veterans or pensioners, and others having military experience sometimes acquired the status of Rockite leaders. The gentleman John Busteed of Tralee summarized the situation with regard to military pensioners in the south at the end of 1821: "Perhaps there are not altogether less than 5,000 pensioners [military and naval], if so few, in the three counties [Limerick, Cork, and Kerry], out of which there is not probably one in ten who cannot do local duty and fight. . . ." Busteed saw this large number of former soldiers and sailors as potential Rockite recruits and trainers of the agrarian rebels. He maintained that "the pensioners are themselves very disorderly, vicious characters whose habits have a great tendency to infect the other classes of the community."[187] The pensioner William McDonald, a tailor by trade who had spent years in the British army, was said in 1822 to be constantly conducting military drill for Ribbonmen-Rockites in the mountains above Borrisoleigh in Tipperary, with at times hundreds participating in these exercises.[188] One military deserter named John Horan, taken into custody in 1822 on charges of being "deeply concerned in the plans which have been going forward to promote a rebellion," had reportedly been "in close communication with several opulent shopkeepers in the town of Nenagh, some of whom, from the apprehensions they entertain that he will betray them, are making preparations to be off at a moment's notice."[189] Distinctly higher in social status was the military man whom the magistrate Jemmett Browne believed to be the Captain Rock of his own district near Cork city. This individual, Browne remarked in 1823, "purchased some time since a majority [i.e., a commission] in the S[outh] American patriot service & has been an idle, disorderly, suspected character in the country ever since his return."[190] Well, perhaps not exactly idle!

As in certain earlier agrarian movements, there was a tendency for persons of elevated social status (yet well below the rank of gentleman) to be recognized as leaders of Rockite bands whose members were drawn from different layers of society. Thus Robert Cussen, found guilty of acting and dressing as the captain of a party of Rockites who had attacked a gentleman's house in County Limerick, was, said a police officer, "a person of the better class, having been bred an apothecary."[191] The three sons of a man named Daly were ringleaders of disturbances in the Castleconnell district of Limerick, according to a local gentleman early in 1822. They had been "for years past the terror of the country," guilty of every kind of crime, murder not excepted. One of them had been transported, but the other two, previously fugitives, had now returned and were promoting the Rockite cause. Their father was the former tenant of a large mill near Castleconnell.[192]

Perhaps the most interesting example of this type concerns David Leahy and his son Daniel, both of whom ranked high among the Rockites in the Abbeyfeale district of Limerick. David Leahy was "a very wealthy man of great influence." He was postmaster of Abbeyfeale, an innkeeper there, and the land agent of Master Thomas Ellis.[193] In his capacity as an innkeeper Leahy had rented one of his buildings to the government for use as a military barracks. But as soon as the troops had moved to other quarters, a party of Rockites led by his son destroyed the building. Seeking to profit from this seeming loss, Leahy claimed compensation amounting to £800, though the real value of the former barracks was said not to exceed £200.[194] Eventually, the authorities secured sufficient evidence of this gross fraud, and at the Limerick assizes in 1824 father and son were convicted of appearing in arms at night and destroying the building.[195] But the Leahys did not confine themselves merely to fraud. According to an accomplice who later became an informer, Daniel Leahy, the son, was a leading participant in the attack on Abbeyfeale church in February 1822; it was he who procured from his father's house the ladder used by the Rockites to strip the church roof of all its lead for the purpose of making ball cartridges. Besides numerous farmers or their sons, this band allegedly included at least four laborers, two subtenants (probably cottiers), a dairyman, a weaver, and a tailor.[196]

Another revealing indication of the somewhat elevated social status of many Rockite leaders was their connection in some cases with members of the landed elite. Cornelius Murphy, alias Captain Starlight, a principal in the attack on Newmarket during the insurrection of January 1822 and "well known in Duhallow as a notorious Whiteboy leader," took the oath of allegiance to the king—a possible protection against arrest—from the magistrate Jeremiah McCartie on the day after the attack. McCartie, it was said, could not have been ignorant of Murphy's character, since a local priest had earlier denounced him before McCartie as "the ringleader of all disturbances" in the district.[197] Timothy Cotter, even more famous as a Rockite commander in northwest Cork during and after the insurrection, long escaped apprehension; he did so at least partly because of the benevolence or restraint of more than one local magistrate.[198] An irate and sarcastic Kanturk gentleman complained in 1824, shortly before Cotter's capture: "In a letter to Judge Torrens last assizes I mentioned that to the disgrace of our *authorities*, Cotter was known to remain at his ease within a few miles of this town, and had spent last Patrick's day coursing with the eldest son of one of our active magistrates. . . ."[199] Unlike Cotter, John Halloran did not long remain at large after the authorities concluded early in 1823 that he was a chief leader of Rockite operations in the North Liberties of Limerick city. Tried and convicted, he was sentenced to be transported. But Halloran too had influential friends among the elite. Described as "a man in very comfortable circumstances," he boasted "when going off that he would soon be back, as he had interest sufficient to bring him home again."[200]

Lastly, there is the fascinating and somewhat mysterious case of the celebrated Walter Fitzmaurice. Finbarr Whooley has described Fitzmaurice as a "prominent under-world figure," "apparently a highwayman with his own group of followers," and as "an outlaw turned Rockite."[201] His name was associated with many different exploits, including the abduction of the teenage heiress Miss Honora Goold, the crime for which he was eventually sentenced to be hanged.[202] One gentleman went so far as to claim in late March 1822 that Fitzmaurice was "well known to have instigated all the murders & atrocities which have been committed" in north Kerry, west Limerick, and northwest Cork during the previous six months.[203] This assertion was of course remote from reality. But even a judicious police official could say that Fitzmaurice had been "the most active agent and ringleader of the Whiteboys" during the winter of 1821–22.[204] Soon after surrendering himself in July 1822 to a County Limerick magistrate, apparently in hopes of gaining his protection, Fitzmaurice became the object of strenuous efforts at intercession by persons of political influence, who tried to prevent his transmission to County Cork, where capital charges had been laid against him.[205] These efforts were unsuccessful, however, and at the Cork summer assizes he and another prisoner were condemned to death for their roles in Miss Goold's abduction. Actually, Fitzmaurice's case never went before a jury because he pleaded guilty and threw himself on the mercy of the government, though he was warned by the court not to hope that such a plea would save him from the gallows.[206] Yet pardoned he was, without explanation, much to the chagrin of many members of the southern gentry,[207] but to the satisfaction of at least a few members of that same class.

Playing prominent roles along with Fitzmaurice in the Goold abduction were several farmers. It was allegedly a farmer named John Browne who abducted the fourteen-year-old Honora Goold early in March 1822 from her family's house at Ballinla in northwest Cork.[208] Richard Goold, the girl's brother, later testified that Browne had been a leader of the abduction party and had raped his young sister.[209] Browne may have been led into this escapade by a desire to recoup his reduced fortunes through the acquisition of Ms. Goold's dowry; he had tried unsuccessfully to obtain a license to marry her, according to one report.[210] Among the first seven persons to be capitally convicted in the Goold abduction case (in July 1822) were four Leahys: Maurice, James, David, and David, Jr.[211] Maurice Leahy was described at the time of his conviction as "a very opulent farmer."[212] His relatives were apparently men of substance as well. As previously noted, David Leahy was a well-to-do innkeeper, a land agent, and the Abbeyfeale postmaster.[213] Also convicted of the abduction was Daniel Ready, whose family had earlier been characterized as "the most active and notorious Whiteboys in the county [of Cork], intimately associated with the gang of the celebrated Walter Fitzmaurice."[214] It must have been a keen disappointment to the authorities when three of the Leahys capitally convicted in the Goold abduction case were set free in July 1823 because of a small technical error in the form of their original indictment—the omission of the phrase "then and there."[215] They benefited from having the illustrious Daniel O'Connell as their defense counsel.[216]

Summary of the Arguments

While emphasizing the complexity and diversity of the social composition and leadership of the Rockite movement, this chapter has advanced certain arguments about these important aspects of the upheaval. First, the severity of the economic downturn of 1819–23 aggravated grievances and pressures at all levels of agrarian society, and below the landed elite (itself badly scalded); hardship in varying degrees was experienced all along the economic spectrum, from once-wealthy graziers and formerly comfortable farmers down to the always deprived cottiers and laborers who dominated the prefamine population. This meant that middling and large farmers, when confronted by landowners who refused to grant generous abatements or tried to collect arrears of rent through distraint, either actively joined the rebellion or abetted lower-class Rockites in a variety of ways. Violent opposition to eviction, or the punishment of those who disobeyed the Rockite ban on land grabbing, also strongly engaged the sympathy of many substantial farmers, even though numerous such farmers became the very "land canters" whose transgressions the Rockites so often sought to check through intimidation and violence. Secondly, the grievance of tithes usually transcended the sharp social divisions afflicting prefamine rural society; opposition to the tithe system had raged for decades and had long generated resistance at all levels of society, though its quietest opponents (up to 1824) were probably the graziers and its most vociferous enemies were tillage farmers and cottiers. Admittedly, the passage of the Tithe Composition Act in 1823 had the effect of exacerbating the very social divisions that this grievance had previously transcended, for the law unwittingly highlighted the privileged position enjoyed by those at or near the top of the agrarian class structure. Lastly, the bulk of the activists in the Rockite movement of the early 1820s were drawn from the huge armies of impoverished laborers, cottiers, and small farmers who formed the overwhelming majority of the rural population in the south and the west. Urban artisans, as noted earlier, played a significant role in Rockite operations occurring within the vicinity of cities and towns, but elsewhere the various layers of the rural poor furnished most of the rank-and-file members of the movement. Perhaps surprisingly, they also furnished a very high proportion of the leadership as well. While often ready to accept the direction of persons of higher social status in their communities, the poor were certainly capable of generating their own leaders in great numbers.

Six

The Issue of Tithes

Complaint against the tithe system became deafening in the south of Ireland during the early 1820s. When the Rockite movement revived after the harvest of 1822, tithes were the issue around which popular protest centered, and overall this grievance was the most salient of those pressed by the agrarian rebels of 1821–24. At the popular level the chorus of denunciation was nothing new. But for the first time supporters of the system found themselves on the political defensive as large numbers of Protestant aristocrats and landed gentlemen broke openly and decisively with the hierarchy and clergy of their own church, thus clearing the way to the first tentative attempts at legislative reform. Rockite opposition to tithes was so bitter, so extensive, and often so effectively waged that it drove home the simple lesson taught by the long history of intermittent agrarian warfare over this issue. As Viscount Clifden trenchantly remarked to Pierce Butler in December 1822, "the present mode of paying the ministers of the church establishment in Ireland has been the source of bloodshed, insurrection, and endless executions for half a century; if this is not a practical proof that it is not suited to Ireland, I do not know what can prove anything."[1]

Abolition or Reform

In earlier agitations agrarian rebels had usually aimed not at the total abolition of tithes but rather at reductions in tithe rates, and this less sweeping objective was sometimes voiced by the Rockites as well. In the counties of Tipperary and Kilkenny notices were delivered to farmers or publicly posted, ordering that tithe owners be paid at the rate of only 1s. 6d. or 2s. per acre, regardless of whether the land was tilled or under grass.[2] These notices and similar ones elsewhere represented an effort to eradicate a glaring injustice of the Irish tithe system—the virtual exemption of grassland from liability to tithe—by spreading the burden of payment equally over every type of agricultural enterprise instead of restricting it to those who produced corn or potatoes. As matters then stood, the heaviest burdens fell on tillage farmers and especially on cottiers and laborers, whose very food was taxed, potatoes being titheable throughout Munster and in parts of Leinster; dairy farmers and graziers, on the other hand, paid little or nothing in tithes.[3]

In most districts, however, the Rockites intended that the incomes of tithe owners should, if possible, be reduced to zero. As the Reverend John Chester of Mallow grumbled half humorously in January 1822, "My parishioners are trying to make me live on air whilst the Whiteboys are endeavouring to kill me by exercise."[4] Evidence of the widespread aim to get rid of tithes altogether (and rid of county cess as well) piled up in Rockite notices, in their oaths, and in their reported words to victims throughout the south. Placards posted in November 1821 at the fair of Templemartin near Kilkenny city proclaimed, "No tithes! No taxes! Sixty percent! Reduction in rent!"[5] Among the laws of Captain Rock set forth in a notice attached to the church gate in the west Cork parish of Kinneigh in January 1822 was one declaring flatly that tithes were to be "utterly abolished."[6] And Rockites conducting arms raids in the districts of Clonmel and Caher in December 1821 told one of their victims that "they would soon swear him to pay no tithes or taxes," while they "announced themselves to others as volunteers determined to abolish tithes and taxes."[7]

The thoroughgoing opposition of most Rockites to tithes is to be explained largely by the extreme severity of the harvest failure of 1821 and by the general collapse of corn prices after 1820. When proctors went on their customary rounds in the summer of 1821, they found tithe payers ready enough to enter into bargains and to pass promissory notes, even at high rates, because the fine weather seemed to herald "the most abundant crop ever witnessed."[8] But after this prospect was blasted by the destructive rains and floods that began in mid-August, the proctors did not return to assess the damage and lower their valuations. One gentleman summoned before a consistorial court in the diocese of Ossory in April 1822 for subtracting tithes due on the previous harvest bitterly declared that "the tithe proctors were setting the country in a flame by their oppressions." A proctor examined in this case admitted that he had never viewed the crops after "the wet weather"; he therefore could not have known that most of the people to whom this gentleman had rented conacre plots had left the potatoes in the ground because they were "not worth digging."[9] Because of the expense of resisting inflated claims in the ecclesiastical courts, ordinary tithe payers did not bother to do so, but the experience of this wealthy Protestant was similar to their own, and like him, they were convinced that natural disasters invalidated whatever bargains they had made.

The repudiation of tithe agreements was one thing, but the eradication of the entire tithe system was obviously far more ambitious. If popular hope soared, it was largely in reaction to the refusal of so many tithe owners to adjust their charges to the postwar deflation of agricultural prices. Tithe rates had risen sharply in many areas of the south during the wartime boom between the early 1790s and 1812.[10] In the united parishes of Thomastown and Collumbkille in County Kilkenny those for wheat and potatoes doubled from 6s. per acre before 1790 to 12s. by 1813.[11] Though this may be an extreme example, it should also be remembered that the incomes of tithe owners grew rapidly during the French wars because of the considerable expansion in the land area devoted to tillage and hence subject to tithes.[12] What incensed tithe payers in the time of the Rockites was the lack of anything resembling an adequate response to the dismal conditions of the early 1820s. In a memorial of January 1822 to the Reverend William Power, the landholders of Aglish parish near Dungarvan pointed out that their wheat, once worth from £2 10s. to £3 per barrel, now realized only 15s. to 20s., and that their barley, once sold at 25s. to 31s. 6d. per barrel, currently brought only 7s. 6d. to 10s. 6d. Despite this enormous fall in prices, they complained, tithe rates in their parish had not been reduced a whit; on the contrary, the proctor's exactions for vicarial tithes had actually increased "of late years."[13]

Of course, some tithe owners, either voluntarily or under pressure, made concessions, but these were generally modest, rarely exceeding

20 or 25 percent.[14] The tithe payers of the united parishes of Aharney and Attanagh, meeting at Lisdowney, Co. Kilkenny, in January 1824, adopted a petition to parliament in which they asserted that while farm prices "in most years since the peace were as much as two-thirds below wartime levels, in no year since the peace had tithe levies fallen by more than one-fifth below the highest rates charged during the late war."[15] Concessions by tithe owners were neither so general nor so substantial as concessions by landlords.[16] In his testimony before a parliamentary committee in May 1824, Major Richard Willcocks, then chief inspector of police in Munster, was asked whether in the counties of Limerick and Tipperary the abatement in the demand for the tithe of potatoes had been commensurate with the reduction in rent. Replied Willcocks:

> I think not, and I had a conversation with two respectable clergymen on that subject . . . , and they admitted to me that the proctors valued the land at the same acreable sum for tithe of corn and potatoes that they did in the year 1812, which was the time when the produce of the land, I believe, was at the highest, and I argued with them upon the unreasonableness of that demand. . . . They said they took the good and bad [years] together.[17]

In one telling case in County Limerick a parson agreed at the start of the disturbances to an abatement of 50 percent across the board, taking only 6s. an acre for wheat and potatoes and 5s. for oats and meadow. But with the coming of troops to his district, and with the introduction of the Insurrection Act early in 1822, he promptly restored the old rates of 12s. and 10s. per acre, respectively, an abrupt change of front to which his hardy parishioners not surprisingly responded by withholding payment altogether.[18]

Legislation Favoring Tithe Owners

Since the late eighteenth century the legal resources available to tithe owners facing resistance to their claims had been increased substantially. Until 1814 it was in general still necessary to begin proceedings concerned with the subtraction of tithes in an ecclesiastical tribunal. In such cases that court invariably issued what was called a monition to defaulters, but it was a secular tribunal that gave binding force to the monition by issuing a civil-bill decree.[19] Except where the tithes due from one person did not exceed 40s. per year,[20] double suits in the same cause were thus the rule. This was cumbersome, as was the procedure before 1796 requiring that only judges of assize hear civil-bill cases, including those based on monitions. In that year, however, jurisdiction was transferred to the newly created assistant barristers acting at quarter sessions.[21] At first, tithe owners, like other creditors, could sue before the assistant barrister for any sum not exceeding £20 in an individual case. But a statute of 1799 made tithes recoverable by civil bill in actions on monitions without any limitation as to amount.[22] In part at least, the two laws of 1796 and 1799 were negative responses to outbreaks of popular opposition to the payment of tithes.[23]

So too was an even greater legal boon conferred on harried tithe owners by a statute enacted in 1814. This law greatly increased the number of tithe cases that could be decided summarily by two magistrates, for it extended their jurisdiction in suits for subtraction of tithes from 40s. to £10 a year.[24] A tithe owner could now in effect choose two magistrates out of those residing in his locality to hear and determine his cases.[25] Tithe payers saw this system as an engine of oppression—with some justification. Admittedly, parsons and other tithe owners were forbidden to adjudicate suits in which they were personally involved as claimants, but there was nothing to prevent a lay impropriator or a lay lessee of tithes from sitting on the bench and deciding cases entered by other tithe owners. This practice was common enough.[26] In one egregious case the Reverend Robert Morritt, rector of Castlehaven parish near Skibbereen, the scene of a furious and fatal tithe affray in July 1823,[27] led a neighboring parson and a lay tithe owner, both magistrates, through more than three hundred cases of default in a single day at his own house.[28] On another occasion Morritt asked a small group of local magistrates that included two lay lessees of tithes to hear his cases at Skibbereen petty sessions; he was reportedly preparing to have as many as six hundred summonses issued against his parishioners.[29]

While similar scenes involving armies of defaulters for very small sums (usually cottiers owing less than 10s.) took place before magistrates' courts in other counties,[30] the spectacle at quarter sessions before the assistant barristers was nearly as bad. A mass of tithe business was dispatched amid a great din and in a tremendous hurry.[31] By far the greater portion of the civil bills issued from quarter sessions in the south were for arrears of tithe. Two examples from County

Cork in late 1822 illustrate the general pattern in the time of the Rockites. At Midleton over seven hundred of about a thousand civil bills were for recovery of tithes, and at Macroom, of the more than sixteen hundred processes for debt, as many as thirteen hundred had their origin in unpaid tithes.[32] Especially for small tithe payers, the costs of legal proceedings taken against them for the recovery of arrears were high in relation to the sums owed. Even in the cheapest form of such a proceeding, namely, summary adjudication before two magistrates, the statutory limitation of costs to 10s. did not include the expenses of distraint or of selling the goods distrained, if matters were pursued to this extreme.[33] Where no distress was levied, the costs awarded by the magistrates might be only a few shillings, but in such cases the tithes claimed often amounted to as little as 5s. or 10s.[34] In civil-bill suits before the assistant barrister based on failure to honor a promissory note, the charges were generally somewhat more than £1, again exclusive of the expenses of distraint.[35] And if the suit at quarter sessions was based on a monition issued by a consistorial court, then the legal costs were more than doubled.[36] In one instance a small Kilkenny farmer was presented with a tithe bill of £3 8s. 6d. for 1821, and when he refused to pay it, the proctors threatened him with suit in the consistorial court, where they expected to obtain £3 1s. 8d. in costs alone.[37]

Inherited Tactics of Opposition

Although tithe payers could normally expect only summonses and costly adverse decrees from the courts, they did not lack resources of their own in the battle against the hated exactions of ministers and lay impropriators. Even in seeking to abolish tithes, the Rockites adopted tactics that had been pioneered long ago (when the aim was usually to reduce tithe rates) and that had become ritualized through repeated use. Like earlier agrarian rebels, the Rockites often deprived proctors or their employers of valuation books, tithe notes, and legal processes. They sometimes achieved this with the help of crowds in the daytime, as proctors or process servers were going their appointed rounds, but more frequently they accomplished it through domiciliary visits by smaller bands at night.[38] A process server's papers would be seized and carried off in triumph, or a proctor who had hidden his account books and tithe notes might be sworn to deliver them to an assigned place for destruction.[39] The destruction of the records was of course designed to frustrate legal proceedings for the recovery of tithes.[40]

Again like their predecessors, the Rockites employed intimidation and violence to force tithe farmers, proctors, valuers, and process servers to abandon their activities. Such functionaries were warned to "keep [to] their fireplace for the future,"[41] bound by oaths to have nothing more to do with tithes, and threatened with death or the destruction of their property if they did not quit their work.[42] Intimidation alone was often insufficient because those engaged in tithe business tended to be a hardy breed stiffened by drink and emboldened by the possession of firearms, which they were quick to use if challenged or attacked.[43] Many tithe farmers, proctors, valuers, and process servers were beaten,[44] but it was frequently safer and no less effective to inflict damage on their property. Thus in numerous instances Rockites burned their houses, set fire to their out-offices, or destroyed their crops.[45] Particularly obnoxious and obdurate individuals were made to suffer repeatedly. The west Cork corn merchant Joseph Baker, who had taken a lease of part of the tithes of Skull parish, the largest in the county, was given ample reason to regret his bargain with the absentee rector. In late December 1821 and early January 1822 his stores, full of tithe corn, were burned to the ground, his haggard was pulled down and its contents were scattered, and his sloop in Crookhaven harbor was set on fire.[46] Early in the following October a shipment of his butter was destroyed while on its way to market, and finally, later in the same month, Baker was himself attacked and beaten almost to death.[47]

Those involved in the collection of tithes risked loss of life either in confrontations with crowds of resisting peasants or as a result of premeditated murder. A driver employed by the Reverend Michael Dowling, who farmed the tithes of two parishes near Cahersiveen in south Kerry, was beaten to death in one affray in January 1822.[48] Another driver was killed in July 1823 in a clash near Skibbereen with the parishioners of the Reverend Robert Morritt, the hated rector of Castlehaven parish.[49] A few of the apparently premeditated slayings included the mutilation of the victim as a means of exciting terror among those similarly employed. The murderers of John Corneal, who had gone on a mission to draw home the tithes of a parson in the Rathkeale district in September 1821, severed his head from his

In this cartoon of 1823 the artist portrayed tithe proctors as plunderers of both Captain Rock's followers and the Church of Ireland parsons who hired these functionaries. Many Anglican parsons would have shared this view that proctors fleeced them too, even if they would have strenuously opposed the resort to popular vengeance depicted here. Since the Rockites sometimes used female attire to disguise themselves and perhaps to signal the inversion of the traditional order, the cartoonist's use of "Lady Rock" to administer revenge was full of contemporary meaning. (Nicholas K. Robinson Collection of Caricature, held within the Early Printed Books Department of Trinity College Library Dublin)

body.[50] Equally gruesome was the fate of a process server named Conway, who performed various duties for the two largest tithe owners in all of Kerry—the Reverend Thomas Stoughton and his younger brother, the Reverend Anthony Stoughton, rector and vicar respectively of no fewer than twelve parishes in the northern baronies of Iraghticonnor and Clanmaurice. The Stoughtons, it was claimed, had never encountered resistance to the payment of their tithes in over three decades, but when their agent sent the unfortunate Conway in January 1822 to collect from defaulters near Ballybunnion and Lisselton, he was stoned and bayoneted to death. His ears and nose were cut off and left staked on a bank near the high road a few miles from Listowel.[51] At least four other persons engaged in tithe business were also murdered during the Rockite movement.[52]

Not surprisingly, this onslaught against the agents of tithe owners produced a fair crop of withdrawals from the business. After the killing of Conway in January 1822 the Stoughtons' other agents were "so threatened they dare not ask for tithe money, nor could they get any man to undertake the collection of it."[53] The effects of this slaying extended to the Reverend Thomas Russell, whose parish lay between Tralee and Killarney. From a living usually worth over £500 a year Russell's collectors had managed to extract only £28 in tithes due from the harvest of 1820. Prospects for recovering the arrears looked bleak, since "no man c[oul]d be procured who w[oul]d dare to serve any [processes] for the last sessions."[54] Resistance to the Reverend Robert Morritt's driving of parishioners' cattle for tithe arrears was so fierce that one of his agents, an eight-year veteran, resigned in 1822 because of "the dread he was in of losing his life," while another, a miller, quit the following year because "the people were tardy in bringing him corn."[55] Morritt's difficulties were compounded by the uproar over the Castlehaven tithe affray of July 1823. Two months later, one of Morritt's drivers on that occasion told government investigators that he doubted whether "fifty police constables could drive the parish now," and that he "would not go out with them for £150."[56] Elsewhere, as in the barony of Kells on the Kilkenny-Tipperary border at the start of 1822, dire Rockite threats against all tithe proctors reportedly prompted some of them to resign.[57]

Yet except for those districts where popular passions on this issue were kept constantly inflamed, tithe owners usually found the human tools essential to conduct their business. Though some proctors, drivers, and process servers quit their posts, their withdrawal was not always permanent, and even when it was lasting, other courageous or foolhardy souls stepped forward to fill their shoes. In a country glutted with poor people, even despised jobs were seized; thus the tithe agent, especially in the lower classifications, never was an endangered species. In Kerry and parts of other counties affected by the Rockite movement, tithe proctors and valuers met almost no opposition as they performed their customary tasks of viewing the crops and securing the consent of landholders to bargains in the summer and autumn of 1822. A Kerry squire observed in October: "The tithes have been taken everywhere through the entire of this county, as well from the lay impropriators as from the clergymen. . . ."[58] Whether this tranquility would continue was thought to depend largely on what transpired in Cork and Limerick; Kerry might be reinfected, "especially as the aversion to the payment of tithes as now levied is common to all the southern counties."[59]

Explosion of Incendiarism

What actually happened in Cork—specifically, in three northeastern baronies and some adjacent areas—was the greatest outbreak of incendiarism against tithes ever witnessed anywhere in Ireland. Together with the contemporaneous incendiarism arising from conflicts over rents and evictions in both Limerick and Cork, the tithe-related conflagrations boggled and of course appalled upper-class minds. An editorial writer in the *Dublin Evening Post* commented with some amazement in late April 1823, after about eight months of seemingly endless fires: "Even in the rebellion of 1798 there were not . . . , during its entire continuance, so many houses burnt; and though more property must have been destroyed throughout the whole kingdom, yet certainly no two counties have suffered as severely as Cork and Limerick are doing at this moment."[60]

Beginning in August 1822, as the accustomed time for tithe bargaining approached, notices such as the following appeared in a wide area north, east, and south of Mallow: "A hint to farmers and gentlemen on their perils to take particular notice of the same. That they are not to buy tythes from or bargain for the same from any minister, tythe farmer, tythe proctor, or impropriator, or from any other person whatsoever employed by them, under pain of incurring the severest reprimand from Captain General Rock."[61] In part, the Rockites were reverting to an old technique of agrarian combinations against tithes: instead of consenting to compound with the tithe owner in money, parishioners dissatisfied with the terms offered would serve legal notice on the parson or lay impropriator of their intention to set out their tithes in kind in the field. In theory the tithe owner would then send his agents through the parish to draw home the tenth sheaf of corn, the tenth cock of hay, and the tenth part of each ridge of potatoes. But the theory presumed that the tithe owner possessed sufficient men, horses, carts, and barns to do the job, and this condition was all the harder to meet when most or the whole of his parishioners, by pre-arrangement, set out their tithes and served notice on the same day.[62] In the eighteenth century this type of concerted anti-tithe action had been declared illegal by statute,[63] though even then, decrees of the courts based on the law were largely unenforceable.

By the early nineteenth century, however, legal opinion on this matter had changed radically in both England and Ireland. In September 1822, just as the Rockite campaign was getting under way in County Cork, the *Dublin Evening Post* published a number of letters in which experts offered practical advice and encouragement to disgruntled tithe payers. In one letter the barrister Denys Scully showed in detail how farmers could set out the various tithes in compliance with the law, and insisted that without illegality parishioners might "serve any number of notices . . . for the same hour of the same day."[64] The writer of another letter pointed out that an allegation of illegal combination against tithes could be tried only in a court of common law; he quoted no less an authority than the current lord chancellor of England as having declared, "The judge who would decide that such an act [i.e., the simultaneous serving of multiple notices to draw tithes] was combination should never get leave to decide another case."[65] Of this letter, the *Post* remarked triumphantly: "It shows distinctly that farmers who serve notices to the incumbent, though they should be a thousand on the same day . . . , may defy the bishop's court and all the proctors."[66] For the benefit of those uninitiated in this art, sets of how-to-do-it-yourself directions glared from the pages of certain Irish newspapers of the time and were probably circulated in the form of handbills.[67]

The Rockites, who needed no such directions, wanted tithes to be set out in the field, but their preference for this practice did not derive from any wish to see them drawn home. Rather, it stemmed from the convenience of burning them there with almost no risk of arrest. The only major contribution of the Rockites to the age-old tactics of opposition to tithes lay in the systematic use of the firebrand. Throughout September, October, and part of November 1822, Rockites lighted the night sky of northeast Cork with hundreds of little bonfires of tithe corn and tithe hay.[68] Writing from Churchtown in late September, a gentleman informed the *Cork Advertiser:* "Burning every night quite close to us; it has now the appearance of St. John's Eve; we can see the fires quite plain."[69] Similar reports came at the same time from other parts of north Cork. "From the hill of Kildorrery near this village [i.e., Glanworth]," remarked the Reverend Richard Woodward, "five fires were seen last night."[70]

Once the burnings started, most farmers who had previously ignored Captain Rock's orders by purchasing their tithes and passing their promissory notes to the proctor hurried to make amends. They separated the tithe corn and tithe hay from the rest of their stacks and placed them outside their haggards or drew them back into the fields in order to save the other nine-tenths from destruction. Not only farmers, noted a military officer, "but many of the gentry have through fear, I regret to say, behaved thus dastardly."[71] Those who persisted in disobeying Rockite mandates to divide their corn and hay became the victims of elusive incendiaries. "Tho' the troops and police are constantly patroling [*sic*] all night," said one observer, "they have not taken (that I can hear) one person for these atrocious acts."[72] The greatest sufferer on this account was a gentleman, James Hill of Graig near Doneraile, whose immense haggard was destroyed by fire along with other property in mid-September: "Nine stands of wheat, three of old oats, twelve stacks of various grains, a great quantity of hay, the produce of thirty-five acres of meadow, and a cow house capable of containing twenty-six head of cattle were all consumed." Hill's losses were estimated at over £1,000.[73]

In some districts clergymen and lay impropriators succeeded in drawing home most of their tithes, despite the obstacles placed in their way not only by the burnings but also by the houghing or maiming of horses belonging to country people who had given them on hire for this purpose.[74] Yet it was one thing to draw tithes home and quite another to keep them safe once they got there. Early in October 1822 the Reverend C.P. Wallis, rector of Monanimy parish near Castletownroche, claimed the dubious distinction of being the first parson in County Cork to have any of his tithe corn burned.[75] Wallis did not have long to wait for clerical company. Later in the same month the minister of Wallstown parish near Doneraile lost three large stands of oats to incendiaries; his colleague in the neighboring parish of Clenor suffered even greater injury when firebrands destroyed tithe corn worth over £500.[76] Nor were lay impropriators spared. Early in December a party of Rockites set fire to the extensive haggard of Lieutenant-Colonel Arundel Hill at Clogheen, located midway between Doneraile and Buttevant. Though Hill had succeeded in drawing home the tithe corn of Caherduggan parish, it was all destroyed—a loss of at least £500.[77]

At first, tithe-related incendiarism was confined to the area north of a line from Mallow to Fermoy. Quiet reigned in the huge barony of Duhallow, which had been the scene of so much turmoil in the previous winter. According to a police report early in October, no outrage by fire had yet been committed there,[78] and none was noted over the next few months. Tranquility also prevailed in the southwestern part of the county. In the large barony of East Carbery and the smaller one of Ibane and Barryroe, for example, there was no concerted opposition to the collection of tithes, though the clergy were said to suffer "from the insolvency of their parishioners & the consequent necessity of taking their tithes in kind in most cases."[79] But in late September the burning of tithes began to extend south of the Mallow-Fermoy line into the baronies of Barrymore and Imokilly. In the first of these, the haggard of a rich farmer named James Fitzgerald, containing an "enormous" quantity of hay and corn, was completely consumed, and burnings were also reported from the districts of Youghal and Killeagh.[80] Yet the scale of the incendiarism in Imokilly and Barrymore hardly matched that in the northeastern baronies.

Compared with the autumn of 1822, the following winter and spring were seasons of greatly reduced incendiary activity. During a period of two months between early January and early March 1823, a total of fifty-two outrages were catalogued by the police in the four northernmost baronies of Cork. Eighteen of the fifty-two—the largest single category—involved acts of incendiarism, but these burnings all arose from other grievances besides tithes.[81] Still, a number of old scores were settled either against farmers who had entered into bargains in defiance of Rockite regulations or against clergymen who had managed to draw their tithes.[82] Having suffered on this account in October 1822, the Reverend C.P. Wallis was victimized again in March 1823. On the first occasion much of his tithe corn had escaped destruction, but on the second he lost far more—a barn, a cowhouse, nine cows, and twenty-two bags of wheat.[83]

Yet, while conflict over tithes abated temporarily, the spring and summer assizes of 1823 in County Cork reminded the public of the ravages of earlier incendiarism. At the spring assizes the grand jury received almost one hundred petitions from claimants who collectively sought compensation of over £10,500 for damages to their property by fire and other causes.[84] And at the summer assizes an additional eighty-six claimants filed petitions soliciting about £2,800 in compensation for their losses. Accounts of the first set of claims did not distinguish between losses sustained as a result of malicious fires and those arising from other causes, but a breakdown of the second set showed that as many as seventy-one of the eighty-six petitioners had been the victims of incendiary offenses.[85] There is no reason to believe that the relative proportions of burnings and other types of property damage were much different among the claimants at the spring assizes. From the surviving records it is impossible to determine how many of the fires for which compensation was sought were tithe-related. But of the seventy-one burnings cited in petitions at the summer assizes, thirty-eight took place in the three northeastern baronies, and eleven more of these fires occurred in the adjacent baronies of Barretts and Barrymore.[86] In view of what is known about the geography of unrest over tithes, it seems reasonable to conclude that this grievance accounted for a large share of the incendiary crimes. It must also be stressed that many of the farmers whose corn and hay had been destroyed by fire in the late summer and autumn of 1822 did not claim compensation at the assizes in 1823. By doing so, they would have incurred the wrath of the Rockites, who issued orders that such claims were not to be submitted to the grand jury.[87]

Anti-Tithe Campaign of 1823

With the arrival of harvest time in 1823 came a renewal of conflict over tithes. The discipline that the Rockites sought to impose was generally the same as in the previous year. Again, bargaining for tithes, or monetary compositions, were interdicted; the tenths were to be set out in the fields, and tithe owners were to be given notice to draw. Concerted opposition not only revived in the old strongholds of northeast Cork[88] but also manifested itself in many parishes north of Cork city, where an avalanche of notices fell on the incumbent clergymen—an unsurprising occurrence, thought one observer, in view of "the drilling which the occupiers of land have got from 'A Southern Farmer'" in the press.[89]

But there were certain sharp contrasts with the experience of the previous year. A greater disposition to compromise was now evinced on both sides. In the parishes around Buttevant tenants generally put all their hay into their haggards "with the understanding that the clergyman will receive a compound in money for the tythe according to the valuation." On the other hand, for a time at least, most parishioners declined to compound for their corn, which "has almost universally been stacked in the field."[90] Indeed, as late as the first week in November "scarcely a field of corn" had been cleared of its stacks "in the whole parish of Doneraile, which was the most prominent one for burnings last year."[91] But within the following month numerous agreements were reached. Farmers and gentlemen in the Buttevant district generally consented to purchase their tithes of corn because of what were termed the moderate charges of the neighboring ministers and lay impropriators, a statement implying that they had reduced their usual rates.[92] Terms of settlement were also reached in Doneraile parish. There the tithe owner was a lay impropriator named Giles, who proposed to commute the tithes in return for an annual income of £900, to be obtained by a fixed acreable assessment on all the lands of the parish. Even though the acreable rate was calculated on the basis of his average income over the past seven years, for much of which period corn prices had been substantially higher than those currently prevailing, the new method of assessment lowered the tithes owed by most parishioners sufficiently to induce their acceptance of Giles's proposal.[93]

Implacable Rockites were angered by these retreats from unqualified resistance to tithes and again engaged in incendiarism and the mutilation of cattle and horses in order to punish the backsliders.[94] But there was no systematic burning of tithe corn in late 1823, as there had been a year earlier, and the number of reprisals against those who had purchased their tithes was only a fraction of the figure in the preceding year. Several factors contributed to the reduction in tithe-related violence. First, farmers increasingly opposed the tactic of incendiarism. As much as they hated tithes, they were daunted by the steep increase in county cess that the grand jury had imposed, especially at the spring assizes of 1823, to defray the cost of compensation for malicious damage to property.[95] Presentments for over £100 fell on the county at large, but those for lesser amounts were charged directly to the barony in which the offense had been committed,[96] and farmers' losses on account of Rockite incendiarism rarely exceeded £100 in individual cases. Second, the various instruments of official repression had become more effective. The apprehension of so many alleged Rockites in 1823, including a sizeable number of reputed leaders, some of whom became informers and prosecuted their former comrades, disrupted the movement in northeast Cork and discouraged its adherents.[97] Third, the Tithe Composition Act of 1823 (to be discussed in detail below) held out the prospect of substantial relief from the tithe burden for small and middling landholders, as well as for tillage farmers generally, in parishes having a substantial quantity of grassland, to which a share of the burden could be shifted.[98] The commutation of tithes that the law of 1823 was supposed to facilitate encountered stiff resistance from graziers and dairy farmers, who were often able to prevent its adoption. Nevertheless, parochial negotiations over the terms on which the statute might be implemented absorbed much energy in late 1823 and early 1824.[99] And though the act was not adopted throughout northeast Cork, it was carried into effect in such previously disturbed parishes as Mallow, Mourneabbey, and Rahan.[100]

In other southern counties in 1823 and 1824 resistance to tithes was usually less well organized, more sporadic, and confined to smaller areas. Numerous attempts were made, however, to prevent tithe valuers from viewing the crops. In the parish of Saint Canice near Kilkenny city, for example, a large crowd chased two such functionaries off the ground in July 1823.[101] A month later, not far away, another crowd attacked two valuers and beat one of them so badly that he was not expected to survive.[102] The provision of police protection to tithe agents sometimes led to serious affrays like that at Cloonlahard, Co. Limerick, in September 1823. The tithes of this parish (Kilmoylan) were owned by the Countess of Ormonde. Her agent and his assistants were valuing tithes under the protection of four constables when suddenly they were assailed by 150 men dressed in white shirts. After shots were directed at the police, they returned the fire and wounded two peasants. (Some women who had accompanied these Whiteboys urged them to kill the constables, but without effect.[103]) A similar incident occurred about six weeks later near Milltown Malbay, Co. Clare, where a crowd of country people armed with reaping hooks, clubs, and stones attacked two valuers and a small party of dismounted police. Provoked by volleys of stones, the constables fired into the crowd, killing one woman and wounding two men.[104]

Incidents such as these were not always as spontaneous as they seemed and might have been episodes in a protracted conflict with local tithe owners. This was certainly the case in Knockgraffon parish around the village of Newinn in the Cashel district. A clash in this parish in September 1823 between some tithe valuers, accompanied by police and troops, and a party of Rockites (two of whom were wounded) had a long history.[105] In December 1821 the proctor employed by the Reverend James Butler, rector of Knockgraffon, had been attacked and badly cut. Mentioning this act of violence in a letter of April 1822, Butler lamented, "I have not received twenty pounds from my parishioners, & although I have offered to give them an abatement of twenty-five percent, I cannot get them to come into terms."[106] It was not a 25 percent reduction that they wanted, but rather the low rate of 2s. per acre, whether tilled or under grass, and if sheer persistence would have won it, it would have been theirs. Shortly before the clash of September 1823 a band of ten Rockites, their faces blackened, severely bludgeoned two men valuing tithes in the parish.[107] The struggle had endured for so long that the police officer in charge locally had almost resigned himself to the annual bloodletting.[108] But Butler, who was nonresident, and the resident minister could not resign themselves to the great loss of income. Eventually, they sent out a party of bailiffs to distrain the parishioners' cattle for the arrears. This led early in February 1824 to yet another serious affray between the drivers and a body of country people at Ballydoyle near Cashel. Though "the women and children of the neighbourhood carried away the stock that had been seized," the drivers opened fire (in self-defense, they claimed) and wounded three persons, one of them fatally.[109] In the wake of this melancholy event the crown solicitor Maxwell Blacker could only say that Butler's tithes had been "raised too much in high times," and that there was "a great want of magisterial authority" in the immediate locality.[110]

Heavy Losses for Tithe Owners

That many tithe owners in the disturbed districts of Munster and Leinster were deprived of a large portion of their income as a result of the Rockite movement is indisputable. Though detailed evidence for particular parishes or dioceses is generally lacking, the consensus of opinion among a number of knowledgeable witnesses before parliamentary committees in 1824 was that the losses had been severe. Speaking of Limerick, especially the western part of the county, Major Richard Willcocks declared: "I think in 1822, or probably since that period, the clergymen got very little, and there has been a great deal of mischief arising out of the law proceedings which have taken place to enforce the tithes."[111] Referring mainly to his own neighborhood near Cork city, the landowner Justin McCarthy expressed the view in June 1824 that during "the last two years the clergy have received very little" income, even in instances where their charges were moderate.[112] Francis Blackburne, a king's counsel appointed to administer the Insurrection Act in east Clare and the whole of Limerick, declared more generally, "I believe there are immense arrears due to the clergy in the south of Ireland."[113] His assertion received detailed support in the debates at Westminster over the Tithe Composition Bill of 1823. George Dawson, Tory M.P. for County Londonderry, told the House of Commons in June of that year:

> He had that day received a letter from an individual in [the south of Ireland] . . . , and amongst other instances the writer stated that the clergyman of a parish valued at 800 *l.* a year was now in a state of great distress, having 7,000 *l.* due to him for tithes. In another parish, valued at 1,200 *l.* a year, the clergyman was in the same situation and was now deliberating whether he should or should not throw up the living. The writer of the letter himself, who was the incumbent of a living estimated at 1,400 *l.* a year, had in the last year received no more than 160 *l.* and had several thousand pounds due to him for arrears.[114]

It is probably best to regard such heavy losses as extreme examples of what concerted opposition to tithes could achieve. At the other end of the spectrum it was certainly possible for the most remorseless of tithe owners, by constant distraining, to secure a very high proportion of their traditional revenues. In the three years before the violent tithe affray in Castlehaven parish near Skibbereen in July 1823, the Reverend Robert Morritt managed to collect as much as £2,400 out of the £2,700 due to him (the living was worth £900 per annum). But Morritt's achievement, if such it can be called (he resigned as rector shortly after the affray, his life in jeopardy), was considered something of a miracle.[115] In many southern parishes it seems likely that for two or three years during the early 1820s anywhere from one-third to two-thirds of the tithes due went uncollected.

The Road to Legislative Concessions

The intensity of Rockite opposition to tithes, coupled with other considerations, prompted an influential section of the Irish landed elite to call for reform. There had always been a certain amount of lay Protestant hostility to tithes, not only among tenants belonging to the established church but even among the gentry and aristocracy. The famed resolutions of the Irish House of Commons in 1735, which secured the virtual exemption of grassland from tithes, were warmly endorsed by members of the landed elite, who benefited both indirectly as landlords and directly as demesne farmers often heavily involved in commercial grazing. For many years thereafter, criticism of the Irish tithe system was muted among the aristocracy and gentry. With the reorientation of Irish agriculture toward tillage, a shift that gathered force after 1780, upper-class complaints about the defects and inconveniences of the system sometimes grew loud. But until the 1820s such opposition as there was at the level of the elite had always been thwarted by the hierarchy of the established church in cooperation with the government of the day. Grattan's efforts at reform in the late 1780s had certainly foundered upon this rock.[116]

From the end of 1821, however, a pro-Catholic and reform-minded administration headed by the Marquis Wellesley was in command at Dublin Castle, and its leading members recognized the need for certain measures of conciliation as well as repression if they were to quell the agrarian rebellion in the south.[117] Among the gentry and aristoc-

racy the dangers of reform, at least in the matter of the clerical establishment, no longer seemed as great as they had appeared during the
1790s, when the abolition of tithes had been one of the aims of Irish
radicals and republicans, and when the almost universal upper-class
view was that church and state must stand together or they would both
fall. Lastly, the positive attractions of modifying the tithe system had
now become more compelling. Not only might an effective scheme of
commutation assuage the most persistent and unifying of all popular
grievances, but by reducing one of the charges on the land, it would
presumably make the burden of rent, about which there was great outcry, easier to support. To the established clergy, exorbitant rents that
left nothing for tithes were the chief source of their woes. "It is therefore absurd," declared the rector of Glanworth parish near Fermoy in
September 1822, "for gentlemen at their meetings to talk of fair commutation; they want to deprive the church of their tenth to add it to
their rentroll."[118] But a more accurate assessment would have been that
the gentry, by throwing their weight behind commutation, hoped to
stop or limit the erosion of their incomes under the impact of agricultural depression.

What many of the established clergy regarded as the grand act of
lay Protestant betrayal was the meeting that took place in August 1822
at the Thatched House Tavern in London. There a series of resolutions endorsing a commutation of tithes were adopted by a group of
Irish peers and commoners initially numbering about 60. Within a
short time as many as 124 persons had subscribed their names to the
resolutions. The list included most of the 28 representative peers, a
majority of the Irish M.P.s, and a great portion of what was known as
the absentee interest. Among the signatories were to be found both
Tories and Whigs as well as such warm Anglicans as Lord Louth, the
Earl of Lucan, and Viscount Powerscourt.[119] At the summer assizes of
1822 the grand juries of at least eight counties also urged that tithes
be commuted, and their actions were subsequently seconded by meetings of gentlemen and freeholders in such counties as Waterford and
Kilkenny.[120] At an assembly in March 1823 the Kilkenny freeholders,
under their most respectable chairman Pierce Butler, actually adopted
by acclamation a resolution calling for the abolition of tithes upon the
death of the present incumbents, with an equitable commutation in
the meantime.[121]

In the course of all these proceedings harsh words were written or spoken about the established clergy by leading Protestant laymen. Thus Somerset Butler, a younger brother of the Earl of Kilkenny, declared in a remarkable letter, which he allowed to be published, that without parliamentary reform it was futile to expect any alteration in what he called "our barbarous tithe system." In his opinion "the non-resident pastor, the pluralist, and proctor will continue to live by the sweat from the brow of the unfortunate husbandman as long as the corrupt borough system exists."[122] Some lay impropriators of tithes appalled their clerical brethren by denouncing the current system. One of them, who drew from three parishes an annual income amounting "on paper" to nearly £2,000, observed in April 1823: "Such is the rooted hatred of the lower orders of the people to the very name of tithes that to receive a voluntary payment from them would be next to a miracle." He asserted that because of the present method of recovering tithes, those three parishes lost from £200 to £300 a year "in manor-court costs and bribes to bailiffs."[123]

Their abandonment by the lay elite provoked widespread anger and alarm among the clergy.[124] It was bad enough to be hounded by the Rockites, but to be stabbed in the back by their former protectors was galling in the extreme. And their alarm rose sharply when it became clear that the government was planning to introduce two Irish tithe bills, one described as a temporary measure and embodying the principle of a pecuniary composition in lieu of the usual tithes, and the other, dubbed a permanent scheme, designed to secure to the incumbent, through state-assisted purchase, a certain amount of land, the income from which would replace that now derived from tithes.[125] The hierarchy in particular, which had not been closely consulted by the administration, was outraged by what it regarded as a serious invasion of property rights. In February 1823, a few weeks before the government formally announced its proposals at Westminster, the primate, Lord John George Beresford of Armagh, blasted them as "in principle sinful and unconstitutional, and in operation . . . irritating, vexatious, and impracticable."[126] The bishops seemed oblivious to the fact that many tithe owners in the south, both clerical and lay, were disposed at least to consider ways of placing their precarious incomes on a firmer foundation.[127] The *Dublin Evening Post* could find only amusement and a subject for ridicule in the unrestrained episcopal reaction. Though it was later to pronounce the legislative result as "not worth a straw,"[128] the *Post* remarked irreverently at the beginning of May: "The bishops, without consulting their clergy, clapped their mitres on their noodles, voted an address condemnatory of the [tithe] bills to the lord lieutenant, and dispatched a tantamount petition to parliament. Talk of the pope of Rome and the scarlet lady. This out-Herods his holiness and puts her ladyship to the blush."[129]

Tithe Reform at Last: Cui Bono?

The Tithe Composition Act of July 1823[130] possessed numerous defects, but it was arguably the first significant social reform conferred on Ireland by the British parliament since 1800. The Whig magnate Lord Holland was not too far off the mark when he described the measure in the upper house as "the only miserable pittance which, during twenty-three years, the wisdom and justice of parliament had condescended to give to the people of Ireland."[131] The chief social benefit of the new law was its provision that in assigning shares of the composition to individual occupiers within any parish where the act had been adopted, the tithe commissioners were to include in the applotment the holders or owners of land used to feed dry cattle and milch cows. In other words, in such parishes the abolition of the so-called tithe of agistment under the resolutions of the Irish House of Commons in 1735 was henceforth to be null and void, and pastureland was to bear the same acreable charge as tillage ground.[132] As a result of this extension of liability over the surface of the parish, the cottier or tillage farmer who had previously paid at the rate of 12s. or more per acre for the tithes of potatoes and wheat might be assessed as little as 10d. or 1s. per acre as his contribution to the parochial composition.[133] The biggest boon, in relative terms, was to be conferred on the smallest tithe payers, that is, those holding a few acres or less, who in many parishes constituted a high proportion of the total. "This was generally the case in the southern and western parts of Ireland," declared the Irish chief secretary Henry Goulburn in the House of Commons in May 1823. "He had mentioned a parish in which, out of 2,000 persons who paid tithes, 1,200 paid less than a pound; and he could name cases without end of the same description."[134] Adoption of the act of 1823 would also remove those annual occasions for conflict and violence given by the valuing of crops, the bargaining for tithes, and the distraining of property for arrears. The need for tithe valuers and pre-sumably for drivers as well would be eliminated, and the duties of proctors would be transformed and drastically simplified.

Initially, clergymen and lay impropriators expressed great interest in negotiating about composition with their parishioners, but partly because the expectations of both parties were extravagant and also because the defects of the new law were serious, progress was slow. By mid-February 1824 the lord lieutenant had received applications from 1,033 parishes (there were some 2,450 parishes in Ireland), requesting that orders be issued for convening special parochial vestries to con-sider whether or not to adopt the act. But in only 240 parishes were the decisions positive,[135] thus indicating that the shortcomings of the law had become glaringly apparent. First of all, it was a voluntary measure requiring the mutual agreement of both tithe owner and se-lect vestrymen on the precise amount to be paid as a composition in place of the customary tithes. No rector, vicar, or lay impropriator could be compelled to accept an income that entailed a reduction in his previous level of support. In fact, a tithe owner could boost his in-come (that is, if his parishioners would allow it) by taking advantage of certain provisions of the statute, and many seemed to want to do just that. The *Dublin Evening Post* claimed in late September 1823 that thus far, incumbents were seeking one-third more than the amount for which their tithes had been set during the current year.[136] At this early stage, of course, most tithe owners serious about composition fixed their sights high in hopes of eliciting better offers from their parish-ioners. The rector of Glanworth parish in northeast Cork, for example, originally sought £2,000 a year but eventually settled with his parish-ioners for £1,200.[137] But even when tithe owners might have been will-ing to sacrifice a portion of the income that the new law entitled them to claim, they worried that its provision for triennial revision of the composition at quarter sessions, to be regulated by the average price of wheat or oats during the three years since the first agreement,[138] would result in a further reduction.[139] Some bishops refused to sanc-tion certain compositions for precisely this reason, even though the rector or vicar concerned was prepared to assume the risk.[140]

On the other side of the bargaining table the obstacles to consent were even greater. Though the amount of the composition could be de-termined simply by agreement, without strict regard to either recent tithe income or recent agricultural prices, the act of 1823 prescribed another method, highly advantageous to tithe owners, which could also be used to fix the amount of the composition. Under this procedure the parochial tithe commissioners were to take an average of "all the sums paid, or agreed for, or adjudged to be paid . . . on account of tithes" over the seven years prior to 1 November 1821. This language allowed incumbents to claim the sums that parishioners had promised to pay in their tithe notes, which would ordinarily have been greater, sometimes much greater, than final receipts. Moreover, if the commissioners concluded that an incumbent had abstained from charging the real value of his tithes in the years 1814–21, they were empowered to increase by up to 20 percent the amount of the composition that resulted simply from averaging what was paid or payable.[141] By stipulating this particular seven-year average and by creating the possibility of a substantial addition to it, the government demonstrated ample solicitude for clerical interests. In its defense it could be said that to have taken the immediate past as the standard of future compensation would have simply perpetuated the serious financial injuries inflicted by agrarian combinations. As Goulburn told his parliamentary critics in March 1824, no doubt with exaggeration, "During two years before the passing of the [composition] bill [of 1823] the income of the clergy had been reduced almost to nothing, and to fix their incomes according to the rate of those two years would be little less than fraud."[142]

But if the years 1821–23 were an improper standard in the eyes of tithe owners, the period 1814–21 was an unjust yardstick in the eyes of tithe payers. It was notorious that the high tithe rates generally prevailing between 1814 and 1821 reflected the bygone prosperity of wartime or the accident of harvest failure, and that they could no longer be justified in view of the extraordinary decline in corn prices that had taken place since 1819. The average price of wheat, compiled from figures published in the *Dublin Gazette*, was almost 39s. per barrel during the statutory period 1814–21,[143] whereas the price of wheat at Cork in September 1823, shortly after the composition act came into operation, was only about 20s.[144]

It is therefore scarcely surprising that proposals by tithe owners to make the income yielded by the rates of 1814–21 the basis of composition were invariably rejected out of hand by their parishioners. Thus in the parish of Kilmeen in west Cork the rector's average income dur-

ing the statutory period was calculated at £1,000 a year, but he was offered only £600 as a composition, and no agreement could be effected.[145] In another parish near Cork city the rector's receipts
reportedly ranged from £1,200 to £1,500 per annum and in one year
amounted to as much as £1,800, but the members of the special
parochial vestry proffered a mere £700 by way of composition and negotiations collapsed.[146] In the parishes of Ballyclogh and Dromdowney,
located north and northwest of Mallow, the select vestries were prepared to pay as a composition only a little more than half the average
annual receipts, not of what was due, between 1814 and 1821.[147] Significantly, in some parishes where the vestries did agree to enter into
composition, the figure upon which they settled was 30 percent or
more below the seven-year average.[148] This seems to have been the
case in Mallow parish, where the rector had suffered heavily as a result of the Rockite movement. Apparently, his tithes were nominally
worth about £1,000 per annum, but in October 1823 one of his parishioners expressed the belief that the rector had not received even £300
in any of the last three years and "contended he would now live more
comfortably on six hundred than on ten in wartime." In the end a sum
of £650 proved mutually acceptable; a well-secured income of £650
was vastly superior to £300.[149]

Shortchanging the Poor

The most serious defect of the Tithe Composition Act of 1823, however, was the scope it gave to graziers and dairy farmers to defeat the
best intentions of those officials and legislators who wanted to lighten
the tithe burden of conacre men, cottiers, and tillage farmers. For on
the side of the parishioners the power of decision on the crucial question of whether or not to enter into composition was entrusted to the
select vestry; the legal right to attend and vote at meetings of this
vestry was confined to the twenty-five inhabitants who in the previous year had paid the highest amounts of county cess.[150] Those included in this small group who had paid from £1 to £10 were entitled
to only a single vote, but extra votes were given for each additional
£10, up to a maximum of six votes to any person paying over £60 in
county taxes.[151]

Thus the special vestry comprised only the largest landholders and
landowners, and their predispositions heavily depended on the rela-

tive proportions of tillage and pasture in any given parish. Where the growing of corn and potatoes predominated, economic interest inclined the vestry toward acceptance of a composition, provided that the demands of the tithe owner were not outrageous or unusually excessive. But where the feeding of cows, dry cattle, and sheep was in the ascendant, economic interest ordinarily pulled the vestry in the direction of opposition. In parishes where the two basic forms of agriculture were closely mixed, the extra votes possessed by large resident landowners, charged to grand-jury cess for their demesnes, could tip the balance in favor of the tillage interest. For a combination of reasons, perhaps most proprietors were disposed to see the provisions of the act adopted, but some, especially occupiers or lessors of grassland, allied themselves with the pastoral interest.[152] It is a cause for wonder why the framers of the law, given their sincere desire to relieve the poor and the tillage farmer, created a mechanism with such potential to cripple the operation of their scheme. In large part, the answer is that they considered it both economically unjust and politically unwise to ride roughshod over the vested interests of the most privileged members of Irish rural society.[153] It was thought better to secure their participation in the process and to convert them gradually to acceptance of the act by its provision allowing them to deduct their share of the parochial composition from the rent whenever they took land after the law had gone into effect in their parish.[154]

Graziers and dairy farmers in general were quick to throw their weight against the Tithe Composition Act of 1823, not only openly by their negative votes in the special vestries, but also by exerting "all their influence over their fellow parishioners to prevent indirectly those measures which they could not oppose in fair discussion."[155] The atmosphere that surrounded the defeat of composition (by one vote) in the parish of Swords, Co. Dublin, was replicated in numerous southern districts. Of the Swords contest one observer remarked, "This was a hard-fought battle between the graziers of the parish and the vicar, both parties having for several days previous canvassed the voters as actively as on the occasion of electing a member of parliament."[156] By a margin of nine to six the vestry of Ringcurran parish near Kinsale also rejected composition, the minority consisting of a gentleman and five other persons described as cottier tenants.[157] And the account of still another repudiation—said to be unanimous—by the vestrymen of Gaulskill parish in south Kilkenny explained their reasoning thus: "They seemed to conceive that the burden of tithes would be taken from the mountain farmers . . . and laid upon those who reside on the low grounds, so as to make the latter pay nearly twice as much as the former."[158] This widespread pattern of opposition put a rather different complexion on the ancient tithe question and helped to widen social cleavages within rural society.[159]

Amending the Law of Tithe Composition

Despite the repeated defeats of composition by the pastoral interest in the special vestries, the government refused to change in any drastic way the mechanism by which parochial decisions were reached. When announcing in March 1824 a list of proposed amendments to the Tithe Composition Act of 1823, Henry Goulburn frankly admitted the serious impediment raised by the occupiers of grassland, but he insisted that a solution must await "the operation of that part of the act which, on the granting of new leases, threw on the landlord the burthen of the tithes" in compounding parishes.[160] In fact, however, the amending act of the following June did modify the membership of the select vestries by admitting to both attendance and voting local justices of the peace and those possessed of a freehold estate in the parish worth at least £50 a year.[161] In those parishes—relatively few in number—where fewer than twenty-five persons paid county cess in amounts exceeding £1, the remainder of the quota could now be filled from the ranks of the smallest taxpayers.[162] The effect of these two provisions, and especially of the first, was to strengthen marginally the tillage interest in the special vestries.

In the previous session of parliament some M.P.s and peers had called loudly for making the whole business of composition compulsory rather than voluntary, but when the question was put to a vote in both houses of parliament, it had been soundly defeated by a convergence of forces from left and right.[163] In the amending act of 1824, however, a small step in the direction of compulsion was taken. In any parish where the tithe owner and the vestry agreed to proceed under the law, but subsequently one or the other of the two parties failed to appoint its tithe commissioner, the lord lieutenant was authorized to select a commissioner to act on behalf of the defaulting party.[164] Thus it became possible to override the negative second thoughts of the tithe owner or the vestry. Only to this quite limited extent was compulsion to be employed. By another amendment the power given to tithe commissioners in the act of 1823 to raise a mutually agreed composition by the amount that it fell short of average tithe income during the years 1814–21 was repealed.[165]

But this modest concession to tithe payers was counterbalanced by a much more significant one to tithe owners. Whereas the measure of 1823 had provided for the triennial revision of compositions at quarter sessions in accordance with fluctuations in the price of wheat or oats, the amending act of 1824 stipulated that all future compositions could be revised only after seven and again after fourteen years.[166] This provision effectively laid to rest the fears of the hierarchy of the established church and of many clerical tithe owners that, having already submitted to some sacrifice of income in the original composition, rectors and vicars would have no choice but to acquiesce in a second surrender three years later.[167] The recovery in grain prices that became apparent in 1824 helped to consolidate the impact made on the clerical mind by the provision for septennial revisions.

Though the expectations confidently expressed in some quarters in 1825 that composition would be all but universal within a few years were not realized, rapid progress was made in the late 1820s. Fewer than 400 parishes out of a total of 2,450 had agreed to compound for their tithes by early 1824, but more than 1,500 had done so by early 1832. Over that period the amount of income derived by tithe owners from compositions increased from about £111,000 a year to £442,000, thus reducing the sum due in the form of customary tithes to perhaps £262,000.[168] The early compositions were nearly always the result of haggling and compromise with the special vestries, the device of the seven-year average being almost totally disregarded.[169] And this continued to be the general practice after 1824 as well. Tithe owners were nevertheless increasingly eager to come to terms, the established clergy even more than the lay impropriators. In his diocese, said the Protestant archbishop of Cashel in March 1825, the clergy "have been all very willing, ready indeed to make almost any sacrifice for the sake of an arrangement" under the composition acts.[170] Pressure from the poor, tillage farmers, and landowners gradually eroded the opposition offered by graziers and dairy farmers. In the diocese of Cashel and Emly, for example, where resistance from the occupiers of grassland was initially so strong that only 14 of 152 parishes had compounded by March 1824, the acts were implemented in as many as 104 by early 1832. Over the same period the corresponding figures for the diocese of Ossory (138 parishes) rose from 6 to 70, and for the diocese of Waterford and Lismore (108 parishes) from 3 to 61.[171] Yet there was a bitter irony about the successful working of the Tithe Composition Acts of 1823 and 1824: while they conferred a much appreciated boon on tillage farmers and the poor, they also helped unwittingly to sow the seeds of the tremendous onslaught against the tithe system in the early 1830s. Most graziers and dairy farmers who were subjected to the unwanted novelty of composition, after escaping for so long any significant liability to tithes, became more than ever the inveterate enemies of the entire system. From their ranks many of the ablest and most vigorous leaders of the tithe war of the 1830s would be drawn.[172]

Seven

The Issue of Rents

Complexities of the Land System

Confronted with a punishing economic downturn in the early 1820s, the Rockites were as fully determined to lower rents and control the occupation of land as they were bent on reducing or abolishing tithes. This book began with the story of the violent campaign against the collection of the customary rents and huge arrears on the Courtenay estates around Newcastle West in County Limerick. That originating conflict had many special features, as we have seen, but so too did the much wider struggle that followed in its wake. A proper understanding of the innumerable conflicts about rent that punctuated the Rockite movement of 1821–24 must take into account the central complexities of the land system in prefamine Ireland. In between the proprietor of the land at the top of the rural social hierarchy and the occupying tenant or cottier near the bottom were typically a series of middlemen. It was not uncommon to find from three to five layers of landholders of some kind between the highest and lowest rungs of the tenurial ladder. These middlemen might be Protestant gentlemen, members of the minor gentry who, besides farming on some scale themselves, rented out considerable tracts to middling or extensive farmers or, more usually, to large numbers of small tenants; parts of their demesne lands, or the holdings that they farmed themselves, might be let in potato gardens to conacre tenants for the growing season. Thus, while owing rent to the proprietor or head landlord, these gentry middlemen needed to collect rent themselves from a variety of tenants below them. Many large farmers or graziers, usually Catholic in religion in the south of Ireland, occupied a somewhat parallel position in relation to the smaller tenants, cottiers, and bound laborers to whom they regularly rented or sublet land in differing quantities. In the aftermath of the French revolutionary and Napoleonic wars, when the former agricultural boom gave way to something like a bust, landed proprietors in increasing numbers displayed a keen interest in eliminating middlemen and in pocketing the "profit rents" that the middlemen were in the habit of extracting from those to whom they had sublet most or all of their land. Where elimination seemed inadvisable for managerial or political reasons, then landowners or their agents might seek to clip the wings of gentry middlemen by reducing the extent of their holdings.[1]

Although the process of eliminating such middlemen was a gradual one in the prefamine decades, it was gathering momentum during the economic downturn of the early 1820s. The capacity of this development for generating social conflict mostly derived from its association with the removal of bankrupt subtenants or those who had allowed heavy arrears of rent to accumulate. Large Catholic farmers with subtenants, or with unbound laborers renting conacre plots, were often inclined to be as harsh and unforgiving as any Protestant landowner or middleman when faced with heavy arrears of rent or demands for drastic abatements. The popular response to such farmers if they engaged in evictions or took over evicted holdings was frequently even more vengeful than the treatment meted out to members of the landed elite. Large Catholic farmers and minor Protestant gentlemen living in the midst of the perennially poor or the suddenly impoverished made relatively easy targets for the Rockites.

Popular Outcry over Excessive Rents

Many contemporary observers insisted that Irish landlords in general had been exceedingly slow to adjust their rents in the wake of the first postwar depression and the second that commenced in the years 1819–20. Criticizing the numerous proprietors who had clung to high rents despite the accumulation of huge arrears, the *Dublin Evening Post* predicted that the year 1820, which saw livestock and corn prices steeply decline, "is destined to teach them a lesson which they can never forget."[2] But Irish landowners were apparently painfully slow learners.[3] More than two years later, in November 1822, when the economic depression had taken a further downward plunge, one observer claimed that "the prices of cattle and grain are nearly at the standard of 1760, when land set at 7 to 15 shillings per acre." Nevertheless, as he pointed out, rents of 40 or 50 shillings an acre were still quite common.[4] This widespread lack of adjustment to economic realities greatly helped to fuel the Rockite movement.

To judge from their threatening letters and from the anguished cries of landowners and middlemen, it is clear that the Rockites demanded fairly sweeping rent reductions or abatements to compensate for the heavy losses incurred through falling prices and crop deficiencies; they also believed that as a matter of justice, arrears should be forgiven or overlooked for as long as the economic downturn lasted. Some early threatening notices in County Tipperary demanded reductions as large as 60 percent.[5] And one notice posted at or near Dromcolliher in Limerick in February 1822 ordered the tenants of Luke White not to register as 40-shilling freeholders (county voters) unless they first received abatements of 50 percent.[6] Other notices made it clear that the followers of Captain Rock would resist efforts to collect arrears of rent from defaulting tenants by the customary process of seizing crops or livestock. One notice posted on the chapel door at Rockhill near Bruree in County Limerick in November 1821 "cautioned all persons from paying rents" and warned that "no landlord [is] to attempt distraining cattle or goods for rent, [or] if they do, they are to suffer the death of Major Going."[7] And a lengthy and very literate Rockite notice posted on the chapel gate during Sunday Mass in Kinneigh parish in west Cork early in January 1822 decreed that a whole range of financial exactions and hostile legal actions be abandoned forthwith: the collection of "all rackrents and backrents"; the execution of writs or processes against defaulting tenants or tithe debtors; the distraining of crops or livestock for arrears of tithe or rent; and bidding at any public auction of distrained produce that would put cash into the hands of pressing landlords or tithe owners. Violators of these strict injunctions were warned that they would "suffer capital punishment."[8]

Numerous landlords and other elite observers testified to the widespread enforcement of such notices. Colonel James Crosbie of Ballyheige in north Kerry declared in January 1822: "I hear of nothing but parties being out every night . . . trying to deter every person from collecting rent or executing any order of law"; he cited the case of one of his subagents, who was warned by a visiting band of Rockites that "if he attempted to distrain for rent . . . , he would be put to death."[9] Another Kerry landowner, citing attacks on two "canters" of small farms near Killarney, angrily insisted that "the lower classes have the lands nearly on their own terms at present."[10] And writing from the sybaritic security of Bath in England, the Earl of Clare could complain in January 1824, following the murder of one of his cattle-seizing drivers in County Limerick, that "the chief baron [a legal officer] is quite right when he says that tho' the landlords may have the title deeds of their estates, the lower orders are in possession of the country."[11]

Rents Largely Unpaid

Under these circumstances the usual rents were by all accounts extremely difficult to collect. Toward the end of 1821, when the Rockite movement was only in its early stages, the *Dublin Evening Post* declared: "All our country letters agree on one point, that there is no such thing as getting rents from the peasantry."[12] The same newspaper again drew attention in the following summer to the desperate general financial condition of landlords and to their unusual stratagems in certain counties: "It is true they do not receive more than one-half of their rents, on an average; but it is equally manifest that they get all that is to be had—that in some cases, as in Clare and Mayo, they have become factors [middlemen in grain sales] and have brought their corn to the best market."[13] Reporting on County Kerry in the fall of 1822, the landowner Daniel Mahony observed that "there is certainly a general disposition to pay little or no rent, which must bring ruin on many of the gentry of the country who are in any way encumbered [with debts]."[14] At about the same time a well-informed observer, writing about the large baronies of East Carbery and Ibane and Barryroe in Cork, firmly asserted that three-quarters of the farmers of this area were "absolutely bankrupts" and could simply not pay the current rents.[15] Even farmers who had formerly been living "in most comfortable circumstances" had sometimes been reduced to "the utmost distress" by the end of 1822, with crippling consequences for their landlords. At that point one commentator asserted that some landlords had gone without rent "for the last two years."[16] The dire financial position of landlords and Protestant ministers had not much improved by the summer of 1823. It was then said of counties Cork and Limerick that "scarcely any rent has been rendered, and not more than one-third of the tithe has been collected."[17] This situation prompted an editorial writer in the *Dublin Evening Post* to remark sharply: "The peasantry are doing their business much more effectually than if they took the field in the open day."[18]

Some landlords did respond to the severe economic downturn and to the extremely widespread expressions of popular discontent by offering abatements of various kinds. For example, it was announced in December 1821 that the Earl of Charleville had forgiven all arrears on his Limerick estate up to 1 November of that year. And his fellow Limerick proprietor R.J. Stevelly reportedly reduced his rents by as much as one-third, retrospective to 1818.[19] But contemporary newspapers offered very few accounts of such indulgence on a systematic scale across entire estates. Under the pressure exerted by the Rockites, numerous landlords felt compelled to make economic concessions that would otherwise have been unthinkable, as the Limerick proprietor Thomas Studdert explained to the undersecretary William Gregory in May 1822:

> I am acquainted with many gentlemen who, to avoid those evil consequences arrising [*sic*] from ejectments, were forced by the exigencies of their circumstances to remit considerable arrears due on their lands, and in addition had to pay some money to obtain a peaceable surrender, thereby to qualify themselves to select honest tenants likely to pay them in future, who would not on any other conditions become responsible for rent or subject themselves to hostility from the old tenants through the agency of unknown persons, such as Captain Rock & his followers.

Studdert acknowledged that he himself had suffered considerable loss by adopting this approach, and he realized only too well that his actions represented a capitulation to "the Whiteboy system." His excuse was simply that would-be new tenants insisted on his making amicable settlements with the old occupiers before they would risk taking the tainted lands.[20]

Yet such sacrifices of income hardly guaranteed success. For example, the Clare landlord Richard Creagh of Dangan forgave "an immense arrear" to a group of cottier tenants on the lands of Gorteen in the barony of Upper Bunratty; his hope that there would be no reprisals against the new tenant whom he had installed proved vain.[21] And a County Cork landlord named Roberts had a similarly exasperating experience. For some thirty years he had let a farm in the Shanballymore district to the two brothers David and James Regan and had allowed them to retain possession even though they did not pay their rent very regularly. In about 1820 Roberts forgave arrears of some £90 owed by the surviving brother James Regan in return for his surrender of the farm. Roberts kept the land in his own hands until the beginning of 1824. He then let it to a new tenant for whom he was building a new house; Rockites promptly destroyed the dwelling.[22]

The extraordinary conduct of Richard Creagh in Clare demonstrated the leeway that was sometimes extended by landlords even to the most refractory of tenants. Six of them were to be prosecuted under the Insurrection Act at Sixmilebridge for tumultuously assembling and rescuing cattle distrained by his agent—a crime for which they could have been transported if convicted. But at the last minute Creagh decided to let the malefactors off the hook. The prisoners (and perhaps other tenants of Creagh's) paid their rents, the prosecutor stayed away from the court, and Creagh abandoned the case. Very unhappy about this outcome was the crown solicitor; he was incensed because the former prisoners would now be returning to the parish of Feakle, which "had been disgraced beyond any other by its outrages and incendiaries."[23]

Pressing Hard to Collect Arrears

If landlords collected much less than their customary rents, this was usually not because they ceased to press their defaulting tenants. Auctions of seized stock or grain were extremely common in these years as a means of extracting rent from tenants whose financial means had been severely reduced. It was said in December 1821 of the Lansdowne estate in south Kerry, for example, that the canting or auction of the stock of defaulting tenants had occurred at least once a week since the end of the Napoleonic wars; by 1821 the estate was described as a scene of "horror and despair."[24] The grievous loss of income suffered by many landlords stemmed from the violent measures taken by tenants themselves to frustrate or render useless the traditional method of distraining for arrears of rent by seizing crops and livestock. Pound keepers were sworn in effect to abandon their profession, at least for the time being; they were ordered not to receive any cattle taken for nonpayment of rent or tithe.[25] Some pound keepers who remained faithful to their duties were severely punished, to the point of having their houses burned down.[26] In numerous other cases pounds that had been full of distrained cattle and sheep were simply broken open and the livestock rescued. For instance, Rockites assembled in great numbers early in 1822 in the Lixnaw district of Kerry (southwest of Listowel) and leveled a pound near Kilflyn; another band of Rockites in south Kerry destroyed a pound on Lord Headley's property at Rossbehy in the Cahersiveen district.[27] Similarly, a large party of about a hundred men raided the pound of Kilmacthomas in County Waterford at the end of March 1822 and "forcibly took away a great many head of cattle that was there distrained for rent, at the same time declaring their intention of murdering the keepers if they caught them."[28] And a crowd of some four or five hundred people rescued from Ballyclogh pound near Mallow a considerable number of livestock seized for rent by a Cork landlord named Foley; they emptied the pound in February 1823 and destroyed its walls.[29]

Many defaulting tenants did not simply wait for their landlord to levy a distress against them; instead they took prompt action to avoid the possibility of distraint by effectively frustrating the law. This often involved a large-scale collective response. A police official in the Rosscarbery district of west Cork reported early in October 1822 that several landholders had recently assembled "bands of ruffians, sometimes to the number of 200 or 300, with arms and horses," and carried away the whole produce of their farm so as to escape the payment of both rent and tithe.[30] This sort of behavior was not at all unusual. Even a Protestant minister named Kenny residing mostly in Kinsale, who held an interest in a farm in the Macroom district, gathered together a large number of horses, carts, and human helpers and spirited away all the corn, hay, potatoes, and fixtures of the place. As a disgruntled local landlord archly remarked in the following November, "This conduct, though perhaps excusable in Mr. Kenny, is now pleaded by every runaway tenant in this neighbourhood. . . ."[31] A more novel type of dishonesty was practiced by a league of hard-pressed tenants on either side of the Clare-Galway border in October 1823. Accompanying a report that cattle had been driven off a farm at Oughtmama in the

In the face of massive default in the payment of rents during the depression of 1819–23, landowners, middlemen, and their agents frequently resorted to such forcible measures of collection as "driving" the cattle and sheep of impoverished tenants into some local pound. The Rockites, however, retaliated by breaking open the pounds, assaulting the drivers (as well as the "keepers" of distrained crops), and stoutly resisting landlord pressures in other violent ways. "Driving" was a characteristic feature of the Great Famine (as in this sketch of the practice in County Galway in 1849), but famine conditions generally pulverized the capacity for widespread collective resistance. (*Illustrated London News*, 29 Dec. 1849)

barony of Burren in Clare was the following statement: "The lands were held in common with other tenants, and to avoid the payment of their part of the rent, this illegal step was taken" by the friends of the tenants Patrick and Peter Kavanagh of Parkmore, Co. Galway.[32]

In order to prevent tenants from "running away" with the proceeds of their livestock or grain without discharging arrears of rent, landlords regularly resorted to the practice of placing armed keepers on the goods. But in the period of the Rockite movement this was only an invitation to the rebels to collude with the keepers for the recovery of the produce or to punish them severely (even to the point of murder) when they resisted the strong assertion of force. In one case twelve or thirteen keepers reportedly found themselves confined in an adjacent house when under their noses a party of Rockites removed a substantial quantity of wheat that the landlord R.P. O'Shee had seized for rent near Kilkenny city in September 1823. The Earl of Normanton's keepers were equally ineffective at about the same time at nearby Brownstown in the face a crowd of over two hundred country people, who carried off six acres of wheat in an hour.[33] Still other keepers appear to have placed their own safety well ahead of the interests of their employers and to have allowed themselves to be "tricked" or easily intimidated into inactivity when crowds rescued the livestock or crops of defaulting tenants.[34] Some keepers employed by a distraining landlord named Crofts at Streamhill near Doneraile ran off at the first sign of trouble in November 1823: "The keepers acknowledge that they fled thru terror on observing a large party of men enter the field with fire, and they deny having any knowledge of the perpetrators."[35]

Ready Resort to Extreme Violence

Keepers and others who made the mistake of resisting these large crowds or even smaller parties were courting death or serious injury, as an avalanche of cases gruesomely demonstrated. James Egan of Rathlogan in the Kilkenny barony of Galmoy was stoned and pitchforked to death in September 1822 after distraining a subtenant in arrears and seeking to prevent the distress (some corn) from being rescued.[36] A man named Harvey, searching for a distress stolen out of official custody, was beaten and left for dead at Tubber fair in County Clare in September 1823.[37] A party of Rockites "savagely beat" the keepers assigned by a landlord to guard certain tenants' corn at

Ballyea near Nenagh in north Tipperary in January 1824.[38] And in perhaps the worst case of this kind, the two keepers Robert Vallance and Laurence Lisle were both stoned to death in the Mitchelstown district for their efforts; they were employed by a farmer named Michael Roche to secure the produce of two subtenants on the Hyde estate near Fermoy whom Roche had earlier distrained. Shortly before the keepers were killed, one of the subtenants had reportedly demanded of Roche "if he intended to become a tyrant over them" or "to act as C[a]esar did in Rome and be treated so."[39]

The common practice of distraining for rent or executing court decrees in the face of hostile crowds often constituted a provocation that sparked serious violence and even murder. Deaths occurred on both sides in these frequent affrays. Members of the Ballylongford yeomanry corps distraining cattle on a farm near Ballybunnion in north Kerry had to beat off a crowd estimated at over a thousand people armed with scythes, pitchforks, and stones in December 1821. The corps killed three of the resisting peasants (two men and a woman) and wounded several others.[40] In a successful rescue of cattle from constables acting under a court decree against a farmer named Boland near Borrisokane in County Tipperary, the rescuers killed one of the constables in February 1822.[41] The outcome apparently went the other way in a different affray in County Limerick a year later. On this occasion troops prevented three or four hundred country people from rescuing cattle seized for rent at Kilbehy in the Rathkeale district; in this clash the troops shot one of the would-be rescuers.[42] In a different case of violence that occurred just outside Templemore in Tipperary, the police killed a man named Ryan, wanted for the rescue of some pigs, as he was allegedly trying to escape.[43] But there were numerous victims on the landlord side. A cattle driver for a court-appointed receiver of rent on a large property near Killaghy Castle in the Callan district of County Kilkenny was shot while performing his obnoxious duty in March 1822.[44] Another driver named William Green, "employed by gentlemen . . . in cases where rent has not been paid," was also shot within two miles of Tipperary town in August 1823 as he returned home from his dangerous labors.[45] While Neville Payne Nunan was executing a legal process in the Rathkeale district in the following September, a crowd of country people attacked him and his servant; Nunan was hamstrung and suffered several broken bones.[46] In one of the worst incidents of this kind, a farmer named James Gorman retaliated against two brothers named Fogarty (Phillip and William) after they had distrained his cattle on the lands of Copse near Templemore in Tipperary in June 1824. With the aid of a crowd assembled at four o'clock in the morning, Gorman tried to rescue the distrained cattle but was shot dead on the spot by Phillip Fogarty; his brother William Fogarty had his skull fractured and was "dreadfully" beaten by the crowd.[47]

The agents of landlords big and small had much reason to fear that the extreme pressures they applied to secure rental income for their employers would bring the wrath of the Rockites down on their heads. The agent of a landlord named Creed who had evicted some tenants near Fermoy in north Cork was ordered to reinstate them in a threatening letter of January 1822; Creed might now be hiding in Cork city, asserted the writer of the notice, but Captain Rock would find him there if he refused to comply.[48] Rockites threatened in October 1823 to end the life of James Enright "if he did not drop his late undertaking"—that of agent to the Dawson estate in the Rathkeale district of Limerick.[49] Lord Hawarden's agent, a man named Stewart, was nearly murdered as he returned on horseback from Tipperary town to his residence at Dundrum in August of that year; this was at least the second attempt made to kill him.[50] The "land steward" and the gardener of the Limerick landowner George Massey were also the targets of an abortive attempt at assassination as they came back to Glenwilliam in the Croom district in July 1824 after distraining the cattle of some of Massey's tenants.[51] Rockites did murder Denis Browne, the steward of William Cox of Ballynoe House, in the Ballingarry district of Limerick in February 1822;[52] this same fate befell James Condon, the sub-agent of the Earl of Egremont, as he returned in January 1823 from Limerick city to his house at Fedamore.[53] They also killed a "most respectable farmer" named Callaghan as he went home from the fair of Tipperary town in December 1823. In assigning a motive for this killing, the reporter noted that Callaghan had occasionally acted as an agent for his landlord, a gentleman named Brazier.[54] And in the Newinn district of Tipperary, which was wracked by agrarian violence in these years, Rockites shot the steward of Leonard Keating, and in March 1824 they stoned his herd in the head when he sought to prevent some of them from setting loose Keating's cattle.[55] Unusually, the Rockites employed a more subtle approach in their effort to ruin the career of an "understeward" named John Carroll in the Fermoy district in 1823. Exasperated by his actions in pressing for arrears of rent and distraining their cattle, his enemies concealed arms and ammunition in his house and then sent the police there to discover his supposedly criminal behavior—a transportable offense. Carroll was arrested and tried at a special sessions under the Insurrection Act in Fermoy before the ruse was exposed.[56]

Evictions and Rockite Violence

Apart from the extreme difficulty of collecting the customary rents, no single issue gave landed proprietors, middlemen, and large farmers more trouble in the early 1820s than the systematic and persistently violent Rockite response to the eviction of tenants. Of course, the two questions were intimately related to each other. Unless tenants who fell heavily or hopelessly into arrears could be dispossessed, and new tenants found for the holdings they had vacated, it would only make extracting rents from the tenants who remained in possession that much more difficult. In the vital matters of cost and simplicity of procedure, Irish landlords were in a much stronger position after 1815 than they had ever been before. Prior to 1815 landlords wanting to eject tenants had to contend with heavy expenses and cumbersome legal procedures. Action had to be taken in the superior courts in Dublin, and in the unlikely event that a tenant took defense there (that is, contested the landlord's suit), the cost could be anywhere from £50 to £150; even when no defense was taken, securing a decree of ejectment in one of the superior courts cost about £18. But the Ejectment Act of 1816 transformed the traditional situation almost out of recognition.[57] The new law created the procedure known as civil-bill ejectment, under which a landlord might bring suit in a local court against any tenant whose rent did not exceed £50 a year, and if the landlord's suit were successful, the legal cost of the ejectment decree was less than £2.[58] With the Ejectment Act of 1816 at their disposal during the early 1820s, landlords were much more likely, when faced with concerted opposition to the payment of rent, to resort to evictions in an effort to cripple or weaken popular resistance. But by the same token the economic downturn and a greater propensity among landlords to dispossess defaulters steeply raised the stakes for tenants of all kinds and made it almost certain that they would resort to violence on this score more readily than ever before.

In both word and action the Rockites incessantly proclaimed their opposition to evictions and to the letting of the vacated farms to new tenants. They promulgated "laws" that were designed to ensure that no one else took an evicted holding for a long period of time, with three, five, and seven years being the terms most frequently specified in their notices.[59] These injunctions were intended to be retrospective as well as prospective. Thus one notice posted in County Tipperary in December 1821 decreed: "Any farmer or farmers that has taken ground within these five years must surrender it to the former tenant, or if they do not, they will receive a most unmerciful death."[60] Another notice posted early in March 1822 on the door of the Catholic chapel at Killorglin in Kerry warned that "all canters of ground these seven years past" must "surrender their lands to the former owners."[61] So too did a third document written in rougher language and posted in County Kilkenny around the same time: "i am giving notice to every one that tuck [took] ground this seven year . . . , and there is not one of them that i served with notice but will burn to ashes, and my name is Captain Rock. . . ."[62]

The surviving records of the Rockite movement offer hundreds of individual instances of attacks and violent assaults against the persons or property of those who took land from which other tenants had been evicted. Length of possession provided no defense against such violence. The land agent James Hickson reported an attack in January 1822 on the house of a tenant on the Trinity College estates in south Kerry; the tenant was ordered to quit a farm that he had taken two years earlier and "to obey all future orders from Captain Rock to cant no lands."[63] Only weeks later, Rockites in the Ballylongford district of north Kerry raided the house of Patrick Hennessy, who had apparently appropriated the holding of certain subtenants whom he had long ago evicted. Hennessy was sworn in a violent manner "on a prayer book to quit his farm in five days [or] else they would burn his house and family"; he was instructed "to give up the farm to some tenants who had about seven years since run away with their rent from him."[64] Early in October of the same year Lord Glengall reported two similar attacks in the Caher district of Tipperary. In one case a very large farmer with about 500 acres was threatened with murder if he did not quit his lands; "the only cause assigned . . . is his being a stranger in the country, though he has resided here about five years or more."[65] In another case in the same district the occupants of a holding were badly beaten and commanded to depart within two weeks, "they being considered strangers, though they have resided many years in the country." It was the common practice to burn down the houses of those who ignored such warnings or to strike preemptively against landlords who evicted defaulting tenants by destroying the cabins of the former occupiers. The military officer Thomas Arbuthnot informed his superior that there had been several burnings in his district of north Cork since the end of March in 1823—all caused either by ejectments for nonpayment of rent or by "new" tenants taking evicted farms. Arbuthnot commented that the exaction of vengeance in such cases had "become progressively systematic" and now extended to lands from which tenants had been ejected several years previously.[66]

Besides setting time limits for evicted farms to lie vacant (unless restored to the former occupiers), the Rockites sought to impose other restrictions on would-be land canters and on landowners. Those gentry, middlemen, or large farmers who placed dairymen on evicted lands were firmly instructed that to do so was a serious violation of Captain Rock's laws. "Gentlemen and rich churls" were plainly told in a threatening notice posted in the Kanturk district of northwest Cork in February 1822 that they were "to keep no milkwomen, dairymen, or herdsmen but to let the land to honest, industrious tenants for the value."[67] One of the oaths sworn in 1822 by followers of Captain Rock in parts of County Waterford reportedly specified that any tenant "leaving a farm" was "to prevent any dairyman from taking it."[68] The landlord William Stawell, with property located around Doneraile, was ordered in March of the same year to "discharge all herdsmen, dairymen, and milkwomen from the lands of Wallstown, Bally Losseen, and every [other] place where you hold lands"; he was also informed that "in consequence of a new code of laws recently given out by General John Rock, Legislator General of Ireland," it was "unlawful for any gentleman to hold any more lands than that which immediately adjoins his dwelling residence." Stawell was therefore to remove his livestock from these "prohibited lands" and "afterwards to set the same to poor industrious people at a resonable [sic] rent."[69]

This sketch from the period of the Land War of the early 1880s was subtitled "A Visit from 'Rory of the Hills.'" The same type of nocturnal action by armed agrarian rebels using burnt cork to disguise themselves often occurred back in the early 1820s, and this scene transposed to that period might justifiably be labeled "A Visit from Captain Rock." A prime target in the days of the Land League was the land grabber; such an offender the Rockites were most likely to have called a land canter or a land jobber. In both periods the purpose of making "a visit" to such persons was to intimidate them into surrendering any holding they had recently taken in defiance of loudly proclaimed popular norms. (*Illustrated London News*, 22 Jan. 1881)

Rockites carried out a series of attacks in May and June 1824 against two dairymen holding land from the gentleman John Glover of John's Grove, also in the Doneraile district.[70] Reinvoking the localist ethic that was deeply embedded within Irish rural culture and within the Rockites' value system, other bands of rebels told offending new tenants that they must look for holdings only in the place of their nativity, and that "strangers" must always give way to the claims of local people in the letting of any land.[71] Among the threatening letters reportedly raining down in County Kilkenny in April 1824 was one directed against would-be land grabbers, "save and except a person residing within a half a mile of the spot [of the evicted farm]."[72] Of course, under the cover of the popular legitimacy of the Rockite movement, some people looked to achieve private gain. One Kerry land agent sought bidders early in 1822 for a piece of land from which no one had been evicted; he expressed his acute annoyance at the public misrepresentation of the situation: "Captain Rock is interfering with this letting, vowing vengeance to those that bid for it; this daring gentleman knows it's no canting of the former tenant or of any other person. . . . This stroke of generalship in the Capt[ai]n is to get it [the holding] for his friend, the person I told you I did not like [because of] his connections in the neighbourhood."[73]

Gauging Rockite Effectiveness

To what extent did the intimidation and violence so widely practiced by the Rockites against "land canters" and "land jobbers" actually curb the eviction of defaulting tenants? If one were to judge from the pessimistic comments of numerous landowners and agents, the impact would appear to have been substantial. Writing in April 1822, the Kilkenny proprietor John Flood bitterly deplored the landlords' loss of control in the whole area from Callan to the Waterford border. Threats of death and burning, he declared, were having

> such an effect on many [tenants] who have been for a considerable time in quiet possession, & who have expended much money on their farms, that they are about to surrender them to the landlords. . . . I *know* that any ruffian can raise a party to attempt to gain possession of ground even formerly unheld by his family. Such is the system of terror![74]

According to Thomas Studdert of Askeaton, west Limerick was in worse condition: "Large tracts of rich & valuable land are in consequence left untenanted which have been ejected, and are likely to remain so unless restored to the orriginal [sic] holders or leased to some of their agents in fraud on any terms they are pleased to offer."[75] From his residence near Kildorrery in Cork, the landlord and magistrate Andrew Batwell reported in April 1823 that the Rockites controlled the occupation of land in that district: "The insurgents have completely possessed themselves of all the landed property, as no proprietor can venture to turn out a tenant, nor can he procure a bidder for his lands, tho' surrendered voluntarily by the former occupier."[76]

But accounts of the many hundreds of evictions and reprisals recorded in contemporary newspapers and in the State of the Country Papers make it certain that at most the Rockites were able to restrain the frequency of ejectment and never came close to halting it completely. Few landowners faced with the serious loss of rental income were able to avoid evictions altogether. The predicament of General Barry in north Cork was replicated all over the region in which the Rockite movement took root. "Like most of the landed proprietors in this part of the country," noted the military officer Thomas Arbuthnot in April 1823, Barry had "lately found it necessary to eject a tenant from whom he could obtain no rent," and as was not at all unusual in such cases, several subtenants were dispossessed at the same time. Arbuthnot was bracing for the expected violent reprisals.[77] Efforts to collect large arrears of rent through a limited number of evictions and the installation of some new tenants seem to have been the cause of a series of outrages on Thomas W. Foot's estate in the Mallow district in 1823. An armed party of Rockites broke into his residence at Donville in March of that year and compelled Foot to swear that he would forgive the arrears of a certain tenant,[78] but to judge from the turmoil on this property, other tenants must have been in a similar condition of financial distress, along with their landlord himself. The house of Foot's steward, a substantial farmer named John Halloran, had been burned in the previous January after the steward had been permitted to remove his furniture,[79] while two months later, other tenants or employees of Foot, probably installed recently, were locked inside a farmhouse that the Rockites set on fire.[80] The Rockites also destroyed two different houses, one after the other, occupied by a "strange" tenant to whom Foot had let an apparently evicted farm.[81] Landlords in other counties behaved like those in north Cork, with similar results. For example, the County Clare proprietor Thomas Jackson evicted some tenants near Limerick city and relet their farms to at least six others; as soon as they took possession in the spring of 1822, their dwellings were destroyed by fire.[82] Other Clare landlords had the same embittering experience as Jackson. One named Sampson evicted at least five tenants at Sheeaun in the Scarriff district for nonpayment of rent, with one of those dispossessed owing arrears of as much as £150. In April 1824, Rockites burned five houses over the heads of the new tenants whom Sampson had installed.[83] And Mrs. Prudence Nihil was made to suffer repeatedly in 1824 for having evicted at least seven defaulting tenants from lands in Kilchreest parish in the barony of Clonderalaw in May of the previous year. Since then, four houses on her property had been destroyed, and she had experienced other significant losses.[84]

Middlemen as Rockite Targets: The Franks Murders

The widespread persistence of the middleman system gave rise to much of the popular discontent over rents harnessed by the Rockite movement. On numerous occasions the Rockites manifested a general desire to eliminate middlemen and instead to hold their farms under the head landlord or owner of the soil.[85] This attitude represented a clear recognition that middlemen were charging a premium for their status as intermediate managers of land and collectors of rent for the proprietors. Typically, middlemen leased extensive tracts at low or modest rents from the owners and turned around and charged much higher rents from all those underneath themselves, with the subletting small tenants and cottiers commonly burdened with the highest rents per acre. In urging moderation on landlords in general, even the authorities maintained that middlemen in particular needed to lower their rents as a contribution to the restoration of social peace.[86]

The background to the murder of the three members of the Franks family in September 1823 illustrates in an especially striking way how the middleman system provided a conducive framework for Rockite violence over the issue of rents. The father of the family, Thomas Franks, was a middleman under the dowager Countess of Kingston until her death in January 1823; from her he held the lands of Scart (a large townland of just over 500 acres). His relations with the sub-tenants at Scart were acrimonious, to say the least, and the undertenants were eager to escape from the clutches of the Franks family and to become the direct tenants of Lady Kingston. She was widely known as "the good countess" because of her numerous charitable works, the extent of the local employment that she provided, and her town improvements at Mitchelstown.[87] Her reputation for munificence, which stood in glaring contrast to the neediness of the Frankses, seems to have made a possible change in landlords very attractive to the subtenants of Scart. Even in the wake of their gruesome deaths the Frankses were reliably described as "very disreputable characters in many respects; they were ruined in their circumstances, and [acted] under a false pride to keep up the appearance of holding a rank in society higher than [their] conduct entitled them to; they oppressed their tenantry and those in their employment. . . ."[88] Their dwelling near the village of Rockmills in the Kildorrery district was "a miserable thatched one, unfinished, and several rooms were without windows."[89] Since this hovel had been "in a great measure destroyed" by an incendiary fire in December 1822, the Frankses had lived "for safety" two miles away in Kildorrery itself until exactly a week before they were murdered.[90]

So impecunious was Thomas Franks that he involuntarily gave Lady Kingston the opportunity to eject him, probably on the grounds of the nonpayment of his own rents to her. He was made the subject of formal ejectment proceedings and temporarily ceased in practice to be the landlord of Scart. This development obviously delighted the subtenants there, who seemed to be on the verge not only of eluding the grasping hands of Thomas Franks but also of receiving leases from Lady Kingston. But then things went horribly wrong from the perspective of the Scart landholders. The dowager countess died at the beginning of 1823 and was succeeded by her spendthrift eldest son George, the third earl, with whom she had been quarreling for many years.[91] Lawyers for Thomas Franks identified a legal defect in the ejectment proceedings, and his rights as the legitimate leaseholder were restored. Having recovered possession, Franks immediately pressed for his old rents from Scart, and when the arrears went unpaid, he vigorously distrained the subtenants for some three months. They in turn appealed to the Earl of Kingston to interfere, saying that "they could not live under Mr. Franks," but he or his agent declined to take further action.[92]

Not everything is known about why the dowager Countess of Kingston's effort to eject Thomas Franks misfired, but she had her own problems, and these probably contributed directly to the failure and indirectly to the murders. As previously noted, and as confirmed by the extensive landowner John Hyde of Castle Hyde in Fermoy, the dowager countess had been feuding with her eldest son, the third earl, for quite some time prior to 1822. To make matters worse, the chief agent of the vast Kingston estates (they extended to as many as 75,000 acres in the three counties of Cork, Limerick, and Tipperary) was an absentee, and hence the need for middlemen like Thomas Franks to act as substitute managers of the property. These middlemen, however, were apparently of little social consequence or authority. John Hyde told the undersecretary William Gregory in February 1822, more than eighteen months before the murders, that although the population of the Kingston domains was both numerous and "licentious," not a single gentleman of rank or influence resided permanently anywhere on these huge estates. In the whole long tract of country between Doneraile in Cork and Cahir in Tipperary, there were allegedly only two magistrates, and neither was a person of "weight or influence."[93] Around Kildorrery itself there was a very large district of absentee property, and as the crown solicitor Maxwell Blacker noted at the time of the murders, this area contained "a very small number of resident magistrates of influence."[94] Thomas Franks may have been an extreme case of the slovenly, niggardly, and oppressive middleman, but he had plenty of company in the general ranks of this phalanx of large leaseholders whom the Rockites despised.[95]

Sir John Benn-Walsh, Middlemen, and Rockites

The journals of Sir John Benn-Walsh (1798–1881), an absentee landowner who paid the first of a long series of visits to his estates in Kerry and Cork in 1823 and 1824, also shed much light on the plight of middlemen tenants and subtenants in the days of Captain Rock. Benn-Walsh was one of those proprietors who was eager to eradicate the middlemen who were so well entrenched on numerous estates in Ireland.[96] Of the eleven large farms that comprised his Kerry estate, as many as seven were then in the hands of middlemen.[97] One of his first successes came in April 1823, when he and his agent Matthew Gabbett reached an agreement with a middleman named Moriarty who held the lands of Derryvrin in Kilcaragh parish in north Kerry. Moriarty, recorded Benn-Walsh in his journal on 14 April,

is so much alarmed by the combination of his undertenants against him that he refuses any longer to hold the whole farm. We let it accordingly to the undertenants, Moriarty holding only the dwelling house & that part of the ground which he previously held in his own hands. The poor man seemed quite relieved. His tenants were deeply imbued with the Whiteboy spirit, & he has lived for months in dread of his life.[98]

Benn-Walsh had been prepared for his experience with Moriarty by what he had encountered on his journey by coach from Dublin down to Limerick city—a trip that began at six o'clock in the morning and ended at ten at night. He had traveled with "a sharp, shrewd Paddy, a county of Clare man, whom I suspected to be a Captain Rock's man." What's more, on his way to Listowel via Tarbert, Benn-Walsh had "talked a good while with one of the Highlanders quartered at Adare, who told me that the country was still disturbed & that he had seen six fires a few nights since, all blazing in different parts of the country at once."[99]

Benn-Walsh showed no fear. He had already taken steps to recover possession of two other farms on his Kerry property held by middlemen—Wall of Derrindaff in Duagh parish and Hilliard of Furhane in Kilshenane parish. Like Moriarty and perhaps with the same sense of relief, Hilliard had surrendered possession of Furhane without a struggle, but the impecunious Wall had to be ejected. The lease of a third middleman who held the large farm of Fergus (about 950 acres) some fifteen miles north of Cork city, had recently expired, and Benn-Walsh had firmly decided not to renew it and instead to let these and other lands to the occupying tenants (cottiers in the case of Fergus)—but not to all of them.[100] "Some of the poorest & worst have been refused & their lands let to others," Benn-Walsh coldly remarked. "They have kept possession, however, & it requires some firmness & address to get them out. We went all over the farm [of Fergus on 16 April], followed by a rabble of tenants."[101] Benn-Walsh took the same approach in September 1824, when it appeared that James Julian, the defaulting middleman tenant of Tullamore on the Kerry estate, had "no intention of redeeming it." Benn-Walsh had the lands freshly surveyed and valued, with a view to letting them in new divisions to a portion of the former occupiers, just as he had already done with Derrindaff.

"All these arrangements," he declared, "gave birth to an infinity of petty squabbles, extremely difficult to settle or even understand."[102]

But the most interesting experience of a middleman tenant on the Benn-Walsh estates during the Rockite years was that involving John Hawkes, Sr., who held his position against the rebellious subtenants and against the head landlord's desire to clear the ground of intermediaries between himself and the occupiers. Hawkes was the middleman of Grange, described by Benn-Walsh as "a very pretty estate of 597 English acres situated in an improved part of the country about 7 or 8 miles from Cork [city]." Hawkes was a much more decorous member of the group of gentry middlemen to which the Frankses belonged. He seemed to Benn-Walsh to be "a respectable man of the class of small Irish gentry"; he lived in "an excellent mansion house" at Grange in Athnowen parish near Bandon. The senior Hawkes was fortunate in holding Grange under a lease for three lives, with the lives being those of his three sons.[103] This lease protected him from Benn-Walsh's designs for many years, but Hawkes was hardly immune from the Rockites. When Benn-Walsh visited Grange in 1851, toward the end of the Great Famine, old Hawkes recounted for him the compelling story of his endangered situation in the early 1820s:

> "Why, in 1822 the country was in a disturbed state & the people went about attacking the middlemen's houses, who were a bould, intrepid gentry, & we defended your property for you & kept down the people. Perhaps some day next time you will have to do it for yourselves & won't find it so easy. Now, this house was a very strong house & the walls [were] built of the thickness of the ould castle on which it stood, & it was well garrisoned too, for there was my father & his three sons & plenty of arms & ammunition. And we bricked up the windows & barricadoed it below, so that it would not have been easy to get in without artillery, & that hole you see [in the wall of an upper room] was made to command the back door, that we might fire on the people if they tried to break in."[104]

Among the reasons for the state of siege into which Hawkes vividly recalled having put himself was at least one instance of Rockite violence directed against him or a member of his family. According to a newspaper report of 1 March 1823, Rockites had burned down the house and out-offices of "Mr. Hawkes" at Ballyburden near Ballincollig in the previous month.[105] The land in question might have been the middleman holding of Classes, which Benn-Walsh described as "a

compact little farm [of 147 acres] beautifully situated" at the conflu-
ence of two rivers, and lying four or five miles from Cork city and only
one mile from Ballincollig; he believed that it might "be considered
not merely as a farm but as an eligible villa for any of the merchants
of Cork."[106] Whether the Hawkes family held Classes in February 1823
is not absolutely certain, but its members definitely had land in the
vicinity that was the source of Rockite hostility. And it seems very
likely that Hawkes and his sons were Rockite targets on more than this
single occasion, to judge from the elaborate care with which they had
prepared their little mansion at Grange for repelling Rockite assaults.

Large Farmers as Hated Middlemen

In the ranks of despised middlemen were also large farmers whose conduct toward their subtenants aroused the anger of the Rockites. The
infamous Patrick Shea, whose farmstead at Gurtnapisha near Mullinahone in south Tipperary was the scene of the catastrophic burning
to death of sixteen people in November 1821, earned intense local hostility because of his behavior as a landlord and an employer. Shea was
wealthy enough to be the employer of the five laborers and three maidservants who died with him when flames engulfed his thatched house
and out-offices. Commentators on this event pointed out that Shea
had evicted subtenants and cottiers from his large holding for nonpayment of rent, and that he had hired "strange" laborers to dig his
potatoes.[107] Topping off his offenses was his alleged status as a "land-
jobber."[108] The folk memory about the conflagration highlighted one
act of dispossession in particular: Shea had evicted William Gorman
from his smallholding at Tober on the slopes of Slievenamon.[109]
Though "the burning of the Sheas" may not have been organized by an
established Rockite band, there seems little doubt that Patrick Shea's
perceived oppression of poor tenants and laborers made him an object
of popular hatred.

Several other notable murders during the period of the Rockite
movement were also associated with the tensions inherent in the middleman system. One was the murder of John Marum, the brother of
the Catholic bishop of Ossory and an opulent farmer with extensive
holdings near Johnstown in the chronically disturbed Kilkenny barony
of Galmoy. His killing in March 1823 and the near murder at the same
time of his son will be discussed at some length in chapter 8. Here the

focus is on Marum's status as a farmer-middleman who had earned an unenviable reputation for oppressing his undertenants and for land grabbing. The immediate reason for the killing of Marum was reportedly his taking of a certain property (part of the lands of Rathpatrick) that lay under ejectment but subject to the possibility of redemption by the previous leaseholders (the Steele family). Somewhat like the dowager Countess of Kingston, the Protestant Steeles had the reputation of being benevolent and accommodating landlords, in contrast to John Marum, whose reputation was the very opposite. Moreover, the Steeles had held the land in question for over a century before their ejectment and replacement by Marum. When the head of this Protestant family did succeed in redeeming the property two months after Marum's murder, probably by finding somewhere the means to pay large arrears of rent owed to the Earl of Courtown, the tenants of Rathpatrick rejoiced at their good fortune. The *Leinster Journal* considered this outcome as irrefutable evidence that millenarian prophecies of Protestant doom had not been the inspiration for agrarian violence.[110] A different and more plausible conclusion is that subtenants much preferred a Protestant middleman who was willing to forego his rents in hard times to a Catholic who refused to concede the necessity for doing so.

The murder of the three Kinnealy brothers (Martin, Michael, and William) in July 1824 near Newinn in Outeragh parish in Tipperary is another event that appears to have been embedded in the hostilities so commonly associated with the middleman system. Outeragh and Knockgraffon are adjacent parishes in the southeast corner of Middlethird barony, an old Whiteboy stronghold. Part of Knockgraffon parish (over 9,000 acres in extent) was owned by Baron Richard Pennefather (a judge in the Irish Court of the Exchequer), who actually had a reputation for indulgence toward his own tenants during the depressed times of the early 1820s. It was said early in 1824 that even though his direct Knockgraffon tenants owed him two years' rent, he had urged them not to sell their produce until "they could obtain the highest prices." Nevertheless, the Rockites had recently been active on this part of his estate, particularly in enforcing their "laws" against any kind of land grabbing. In February 1824 two unnamed farmers on Pennefather's property in Knockgraffon parish were directed on pain of death to surrender the lands that they had held "for several years," and a third farmer was also told to quit, or the Rockites would burn his house on their next visit.[111] The need for such intimidation on the estate of a highly indulgent landowner appears hard to explain unless we assume that the victims of these attacks were not the direct tenants of Baron Pennefather but rather those of one or more middlemen. And the circumstances surrounding the murder of three of the five Kinnealy brothers lend even more plausibility to this assumption.

The Kinnealys reportedly took a farm in the adjacent parish of Outeragh from the Reverend Dr. Robert Bell, Protestant headmaster of Clonmel Grammar School, after the wife of the previous tenant, a widow named Doherty, had surrendered it. The Kinnealys had then disregarded a Rockite notice to quit. As a result, they were all savagely beaten, and three of the five brothers died of their injuries. The crime was considered the "most atrocious" act of violence in the district of Newinn since "the burning of the Sheas" in November 1821.[112] The Rockites were not at all dissuaded from this deed by the fact that the Kinnealys lived very close to the military detachment stationed at Newinn, and soon after the crime the Rockites served yet another notice nearby "on some men who had taken ground *three* years ago with the consent of the previous tenant."[113]

The district of Newinn, as we have seen, had long been the cockpit of violent disturbances arising from a protracted controversy over the rectorial tithes of Knockgraffon owed to the Reverend James Butler. Butler was a nonresident Protestant clergyman whose parish was under sequestration and whose Catholic parishioners staged spirited resistance whenever they faced any legal demand for tithes from that quarter.[114] Though a body of "Peelers" had for that reason come to be planted at Newinn, they were ineffective in the absence of magisterial authority over almost the whole area. The one magistrate who was resident, a gentleman named O'Meagher, was of worse than no use. O'Meagher, declared the crown solicitor Maxwell Blacker in February 1824, "is subject to attacks of the gout, and . . . [he] is reported to be rather too well disposed to release persons who have been brought before him on insurrectionary charges."[115] No wonder the Rockites of Knockgraffon and Outeragh went unchecked, and their impunity was to be demonstrated again in the case of the Kinnealy killings.

Who carried out the murderous assault on the Kinnealys? The two surviving brothers identified a group of at least three men, including

William Daniel, whose social and economic status put him in the category of "snug farmer."[116] Surprisingly, William Daniel was the superintendent of the Reverend Bell's glebe (church land) and one of the lessees of the tithes of Outeragh parish, or what contemporaries called a tithe farmer. Daniel also had his own cattle grazing on the glebe of Outeragh at the time when the Kinnealys persuaded Bell to let them rent the valuable farm formerly held by the Widow Doherty.[117] Whether or not Daniel and the two other accused persons who stood trial for the murders at the Tipperary summer assizes of 1824 were guilty is uncertain. A series of respectable witnesses testified to their good character, and the jury that heard the case acquitted them in spite of the positive identifications made by the two surviving brothers. Remarkably, Daniel Kinnealy declared at this trial that even after the death of his three brothers at the hands of a band that, in his telling, included William Daniel, he had been extremely reluctant to identify Daniel because "he wished to hold his ground fair and easy" and wanted "to live in a state of peace" with Daniel in spite of his leading role in the killings.[118] This much is clear: William Daniel had a definite motive for vengeance against the Kinnealys. In letting the former Doherty holding, the Reverend Bell had apparently played the tithe farmer Daniel off against the Kinnealy brothers, and it seems rather likely that Bell was a middleman on the estate of the Pallisers, the locally dominant Protestant family, while the other principals were subtenants.[119]

Other farmers who acted as middlemen and behaved in what was considered an oppressive fashion toward their subtenants were regularly targeted by Rockites. Early in 1822, for example, a single Rockite party operating near Tarbert in north Kerry raided the houses of the "most wealthy" farmers of the district on a particular night, administered beatings to them, and swore them to reduce the rents of their subtenants.[120] At about the same time Rockites in the vicinity of Dromcolliher in County Limerick ordered one well-to-do farmer to give his own tenants an abatement and instructed another large farmer to surrender his land, as it must (they insisted) be set in divisions, presumably for the poor would-be cottiers of the vicinity. As one local proprietor complained about these two attacks on farmer-middlemen, a "general confederacy" existed among the followers of Captain Rock to "bring down [the rent of] lands to their own ideas."[121] And in the Macroom district of north Cork an extensive farmer who had seized the very furniture of some of his subtenants under a decree for rent was made to pay the penalty. An armed band of twenty Rockites attacked his house and thrashed him in February 1823; they remained for about four hours, doing all kinds of mischief, and finally decamped with four horses laden with the rescued furniture![122]

Faction Feuds, Middlemen, and Captain Rock

Disputes arising from the middleman system could also be overlaid with the further complicating feature of factions and their feuds. An agrarian crime that occurred on the estate of Lord Lismore in the Clogheen district of Tipperary in late October 1822 reportedly had its origins in a long-standing faction feud between the rival families of the Hennessys and the Lonergans and their respective followers. Within the previous twelve months this feud, which was said by Lord Lismore to have persisted for thirteen years, had been exacerbated when one of the Hennessys was murdered by a Lonergan and members of his party. But the most recent act of violence between the two factions concerned their competition over a farm of as many as 200 acres (part of the lands of Gurtacullin) that had fallen out of lease within the past month. The conflict between the factions had now intensified because of what Lord Lismore called "their mutual apprehensions least [i.e., lest] the one should take the other's grounds."[123] Given the large size of the farm that was out of lease and the nature of the competition for it, there seems little question that the heads of the Hennessy and Lonergan factions were farmer-middlemen whose respective factions included small tenants, cottiers, and laborers.

In ways that remain only dimly visible, faction feuds must have interwoven themselves with the middleman system and its internal conflicts over rents and evictions in other cases as well. Factions pervaded the Kingston estates, or at least part of them, in the early 1820s, and the area around Kildorrery, where the Frankses had been murdered, was also notable as a stronghold of factions. The military officer Thomas Arbuthnot reported early in 1823 that two very numerous factions came into contention at all the fairs in his general neighborhood (he specifically mentioned the districts of Kilworth, Glanworth, Kildorrery, and Mitchelstown). Though he asserted that their quarrels had nothing directly to do with the Rockite movement, he conceded that this faction-fighting did encourage "a lawless spirit" and pointed out that the factionists tended to live in areas where illicit distillation widely prevailed.[124] One especially impressive example of the strength of factions in the areas of Glanworth and Kildorrery occurred in October of the same year, when at least fifteen hundred people gathered near Glanworth village for the funeral of a man named Sheehan, "who was considered the head of a large faction in that part of the country." In a district where large funerals were not at all unusual, this one was bigger than any other that Arbuthnot had witnessed. He thought this funeral worth mentioning to his military superiors in order to show "how prone the peasantry of this country are to support a lawless leader and the facility with which large numbers can assemble."[125]

The Eclipse of the Movement

The Rockite campaign against the collection of the customary rents and against "land canters" finally began to weaken in 1824. In fact, its demise in County Kerry came somewhat earlier. Writing from Killarney in June 1823, the sensitive observer Daniel Mahony noted that in the barony of Iraghticonnor there had even been instances where "the lower class of occupiers" had been deprived of their lands and their stock, and yet the new tenants had remained "unmolested"; in addition, a particular herdsman at Aghanagran near Listowel, who had once been severely flogged, had not been troubled by a return visit from the Rockites.[126] In north Cork, on the other hand, there was as yet little change. The ever-watchful police magistrate Samson Carter reported with regret in September 1823 that "the practice of placarding threatening notices against tythe proctors and persons retaking lands from which the original tenants have been evicted continues in the baronies of Fermoy, Duhallow, [and] Orrery & Kilmore."[127] But as early as February 1824 the military officer Thomas Arbuthnot found distinct signs of the waning of the movement in these areas. The rise in agricultural prices, he noted, was enabling small farmers to pay their rents, "which, generally speaking, they have lately done regularly." Moreover, when landlords found it necessary to distrain the goods of defaulting tenants or even to eject them, these measures were "generally now carried into effect without opposition from the peasantry, & farms from which some [persons] have been ejected are taken by new tenants, which was not the case last year without much risk."[128]

Reports from other counties in 1824 confirmed that many farmers were abandoning their participation in the movement. The landlord John Waller told the crown solicitor Francis Blackburne that he had noticed a great change in the disposition of farmers in the Pallaskenry district of west Limerick. Until recently, Waller maintained in February of that year, the farmers there had been "at best neuter," but now they were much more favorably inclined toward the restoration of social peace owing to improved markets for their produce.[129] The police magistrate George Drought was even more sanguine about west Limerick by the following October. With the single exception of "a disposition to resist a change of tenantry," he declared with unwonted optimism, "I do not see any reason to apprehend a recurrence of outrage, as the great mass of the people appear to me to be more devoted to their several callings than they have been . . . for the last three years in this part of Ireland. . . ."[130] And the yeomanry-brigade major Henry Croker of Mallow, besides drawing attention to the abundance of potatoes in Cork and Tipperary, stressed that corn "of all sorts" was in strong demand, exceeding "in price the most sanguine wishes of the farmers." This, he declared, "in my humble opinion has tended to tranquillize this county (and most others in this province) more than any other measure."[131] The *Dublin Evening Post* was quite correct in pronouncing with satisfaction and relief in November that Irish farmers were "again holding up their heads," and Irish landlords "again receiving their rents."[132]

But even when they engaged in what became their final actions, the Rockites, or at least some of them, wished to proclaim their ultimate legitimacy and respectability. As a group of them insisted in a threatening notice posted near Doneraile in December 1824,

> We are none of the course [*sic*] rank of people that [go about] house breaking or burning farmers or whipping Kerrymen, but we are the Royal Society which is called the Green Levellers [who] will hold out the old law which was enacted in the year 1821, that's to put down land canters and tythe takers.
> We'l [*sic*] consume such people to ashes and hough their cattle. . . .
> John Rock.[133]

Eight

Patterns of Rockite Violence

Investigating the Wave of Murders

By the early 1820s agrarian murders were already a frightening commonplace of the deeply troubled Irish rural scene. What was so striking during the years of the Rockite movement was the scale on which
the crime of murder, to say nothing of attempted murder, was committed. Detailed analysis of this important subject, however, is fraught
with a variety of difficulties. One major problem is the lack of comprehensive contemporary statistical data. Not until the 1830s did the
constabulary begin to collect, collate, and present crime statistics systematically, and specifically agrarian offenses were not separately distinguished from general crime in the official compilations until the
1840s.[1] The historian is therefore forced to rely for information about
agrarian murder on the incomplete and unsystematic documentation
that survives in national and local newspapers as well as in the State
of the Country Papers (SOCP). For the discussion that follows, the
Dublin Evening Post for 1821–24 (complete files), the *Leinster Journal*
for 1822–24 (incomplete files), and SOCP 1, 1821–24, have been carefully examined. The counties covered are Clare, Cork, Kerry, Kilkenny,
Limerick, and Tipperary. Additional sources might have been in-
spected, but probably without any substantial gain in the fullness or
reliability of the picture that can be presented.

It is obviously a matter of considerable importance to know how reliably murders in general and agrarian murders in particular were reported in the sources that have been consulted. Understandably and happily from our perspective, the press and the authorities heavily focused their attention on murders that arose (or were believed to arise) out of the ongoing agrarian rebellion rather than on homicides that stemmed from intrafamily violence, private feuds, or other causes unconnected with the Rockite movement. Yet the line between homicides arising from private feuds and murders related to or inspired by the Rockite rebellion was not always clear-cut, a fact realized by the authorities. They also recognized that private feuds terminating in murder increased in number during periods of intense agrarian turmoil.[2] Deaths in faction fights, however, decreased in such periods because faction fighting itself greatly declined.[3]

Given the lamentable state of the surviving evidence, the question of what proportion agrarian murder bore to homicide in general cannot be answered at all satisfactorily for this period. But it does appear that the proportion was extremely high in County Limerick during late 1821 and early 1822. According to a newspaper report in early March 1822, the county coroner had conducted twenty-five inquests for murder in Limerick since the assizes closed in late September 1821.[4] Of this number, Rockite violence was definitely responsible for sixteen of the killings and perhaps for another four as well.[5] On the other hand, the proportional relationship in County Kilkenny somewhat later was quite different. There the county coroner conducted nineteen inquests for murder between the assizes in August 1822 and late March 1823.[6] But only a single agrarian murder, according to newspaper and other reports, occurred in Kilkenny during that span.[7] This sharply different pattern for Kilkenny seems to have prevailed throughout most of the Rockite years. In May 1824 an official bemoaned the fact that no one had been brought to justice for any of the twenty-two murders committed in three Kilkenny baronies during the previous two years.[8] But the sources consulted for this analysis indicate that there were only six agrarian killings in all of Kilkenny from the beginning of April 1822 to the end of May 1824.[9] Although some doubt must remain as to whether agrarian murders were less often reported for Kilkenny than for Limerick, it seems unlikely that the sources fail to record the sheer occurrence of such crimes to any substantial degree.

Yet in other respects the surviving evidence is often deficient. In a substantial number of cases murders were stated to have taken place in districts highly disturbed by Rockite violence without the assignment of any motive for the killing, or without the adducing of any facts, such as those relating to the occupational status of the victim, from which a motive might reasonably be inferred. In all such cases— nine in number—it has been decided to exclude the victims from the group whose deaths are here taken to have been related to or inspired by the Rockite movement. Clearly, this simple procedure has the effect of removing a number of persons whose murders were very probably connected with the Rockite upheaval.[10] No motive was assigned for the murder of an unidentified man in the Newmarket region of northwest Cork in January 1822, but his death followed by one night the killing of the suspected informer John Sullivan in the same district, and the chief agent in both of these deeds, as well as in the earlier slaying of another suspected informer, was reputedly John Murphy, "one of the most criminal leaders who disgraced that country [around Newmarket]."[11] In the other eight cases in which a motive was unspecified, only the fact that the murders took place in areas that were highly disturbed at the time of occurrence suggests their possible Rockite origins.[12] This geographical and temporal linkage has been deemed insufficient to include them in the group further analyzed below.

Whether or not certain other murders were associated with the Rockite upheaval is unclear, even though a relevant motive either was specified or can reasonably be deduced from the attendant circumstances. For example, at a time when Rockites in north Kerry were engaged in arms raids against local gentlemen and farmers, Major Collis of Kent Lodge near Tralee was mortally wounded at his residence in late November 1821.[13] By some this was adjudged a Whiteboy killing on the assumption that the raiders wanted Collis's arms and that he had resisted their demand. But others fixed on the fact that the attackers had robbed Collis of more than £200, mostly in silver, and concluded that the murderers were simply thieves and not Whiteboys.[14] They could of course have been both at once, for the notion that Whiteboys rarely stooped to the robbery of money or other valuables is very wide of the mark. That this was a Rockite crime is at least plausible not only from the general arms raiding in the district and from the attackers' demand for "money to buy powder,"[15] but also from the contemptuous conduct of one of the three men tried and condemned for the murder at the Kerry spring assizes in March 1822. He listened to the sentence of the court with "the most determined and deplorable obduracy of mind," vigorously protested his innocence, and exclaimed that "many fine fellows died in this way."[16]

Mixed motives, or the probability of a mixture, also appear to have operated in five additional cases of murder. Denis Mahony, a farmer beaten to death near O'Donovan's Cove in west Cork in October 1822, was described as a land grabber, but the half dozen men who invaded his house also stole £30 from their victim.[17] Similarly, the police labeled as Whiteboys the party of men who killed the wealthy farmer James Sullivan at Aghacross near Kildorrery in March 1824, but it was noted that what precipitated his death was his resistance to the Whiteboys' efforts to break open a chest in which his money was apparently secured.[18] In cases where tithe proctors were murdered, as in the killing of land grabbers, a strong presumption exists, in the absence of other confirming evidence, that these were Rockite crimes, but in a third instance this may not have been so. In the hours before his murder in July 1823 the proctor Andrew Nowlan had been settling tithe business for the Reverend William Sutton in Templetenny parish near Ballyporeen, but he was fatally shot on his way home to Kilworth, apparently by a single assailant along the road who demanded his money.[19]

Irresolvable ambiguities also surround two very different murders committed in Tipperary during the Rockite years, one of a peasant woman in March 1822, the other of the head of a leading gentry family in February 1824. The woman was beaten to death near Ardfinnan as a suspected informer, but since the information that she was believed to have given concerned the illegal cutting of trees, the killing may have had no connection with the Rockite upheaval.[20] On the other hand, the murder of the gentleman Michael Hamerton appears to fit into a well-established pattern. A small party attacked his residence, and after he fired on them, the raiders smashed his skull with bars of plough iron. The fact that they then carried off all his arms may indicate that they were engaged only in the typical Rockite hunt for weapons, but as in the case of Collis earlier, they may also have wanted Hamerton's money as well as his guns.[21] These six murders—from Collis's to Hamerton's—thus raise undecidable issues of motivation that

Guide to Rockite Violence
— Provincial boundary
— County boundary
--- Select barony boundary

ATLANTIC
OCEAN

Burren

Corcomroe   Inchiquin

Bunratty
Upper

C l a r e

Ibrickan   Islands

Bunratty
Lower

Clonderalaw

Killadysert

Moyarta   •Kilrush   •Pallaskenry•

Askeaton•

Cappagh•   •Adare

Ballylongford•   Rathkeale•

Ardagh•

Knockane•   Ballingarry •

Newcastle•

Listowel•   West   L i m e r i c k

Abbeyfeale•   Killeedy•

•Tralee   Newmarket•

Liscarroll•

Castleisland•   Kanturk•

•Killorglin

•Killarney

K e r r y

Kenmare•   •Kilgarvan
Templenoe•

•Inchigeelagh

C o r k

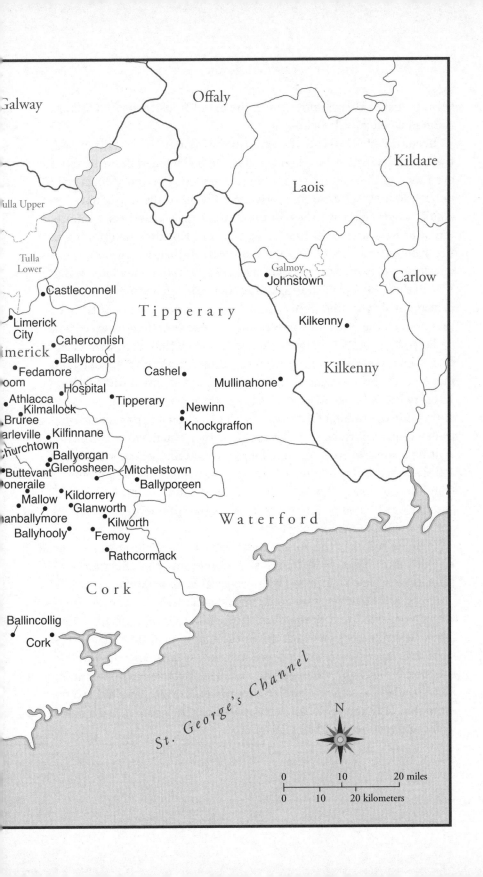

seem to justify their removal from the list of Rockite-related killings examined collectively below.

In contrast to the total of sixteen excluded murders just discussed, there stand a further ten slayings that will be included in the analysis even though their omission could be defended. The ten killings were unquestionably agrarian in motive. Nearly all of them originated in conflicts over rents or tithes. But none appears to have been committed after long premeditation or by bands of Rockites functioning in the usual fashion under cover of darkness. Rather, as a group, these murders were more or less spontaneous reactions to immediate events. Five of them took place during affrays between, on one side, parties of drivers, or drivers and police, sent to distrain for arrears of tithe or rent, and on the other side crowds of peasants resisting the execution of distress warrants.[22] In four other cases the victims were slain by individuals or very small groups, enraged by the victims' specific and immediately preceding acts: a policeman shot to death while forcing an entry into a house to execute a magistrate's warrant; a driver shot dead while distraining for arrears of rent by a defaulting tenant firing from within his farmhouse; a bailiff killed by a "small party" while executing a court decree; and a tithe proctor fatally beaten while serving processes and collecting tithes by a group of perhaps eight men.[23] Only slightly different was the last of these murders, in which a land grabber was bludgeoned to death by a brother of the previous tenant acting alone and in a drunken rage.[24]

In some respects the connections between these killings and the Rockite movement were tenuous. Murders under similar circumstances sometimes happened when no great agrarian upheaval was in progress, and these ten were not in any obvious or direct way the product of concerted and premeditated Rockite activity. On the other hand, the grievances on which the Rockite movement thrived were the same grievances from which these murders sprang, and the Rockite sanction for extreme violence provided a supportive environment for killing under the same or similar conditions. Rockite laws did not require that only active Rockites enforce them; they stimulated a much wider and readier resort to killing.

Rockite Murders: Summary and Analysis

Altogether, then, between mid-1821 and the end of 1824 there were as many as ninety-three persons in six counties whose murder may be said with some confidence to have been related to or inspired by the Rockite movement. During the expansionary phase of the movement from mid-1821 through the spring of 1822, fifty-six persons, or about 60 percent of the total, were killed, although this figure is swelled by the inclusion of sixteen individuals who died in a single incident—the burning of the Sheas in November 1821. After the murder of the servant boy Martin Quane on 11 April 1822,[25] there was an extended period of more than five months during which no agrarian or Rockite-related killings were recorded. This was an interval in which famine or near-famine conditions supervened throughout much of the region previously disturbed by organized agrarian violence, and during this interval the Rockite movement ground to a halt temporarily. But as the movement revived in the late summer of 1822, so too did the train of murders, and from then until the end of 1824 another thirty-seven persons became victims in Rockite-related killings.

The great majority of these ninety-three murders were concentrated in three of the six counties. The other three accounted for a total of fourteen, with nine occurring in Kilkenny,[26] three in Kerry, and two in Clare. By contrast, Tipperary accounted for twenty-nine, Limerick for twenty-eight, and Cork for twenty-two. In other words, 85 percent of the Rockite-related murders took place in these three counties. Among the three there was some shift in geographical incidence over time. Murder was much more frequent in County Limerick during the expansionary phase of the movement from June 1821 through March 1822 than it was in either Cork or Tipperary, especially if the incineration of sixteen people at the Shea farm in south Tipperary is regarded as a statistical aberration. After the movement revived in September 1822, Cork headed the list in Rockite-related killings, with fifteen, as compared with nine in Tipperary and six in Limerick.

In County Limerick, however, what was especially distinctive about the crime of agrarian murder throughout the Rockite years was the narrowly restricted geographical region within which so many of the killings took place. In all of east Limerick (i.e., east of a line drawn from Limerick city through Croom down to Charleville in Cork), fewer than a half-dozen murders were committed, and only one of them (that of Owen Cullinane near Bruree—close to the line—in December 1821) occurred during the expansionary phase of the movement.[27] Of the four others west of the line, one took place near Castleconnell in September 1822, a second north of Fedamore in January 1823, a third near Kilmallock in November 1823, and the last at Bilboa near the Tipperary border in December 1824.[28] What few murders did occur in east Limerick were thus scattered, and they were unconnected with one another. By contrast, as many as twenty-two of the twenty-eight Limerick killings occurred in the western portion of the county, and within it they were heavily concentrated in a handful of specific localities.

The first to acquire a homicidal reputation was the district around the Palatine village of Adare, where there were at least five Rockite-related murders within less than a year, including those of the gentlewoman Susanna Torrance at Mondellihy in June 1821; John Walsh, the fifer of the local yeomanry corps, in the following October; Thomas Murphy and Frederick Petit, the domestic servants of local gentlemen, in December 1821 and February 1822, respectively; and the informer John Neill in March 1822.[29] When Petit was waylaid and shot to death in February, his was said to be the fourth murder committed near the very same spot.[30] To the southwest of Adare, the Rathkeale district soon attained equal notoriety for bloodletting. Here the toll rose to six murders: Major Richard Going, the former police official with Orange proclivities, in October 1821; the farmer's son Michael Gorman, Jr., in the following December; the gentleman Thomas Hare in February 1822; the postboy Henry Sheehan and his dragoon escort a month later; and the police constable Francis McInnes in the following November.[31]

In the area north and northeast of Rathkeale, stretching from Askeaton and Pallaskenry toward Limerick city, there were five additional murders, two near the city and three around Askeaton and Pallaskenry. The latter three all stemmed from conflict over tithes: the killing of the policeman Thomas Manning in a tithe affray in August 1821 was followed in September and October by the murders of John Corneal, a drawer of tithes, and John Ivis, a proctor.[32] To the southwest of Rathkeale lay the Newcastle district, and here at the original epicenter of the movement, murder was also rife, claiming at least four lives in Rockite-related slayings, including not only that of the young

Thomas Hoskins in August 1821 but also those of the informer James Buckley in September, the Palatine farmer Christopher Sparling in October, and the Courtenay estate subagent Robin Ambrose in March 1822.[33] Together with two further murders (one around Ballingarry early in 1822, the other near Abbeyfeale in 1824),[34] this record of extreme Rockite violence in a small corner of Munster was unique in the annals of Whiteboyism during the early nineteenth century.

The connections between some of these killings in west Limerick were more than a simple matter of geography. The murders of Susanna Torrance and John Corneal provide an instructive illustration. Mrs. Torrance and her husband had moved to Adare from the Liberties of Limerick for reasons of security: in March 1821 they had repelled a Whiteboy raid for arms on their house in the Liberties, apparently inflicting casualties on their attackers, who vowed revenge and made their lives there unsafe. But the hand of vengeance pursued them to the Adare district, where in the evening hours of 10 June 1821 two men waylaid them, stabbing Mrs. Torrance to death and almost killing her husband by slitting his throat.[35] The still more gruesome fate of John Corneal near Pallaskenry in the following September at first seemed unrelated to the Torrance murder. Corneal's death appeared to be rooted solely in his connection with tithes. While engaged in drawing tithes home for the Reverend Francis Langford, he was struck down and his head was severed from his body.[36] But Corneal's family had also taken a zealous part in searching for the slayers of Susanna Torrance. This fact emerged in October when the victim's relative Adam Corneal prosecuted an assailant for "striking him with a spade and at the same time saying he would kill him for having gone out at night, assisting the magistrates in apprehending the *honest fellows* who killed Mrs. Torrance."[37] Special mention was made at this trial of "the laudable assistance [Adam Corneal] and his respectable family and friends had afforded to the magistracy" in the Torrance case.[38]

Major Richard Going's murder also did not stand alone. As previously noted, Going was killed near Rathkeale in October 1821 at least partly because of his handling of the aftermath of a fatal clash between over two hundred Whiteboys and a much smaller body of police at Inchirourke near Askeaton two months earlier. In this encounter the Whiteboys shot dead the subconstable Thomas Manning, but the police killed two Whiteboys and mortally wounded a third. These fatalities occurred in the vicinity of the house of the tithe proctor John Ivis, whom the Whiteboys had come to attack and punish. The dumping of the bodies of the Whiteboys (one of whom was rumored to be still alive when buried) in what was insultingly called "a croppy hole," an ignominious interment arranged by Going, was a proximate cause of his murder in October. And his fate was shared in the same month by the proctor Ivis as the train of vengeance played itself out.[39]

Motivations for Murders

This consideration of the motives for murder in a portion of the west Limerick cases leads to a general analysis of motivation in the total of ninety-three killings related to or inspired by the Rockite movement. The leading proximate motive for these murders was the resistance of the victim to a demand for arms or (in a few cases) money, with the money usually wanted for the purchase of ammunition. This motive was present in as many as twenty-seven killings, exclusive of those in which policemen or soldiers lost their lives. Its prominence is a measure of the fierce determination with which the Rockites sought to disarm their adversaries and to arm themselves—an aspect of Whiteboyism that is often slighted. Admittedly, the figure of twenty-seven killings attributed to this factor is inflated by the inclusion of the sixteen people who were burned or shot to death in the farmhouse or out-offices of Edmund Shea at Gurtnapisha near Mullinahone when his premises were set on fire by a raiding party on 19 November 1821. Besides Shea himself, who was a wealthy farmer, the long roll of the dead included his wife, their five children, six male laborers, and three maidservants. Hostility toward Shea derived in part from his allegedly harsh treatment of his cottier tenants and farm workers, but accounts of this gruesome episode, both at the time and at a later trial, agree that the raiders believed Shea's house to be full of arms, and that Shea and the other inmates refused to throw them out, as the raiders loudly and repeatedly demanded.[40]

But even if the heavy statistical impact of this extraordinary episode must be kept in mind in assessing the role that resistance to arms raids played in murders, it should also be emphasized that the figure of twenty-seven killings attributed to this factor does not include any of the police or soldiers who lost their lives for this reason. In addition to the three constables killed in January 1822 in the daring attack on the police barracks at Churchtown in Cork,[41] two other policemen and two soldiers also forfeited their lives to people who apparently wanted their weapons.[42] Leaving aside these deaths and the extraordinary case of the Sheas and their dependents, the murder roll as a result of arms raids included three individuals of gentry rank (Susanna Torrance, Thomas Hare, and Richard Crofts); three domestic servants of gentlemen (Thomas Murphy, Frederick Petit, and Denis Browne); three or perhaps four farmers (Owen Cullinane, Denis Morrissy, Thomas Max, and George Sparling); and Lord Ormonde's woodranger.[43] Almost without exception these killings occurred during the expansionary phase of the Rockite movement between the summer of 1821 and the spring of 1822.

The second leading motive for the murders under review here was the belief that the victim had given or would soon provide damning information to the authorities about Rockite or other criminal activities. This motive was present in no fewer than sixteen of the ninety-three killings. In certain cases it was not the exclusive motive: the slayers of John Corneal, Thomas Murphy, Frederick Petit, and the three Frankses (father, mother, and adult son) apparently had more than one reason for what they did.[44] But having the odious name of informer or potential informer was the primary factor in the deaths of the great majority of these sixteen victims. Most of them met their fate in seclusion: they were waylaid and dispatched along some dark road or struck down suddenly at their own door. But James Buckley was publicly stoned and pitchforked to death in September 1821 as he and several other crown witnesses were returning in a group from the Limerick assizes to their homes around Newcastle.[45] Part of the crowd that joined in the attack actually came rushing out of Ardagh chapel, where they had gathered for Mass. "There goes Hoskins's spies," the crowd yelled as the attack began. About Buckley's fellow informer Crowley, people in the throng were heard to exclaim, "If they did not kill [him], he would hang them all." One witness to the stoning of Buckley and the others later testified that while it was happening, he "knew some of the most respectable farmers in the parish to be looking on" complacently, perhaps even approvingly.[46]

To earn death at Rockite hands, it was certainly not necessary for an informer to go the whole length of appearing in court or even to take the formal step of lodging an information with a magistrate. Some of those slain as informers, like Scully (a laborer) and Sullivan in County Cork or Stack (another laborer) in County Clare, were only suspected of this treachery, though perhaps with good reason.[47] And in the case of two further victims, the domestic servant Thomas Murphy in County Limerick and an unnamed woman at Rathcormack in Cork, their offense was to have openly recognized Whiteboys who had raided their dwellings previously.[48] Nor did the charges laid by an informer have to be capital in law for murder to result. Patrick Drishane, beaten to death near Mitchelstown in January 1823 by a self-proclaimed band of "Captain Rock's men," had given information only about illegal whiskey stills, with which that district in the Galtee mountains abounded.[49] Even if those operating the stills had been convicted on Drishane's evidence, itself an uncertain matter, they would merely have been fined. But such was the popular hatred for informers of every kind that their murder was unlikely to arouse the least sympathy or regret. The case of the informer John Neill, stoned to death near Adare in late March 1822, speaks volumes on this point.[50] After his body was found, none of his numerous family bothered to inquire about claiming his remains, and "when the police were taking his body for interment, the people were convulsed with laughter, and the children shouting and whistling. . . ."[51]

A third set of motives for the murders scrutinized here involved deep resentment against the victim for behavior related to the taking, occupation, or management of land. Such motives were present in at least twenty-seven of the killings and perhaps as many as thirty. Consideration of these murders is clarified by dividing them into three categories that the actions and status of the victims would suggest to be appropriate: (1) persons punished as land grabbers; (2) persons punished for their efforts to collect or to facilitate the collection of arrears of rent; and (3) persons punished for their behavior as the owners, lessors, or managers of land.

The killings in the first category amounted to twelve, and the victims comprised ten farmers and two servants of farmers.[52] Three of the cases were remarkable in different ways. One was a triple murder in which a band of fifteen or twenty Rockites administered savage beatings to the five Kinnealy brothers at Knockgraffon in the Cashel district in July 1824. Battered by gunstocks, stones, and heavy sticks, three of the brothers died of their injuries. The land in dispute was pasture ground held from an Anglican parson. The previous tenant was said to have surrendered it, but as shown in chapter 7, intense local rivalry over this valuable grassland led to terrible retribution against the Kinnealy brothers soon after they took possession. This triple murder occurred close to the chronically disturbed district of Newinn, and through its viciousness revivified memories of the burning of the Sheas.[53]

Remarkable for other reasons were the killings of the wealthy farmer John Marum, whose property was located near Johnstown in the Kilkenny barony of Galmoy, and of Daniel Connell at Bilboa in east Limerick. Both victims had legions of enemies. As a brother of the Reverend Dr. Kyran Marum, the Catholic bishop of Ossory since 1815, Marum was a socially prominent Catholic and also, as recounted previously, an especially grasping one. As early as November 1821 the *Kilkenny Moderator* noted that the Rockites had threatened to "settle" Marum and another land grabber and explained that Marum had long been "obnoxious to the lawless peasantry" of Galmoy.[54] His death came suddenly, and perhaps treacherously, one evening in mid-March 1823 when he was waylaid by five or six men as he was returning to his residence at Whiteswall from the lands of Rathpatrick. The murderers fractured his skull and bayoneted him several times in the chest; they also severely wounded his son, who was traveling with him.[55] Since Marum usually enjoyed the hired protection of armed desperados (indeed, "notorious murderers," according to one account), many people strongly suspected that one of his unsavory erstwhile guardians had betrayed him.[56] The proximate cause of Marum's death was his recent acquisition of the lease of the lands of Rathpatrick, which lay under a decree of ejectment but subject to possible redemption within six months. These lands, it was said, had "all the attending circumstances of the ground which he took [several years earlier] at Eirke, and which did in some degree cause the necessity of the Peace Preservation Force being sent into Galmoy and Upper Ossory in May 1821."[57]

So bitter and pervasive was the detestation that Marum had aroused by his remorseless land-grabbing activities that his murder was greeted with general approbation, indeed in some quarters with unconcealed rejoicing. During the sitting of a court of inquest into his death, "a mob of 500 people were outside, shouting and showing their joy whenever a witness came out after having denied, as they all did,

The Rockites dealt severely with those who persisted in violating their "laws"
against taking farms or even smallholdings from which the previous tenants had
been evicted. Numerous murders (sometimes committed with an axe, as depicted
in this sketch) and many acts of incendiarism were traceable to evictions in the
early 1820s. The large Kilkenny farmer John Marum, a brother of the Catholic
bishop of Ossory, suffered a gruesome death (a fractured skull and multiple bayonet wounds) in March 1823 as a repeated violator of this iron Rockite law.
(William Steuart Trench, *Realities of Irish life* [London, 1868], facing p. 57)

and generally very insolently, any knowledge of the thing." On subsequent days people who gathered in public houses muttered their regret that Marum's son had escaped death, and declared that Marum's brother Ned "must be dropped also." These people expected that the murder would serve as a powerful cautionary tale: "'See what gentleman will dare to turn out poor people now,' said they. 'Galmoy forever.'"[58] As noted in chapter 7, the redemption and recovery of the lands of Rathpatrick by a locally popular Protestant landlord family two months after the killing produced jubilation among the tenants. Presumably, these same tenants had not been too distressed at the murder of Marum a short time earlier.[59]

In the case of Daniel Connell the resentment was not of such long standing, but he probably had almost as many enemies as Marum. Connell's offense was to have taken an 88-acre farm in the spring of 1824 on Lord Stradbroke's estate from which no fewer than twenty-six families (comprising seventy-two persons) had been evicted in a single day—a minor clearance effected a year or two after the expiration of a middleman's lease.[60] Following this clearance in April, acrimony continued to accumulate, spilling over into occasional outrages, but it was not until December that the violence reached a crescendo when a dozen Rockites, armed with swords, scythes, and pitchforks, slew Connell, dashing out his brains and hacking off his head. Other inmates of his house, including a brother, a sister, and several female relatives, were "cut in a shocking manner" but survived.[61] Unlike Galmoy, the east Limerick barony of Coonagh in which this ghastly murder occurred had been comparatively tranquil during the Rockite years.

Killings in the second category (persons slain because of their association with the collection of rent) amounted to at least five. The victims were all either drivers or keepers. Three of them were waylaid and murdered, while the other two died in the direct line of duty.[62] In the cases of three other victims killed in 1823 (a process server, a bailiff, and a driver), it is not possible to tell whether their activities were concerned with rents or with tithes.[63] The sources are simply silent on this point.

For their behavior as owners, lessors, or managers of land, nine additional persons were murdered. Three of them were or had been associated with the administration of Lord Courtenay's estate around Newcastle. Besides young Thomas Hoskins, the chief agent's son fatally wounded at the outset in July 1821, Robin Ambrose, an under-agent, was slain in March 1822, after a new chief agent had been installed but supposedly because of Ambrose's service with the much hated predecessor.[64] A third member of the Hoskins regime, John Hartnett, was no longer employed by Lord Courtenay when he was killed near Abbeyfeale in January 1824. By that time he was working as a sub-agent on the Earl of Clare's estate; he owed his death to having threatened the eviction of two brothers named Flynn on that property whose cattle he had also seized for nonpayment of rent.[65] Hartnett's fate excited little sympathy in police circles. According to Thomas Vokes, Hartnett was "a horrid character who was in company with young Mr. Hoskins when he was murdered, and although he knew the murderers well, he w[oul]d not disclose the names of any of the seven."[66] The other victims in this third category—six in number—included Joseph Condon, an understeward on Earl Egremont's property, beaten to death near Limerick city in January 1823;[67] an "occasional" land agent and wealthy farmer named Callaghan, waylaid and slain near Tipperary town in December of that year;[68] Gabriel Holmes, the demesne steward of the Castle Blunden estate near Kilkenny city, mortally injured in a beating in June 1824;[69] and the three Frankses.

The triple murder of Thomas, Margaret, and Henry Franks (father, mother, and adult son, respectively) near Rockmills, about halfway between Glanworth and Kildorrery, early in September 1823 deserves special consideration for several reasons. It ranks among the half-dozen most notorious crimes of the Rockite years, along with the murder of Major Richard Going, the incineration of the Sheas and their dependents, the burning of Churchtown barracks and the Palatine village of Glenosheen, and the raping of the wives of members of the Rifle Brigade. Newspapers fastened on the lurid event, and the authorities devoted much time and effort to bringing the offenders to justice, though with only indifferent results. Accounts differed as to the number of men who actively joined in the crime. A maidservant who saw the Rockite party leaving the scene testified at the inquests on the bodies that "there were at least sixty" participants,[70] but at a later trial of three of the alleged murderers, an approver swore that there were no more than nine in the party.[71] Whatever the exact number, they did grisly work. The coroner's inquests indicated that Henry Franks had died of a gunshot wound in the chest and of blows to the head with a crowbar, and that his wife and son had died of multiple crowbar wounds to the head.[72] The evidence also disclosed that the skulls of father and son were "literally smashed in pieces," and that their faces were disfigured beyond recognition. The same report, based on the eyewitness testimony of a teenage servant girl, noted that as the victims lay dying, they were stabbed with a three-pronged pitchfork by several of the participants, "each handing it to the other to inflict a blow," and the leader repeatedly shouting to his associates, "*Do your duty.*"[73]

This manifest mutilation, so horrifying to the upper classes, partly accounts for the notoriety that the slayings almost immediately achieved, but the motives that led to the murders also help to explain the contemporary fascination, for they were emblematic of the reasons for a great deal of Rockite violence. Patent facts and close investigation convinced the authorities that there were two principal motives, both of which arose from fairly recent events. First of all, at the Cork spring assizes of 1823, Henry Franks had successfully prosecuted Cornelius Sheehan for having been a prominent member of a band of Rockites who had raided the Frankses' house near Rockmills.[74] The prosecution helped to incite a Rockite vendetta against this family of decayed gentry.[75] Especially because Sheehan had been sentenced to be transported for life, the authorities were certain from the beginning that the active role played by the Frankses in these judicial proceedings constituted one of the reasons for their murder.[76] But the second motive may well have been the more important. As the crown solicitor Maxwell Blacker put it, the fact that Henry Franks had given damning evidence leading to Sheehan's conviction and transportation was used only to increase the size of the murder gang. Blacker's conclusion was that the "most operative cause" of the killings was the family's persistently harsh conduct toward their tenants.[77] As shown in chapter 7, this was clearly a case in which an impecunious gentry middleman (Thomas Franks), himself in danger of losing his lease, had hounded impoverished subtenants to the point of provoking an extremely violent response.[78] Rockites burned the Frankses out of their house near Rockmills in December 1822,[79] and donning women's clothes for the much greater crime, they killed the trio a week after the detested family had returned to it in the first week of September 1823.[80]

Murders arising from tithe disputes were much less numerous than killings stemming from conflicts over the taking, occupation, and management of land. Apart from policemen, whose victimization will be considered separately below, only seven persons are definitely known to have been slain because (or at least partly because) of the functions they performed in the tithe system. It is certainly remarkable that not a single tithe owner, clerical or lay, was killed, though a great many, of course, suffered serious property losses. The seven victims were less exalted and, as frontline functionaries, far more vulnerable. Besides the already mentioned John Corneal, they included two proctors, two drivers, and two process servers. Their deaths were scattered through five counties.[81] The relatively small number of murders in tithe disputes was not a measure of the low intensity of this grievance among the Rockites but rather a reflection of the extent to which the anti-tithe campaign could be waged effectively by other means, especially by acts of incendiarism directed against property.

The last category to be considered embraces the murderous attacks against members of the forces of repression (police, soldiers, yeomen, and magistrates), among whom there were a total of sixteen victims. The magistrates seem to have been extraordinarily lucky, even when due allowance is made for their protected situations. Though many magistrates reported that concealed parties had fired on them, Major Richard Going was the only magistrate to forfeit his life for acts taken in this capacity.[82] Similarly, only one yeoman was slain: the aforementioned John Walsh, fifer of the Adare yeomanry corps.[83] Even more remarkable was the fact that only four soldiers were killed. During their short-lived insurrection in the winter of 1821–22 the Rockites inflicted a single death on the military forces opposed to them: a soldier belonging to Lord Bantry's party in the Inchigeelagh district of west Cork in January 1821 was first wounded and then beaten to death, and his severed head was carried about by the rebels on a pole.[84] A second soldier was fatally wounded near Rathkeale early in March of the same year when Rockites robbed the mails and shot the postboy to death—in the teeth of his escort of dragoons.[85] The other two soldiers to die were slain apparently for their weapons while riding along the roads either alone or in a small party.[86]

Only the police suffered what should perhaps be regarded as substantial casualties, with their dead amounting to ten in all. The motives for these murders varied, but the leading cause was a desire to capture their firearms. This motive prompted the bold and determined mass attack by perhaps seven hundred Rockites on the police barracks at Churchtown in Cork at the end of January 1822. Though badly outnumbered, the police did not yield without a fight and inflicted some casualties on the attackers. But they were soon forced to surrender when the rebels set fire to their thatched buildings. In the confused aftermath three of the disarmed constables were piked or shot to death, apparently in reprisal for rebel losses.[87] The seizure of their weapons also seems to have been the reason for fatal attacks on two additional policemen in other counties later in 1822.[88] Of the five remaining police victims, two were killed in tithe affrays,[89] a third while assisting in the seizure of cattle for nonpayment of rent,[90] and a fourth while executing a magistrate's warrant.[91] The last constable slain, John Orpen, was reportedly mistaken for Cornelius Shine, an informer against the Rockites of Duhallow barony in northwest Cork. Ten or fifteen men dashed out Orpen's brains near Kanturk in December 1823.[92]

Wounding and Mutilation

Besides these ninety-three deaths recorded in the sources, a large number of other Rockite victims must have had their lives shortened by the severe beatings and gunshot wounds that they received. No systematic effort was made to record and analyze the hundreds of beatings noted in newspaper accounts. The stamina of the author was not equal to this herculean task. At a guess, for every person murdered, at least ten others must have been beaten or otherwise injured. Of numerous victims it was said that their injuries were so serious that they were not expected to recover, or to use the stock phrase of the time, their lives were "despaired of."[93] Such descriptions of the condition of victims were employed so often as to raise doubts about their general reliability. But there is not much question that a delayed death was the fate of a substantial number of those assaulted by the Rockites.

Intensifying the terror that the Rockites spread by their murders was the mutilation that they sometimes inflicted on their victims. As already noted, Thomas and Henry Franks were so battered as to be scarcely recognizable, and the heads of John Corneal, Daniel Connell, and a fallen soldier were severed from their bodies. These cases do not stand alone. In further instances the inexact evidence at least points toward mutilation, while in still others the description of the remains puts the question beyond doubt. Whether the murderers of the three Kinnealy brothers intended mutilation cannot be proven, but the skulls of their victims were pounded into a pulp by the weapons they used.[94] The condition of the soldier slain between Mallow and Kildorrery in April 1823 was passed over in silence, but it may well have been similar to that of his police comrade: he apparently survived the brutal attack but had one of his cheeks and both of his ears cut off.[95]

Among the indisputable cases of mutilation were those of a process server for tithes named Conway, whose ears and nose were cut off and left staked to a bank next to the high road after his murder near Listowel in January 1822;[96] and the mailcoach agent William Brereton, killed at Shinnagh on the Cork-Kerry border near the end of the same month, one of whose arms was cut off and whose whole body was covered with wounds.[97] Three victims believed to be informers also suffered in similar fashion. Jeremiah Scully was burned on "a flaming pile" after being shot near Kanturk in November 1821.[98] Two years later in the same district, the killers of the subconstable John Orpen, confusing him with a hated and hunted informer, not only knocked out his brains but also treated his body "with the greatest barbarity."[99] And John Neill, when found murdered near Cappagh in west Limerick in late March 1822, "was desperately mangled, the ribs broken, skull greatly fractured, right arm smashed, the face quite disfigured, and some stabs of a bayonet in various parts of the body."[100]

As vicious and repulsive as these deeds were, an explanation based on mere ghoulishness does not take us very far toward understanding them properly. No doubt the black hatred felt by the Rockites for many of these victims allowed them to engage in such desecrations of the body. But probably even more important in their psychology was an intention to magnify the cautionary and intimidatory effects of the murders. They must have wanted to ensure that such ignominious and horrifying deaths would be remembered, as one threatening notice had it, by the smallest youth in the parish.[101]

Wielding the Weapon of Incendiarism

As frequently as the Rockites resorted to the extremity of murder to enforce their laws and punish their enemies, incendiarism was the weapon that above all others distinguished their movement, marking it off from earlier and later agrarian rebellions. Whiteboys before and after the Rockites also employed this weapon, but never so systematically, so often, or against such a wide range of victims. It is more than surprising that a distinguished historian of popular protest who has much to say about government repression of the Rockites and other Irish agrarian rebels could make the mistake of stressing the weakness of the incendiary tendency among rural protesters in early nineteenth-century Ireland as compared with their English counterparts.[102] This error arises from an excessive preoccupation with the offenses for which Rockites and other rebels were punished by the state when they could be caught and convicted, rather than on the full range of crimes that they actually committed, for most of which they were rarely caught, much less convicted. To put the matter another way, what is concealed about incendiarism in the records of courts and their punishments and in documents relating to transported felons is spread over the pages of contemporary newspapers and correspondence. Among Irish Whiteboys, the Rockites were the incendiaries par excellence.

An artist drawing a single picture of the mythical Captain Rock could not do better than to depict him slinking through some field toward a nearby farmstead, his hand tightly grasping a kettle full of red-hot sods of turf or lumps of coal. This very simple device, usually carried along by Rockites pursuing an incendiary mission and just as often left behind at the site of their target,[103] commended itself for numerous reasons. Among these were the ubiquity of access to fire, the minimal manpower needed for its application (a single individual was frequently enough), the silence and surprise of the attack as well as the ease and safety of withdrawal, the sure infliction of costly damage on victims, with little effort by the incendiaries and even less risk of capture, and the extreme vulnerability of all those who lived in thatched dwellings (the vast majority of the rural population) or who had corn or hay stored in their haggards and barns. Besides its attractions as an effective form of punishment, fire also had an enormous capacity for sheer intimidation: the fear of being "consumed to ashes"—of watching helplessly as flames engulfed one's house or outoffices—easily induced a far-reaching social paralysis against resistance to Rockite demands. Only the rain and damp of the national climate sometimes frustrated the intentions of the incendiaries, but what could not be done on one night that was wet could be accomplished on another that was dry.

Paradoxically, the flooding of chronically disturbed districts with troops or armed police only heightened the tendency of the Rockites to resort to burning as a weapon of aggression. Large parties of Rockites traveling to their targets along the roads on horseback or on foot could more easily be interdicted by military or police patrols. In conjunction with the strict nighttime curfew imposed under the Insurrection Act of February 1822, intensive patrolling of the roads was the principal official defense against the operations of the agrarian "nightwalkers." But once the Rockites abandoned the roads for the fields, once they began to travel in small parties or even in ones and twos, and once they exchanged their loud guns for small, silent kettles, they were almost always beyond the reach of the forces of repression, as the authorities themselves bitterly confessed.[104] The more that police and soldiers scoured the highways, the more the Rockites took to the fields with their little portable fires—from which they soon made much bigger ones.

Incendiary crimes did not reach full stretch until after the revival of the Rockite movement in the late summer of 1822, but they were far from absent during its expansionary phase between mid-1821 and the spring of 1822. Rockites seeking to spread their fledgling protest from west Limerick into adjacent parts of Kerry in early October 1821 set fire to the residence of the magistrate Edward Hartnett of Sandville near Castleisland. His was the first such house to be burned by Rockites, and one of the relatively few gentry-occupied residences to be so treated during the entire period of the movement. The attacking party in this instance consisted of about sixty men (forty on horseback and twenty on foot) who had come down from the mountains around Abbeyfeale into the Castleisland district.[105] Whether Hartnett had drawn their ire by rejecting a demand to hand over his arms is unclear, but it was a standard Rockite practice at this point and later to use the threat of burning to compel surrenders of weapons. As the Kerry proprietor Lord Headley lamented about two weeks after the attack on Hartnett, "now people are forced to supply arms and powder under the penalty of having their houses burnt over their heads. . . ."[106] His complaint was often echoed by others elsewhere in the months ahead.[107]

Incendiarism was also a feature of the insurrectionary phase of the movement in January 1821 and of its immediate aftermath. A huge body of insurgents, said to number over five thousand, reportedly participated in burning the house and offices of the gentleman George Bond Low at Sally Park near Liscarroll in north Cork during the short-lived uprising. Low and his family were not in residence at the time; for the previous two months troops had used the premises as a military barracks, thus explaining why the rebels destroyed them. The same insurgent force attacked a house in Liscarroll and a set of out-offices at nearby Altamira belonging to different members of the Purcell family, destroying the former and setting fire to the latter; the Altamira property was said to be the last one in the district where soldiers could be accommodated, a possibility of which the incendiaries were no doubt mindful.[108] In a few other cases in this early period Rockites had recourse to fire against police or military targets: the stables used by the police for their horses in Tipperary town were burned in late November 1821,[109] as was the barracks at Ballynagranagh in southeast Limerick in late February 1822,[110] not to mention the blood-stained burning of the police barracks at Churchtown in north Cork a month earlier.

Church Burnings

Among the most conspicuous acts of Rockite incendiarism during the expansionary phase of the movement were the burnings or attempted burnings of six Protestant parish churches, three in County Kerry (Knockane, Templenoe, and Kilgarvan) and another three in County Limerick (Killeedy, Ballybrood, and Athlacca). Of these six churches, two (Kilgarvan and Templenoe in Kerry) escaped extensive damage, but the other four were either partly or entirely destroyed.[111] In a seventh incident of this type, involving the parish church of Abbeyfeale in west Limerick, local Rockites (perhaps about forty in number) set out with the intention of burning it but changed their minds, allegedly when one member of the party pointed out that as taxpayers, they would be forced to help shoulder the costs of building a new one. Instead, they demolished everything inside the church and pragmatically stripped all the lead off its roof for the purpose of making ball cartridges.[112]

What explains these incendiary attacks on Protestant places of worship? In chapter 2 it was argued that they were motivated in part

269

by the sectarian animosities of Catholic Rockites, influenced by the anti-Protestant millenarianism so prevalent at the time.[113] Yet in most of these incidents more immediate grievances and mundane objectives (including lead from the church roof at Abbeyfeale) also played a significant role. Opposition to the levying of church cess was evidently a factor in the south Kerry parish of Knockane, since a few days prior to the destruction of the church there in late November 1821, the churchwardens had been warned to stop collecting this ecclesiastical tax.[114] Perhaps the same grievance or that of tithes was operative in the parishes of Kilgarvan and Templenoe. Early in January 1822, soon after Rockites attempted to burn these churches, two local Protestant gentlemen scouted the prevalent notion that any of the three incendiary incidents had stemmed from anti-Protestant animus: "We are both completely satisfied that nothing of the kind exists, particularly among the better order of the laity and the whole clerical body of Catholicks. . . ."[115]

Practical grievances were also advanced to explain two of the three church burnings in County Limerick. In Killeedy parish near Newcastle the Protestant parson was believed to have given information against one man who was later hanged and to have been the cause of the conviction and flogging of another; the popular rage against the minister was attributed to these circumstances, which might also help to explain why Rockites burned his glebe house and out-offices as well—a unique event among these cases.[116] In Ballybrood parish near Cahirconlish, as the Reverend George Madder noted ruefully on the day after his church had been badly damaged by fire, the parishioners were disgruntled about tithes, and since two soldiers constantly guarded his house, the burning was the safest means that his enemies had to strike at him.[117]

There also appears to have been an element of imitation or contagion in these church burnings. Knockane and Templenoe are contiguous parishes in south Kerry, and Kilgarvan lies only one parish to the east of Templenoe. It is possible that some of the same activists were involved in more than one of these incidents. In addition, a mere six weeks separated the destruction of Knockane church (23 November 1821) from the attempted burning of that of Templenoe (4 January 1822), and the effort to destroy Kilgarvan church occurred within days of the attempt on Templenoe.[118] A temporal linkage is also a striking feature of the Limerick cases. The attacks on the churches of Killeedy and Abbeyfeale (adjacent parishes in the southwestern corner of the county) took place on the same night (7 February 1822), and they were separated from the burnings of the Ballybrood and Athlacca churches in east Limerick by approximately two and three weeks, respectively.[119] As conspicuous as they undoubtedly were, the church burnings were also a phenomenon that quickly passed away. With only a single known exception (the attempt to destroy by fire the Protestant place of worship in Whitechurch parish near Cork city in September 1823),[120] no further incidents of this kind occurred in any of the disturbed counties throughout the rest of the Rockite years.

Burning Houses and Out-Offices

While church burnings were few in number and confined to the expansionary phase of the movement, the same can hardly be said of the destruction of houses and out-offices by fire. These incendiary crimes were legion (altogether, hundreds of dwellings or outbuildings were destroyed), and they were even more prevalent after the revival of the Rockite struggle in September 1822 than they had been during its initial phase. All of the counties in which Rockites were active were affected by incendiarism, though not equally and with some significant variations in timing. The social status of victims extended across the spectrum, from landed gentlemen and Protestant ministers at one end to farm servants and laborers at the other. But farmers constituted by far the largest category of victims in their various capacities as landlords, land grabbers, tithe payers, and holders of arms. The better-off farmers were especially liable to be attacked in this way.

The first wave of house burnings occurred in the late winter and early spring of 1822. Accounts appearing in the *Dublin Evening Post* from mid-February through mid-April indicated that a total of about sixty dwellings had been destroyed by fire in the four counties of Cork, Kerry, Limerick, and Tipperary. Of this total, Limerick alone accounted for almost forty of the reported house burnings, Kerry for eleven, Tipperary for seven, and Cork for five.[121] Undoubtedly, these figures understate the real number of such incendiary crimes during this period. At the Tipperary spring assizes of 1822 "the petitions for compensation on account of malicious burnings were more [numer-

271

ous]," observed a police magistrate, "than in any year since the rebellion of 1798."[122] Although this comment refers to all kinds of property maliciously damaged by fire, and not to dwellings alone, it strongly suggests that the figure of seven burned houses derived from newspaper accounts falls far short of the actual number destroyed. Even the much higher figure for Limerick probably represents substantial underreporting. So concerned about the number and amount of claims for compensation presented to the Limerick grand jury was one gentleman that he advocated the establishment of a special commission to investigate the claims, which he considered excessive. Early in March 1822, as this epidemic of incendiarism was still gathering momentum, he conceded that "many thatched houses have been burned and much property destroyed," but he insisted that unless some claims were rejected and others lowered, "many parishes would be at once reduced to beggary."[123]

As with murders, so too with incendiary crimes, west Limerick stood out prominently as the great cockpit of such activity during the late winter and early spring of 1822. Districts associated with killings were also conspicuous for burnings. Among these were Newcastle, Cappagh, and Rathkeale. Two newspaper reports at the end of February noted the destruction of "some houses" near Newcastle, of "several others" in Cappagh village, and of "six houses" between Newcastle and Rathkeale.[124] "This moment," exclaimed an inhabitant of Newcastle whose letter was quoted in one of these reports, "the town is illuminated by the blaze from several houses in the line of Camus, which are now on fire; all is bustle and alarm."[125] A second letter writer, residing near Rathkeale and quoted in the other report, declared, "We are now at this moment looking out of the windows and are illuminated with houses on fire all about the country."[126] At the same time that they were setting fire to so many dwellings, Rockites in west Limerick were also burning "a considerable quantity of corn in stacks" and numerous ricks of hay.[127] The bewildering frequency of these events left observers with little time to inquire into their specific causes, but to judge from those accounts in which the motives of the Rockites in this region were made plain, opposition to tithes occupied a distinctly subordinate place. Instead, antagonism to the collection of arrears of rent and hostility to land grabbers provided the principal reasons for this profusion of incendiary attacks.[128]

Burning Tithes in Kind

When, however, the Rockite movement revived in the late summer of 1822, opposition to tithes became the main focus of the incendiaries, and the chief theater of their operations was now north Cork rather than west Limerick. On the orders of Captain Rock farmers carefully separated their tithe corn from the rest of their grain and left it out in the fields, ostensibly for collection by the agents of the tithe owner, but in fact to be burned by Captain Rock's adherents.[129] The destructive little fires that soon consumed the neat stacks of tithe corn became so numerous over the next few months as to defy all attempts at detecting the incendiaries.

Such was the fear of the firebrand that not only farmers but also "many of the gentry" were behaving in the same "dastardly" fashion by removing their tithe corn from their haggards for its quick destruction by the unstoppable agrarian rebels of north and northeast Cork. "Tho' the troops and police are constantly patroling [sic] all night," complained Robert Eames from Buttevant in late September 1822, "they have not taken (that I can hear) one person for these atrocious acts."[130] A well-informed writer in the *Dublin Evening Post* declared on 1 October that the style of this general combination against tithes—serial burnings of tithe corn by gangs or individuals acting under cover of darkness—was "more artful in design and more efficient in its operations than any previously adopted."[131] These incendiary crimes were at first concentrated in Fermoy barony as well as in part of Orrery and Kilmore. The police magistrate Major Samson Carter reported in October that no outrage by fire had yet been committed in Duhallow,[132] and though its undisturbed condition was not permanently maintained, the burning of a farmhouse in February 1823 was said to be the first serious outrage in that barony in a long time.[133] For both incendiary and other offenses this geographical pattern was impressively demonstrated by a rare police analysis of crime in the four northern baronies of Cork during (roughly) the first two months of 1823. Of the total of fifty-two outrages reported, as many as thirty had occurred in Fermoy barony and another fifteen in Orrery and Kilmore. In all of Duhallow there were merely two outrages, and in Condons and Clangibbon just five. Incendiary crimes constituted the largest single category of offense (eighteen out of the fifty-two), and almost without exception they took place either in Fermoy or in Orrery and Kilmore.[134]

273

Targeting the Gentry

The victims of incendiarism in the middle baronies of north Cork during late 1822 and early 1823 included others besides Protestant parsons, tithe functionaries, and farmers who had violated Rockite regulations concerning tithes. The area was remarkable for its heavy concentration of gentry seats, and their owners sometimes incurred Rockite wrath by their own repressive activities or by the assistance that they rendered to the army and the police. Wallstown Castle, the seat of the Stawell family, was the target of a fire set by Rockites in mid-November, soon after its owner had proposed to lease it to the government for use as a military barracks.[135] Although this attempt at burning proved abortive, the attacks directed at Colonel Hill and John Dillon Croker in December were far more successful. The former, known endearingly to local Rockites as "Bloodhound Hill,"[136] lost over 100 tons of straw and hay as well as the equivalent of 180 barrels of wheat and oats in the burning of his great haggard at Clogheen, midway between Doneraile and Buttevant.[137] This incident was one of a whole series of attacks on members of the Hill family.[138] Croker, son of yeomanry-brigade major Henry Croker, suffered the complete destruction of his extensive corn mills, valued at £4,000, in an incendiary fire near Mallow.[139] Another gentry family assailed repeatedly, like the Hills, were the Croftses. The Reverend Christopher Crofts, military chaplain to the troops headquartered at Buttevant, had provided a house at nearby Ballyhoura for the accommodation of soldiers, and as a result, observed their commanding officer, "the disaffected have vowed vengeance against the whole family."[140] In late February and early March 1823, Rockites not only burned the outbuildings at Ballyhoura but also destroyed a lodge and a set of out-offices belonging to the chaplain's brother Thomas Crofts at Ballyhay in the same district. In addition, twenty-five head of cattle owned by the two men were houghed.[141]

But these outrages against the Croftses were overshadowed by what the local gentry considered a far greater calamity. This was the destruction of Carker Lodge, along with a whole range of out-offices, on 7 March 1823 near Doneraile. Three armed men entered the kitchen of the lodge, took from the servants a few sods of lighted turf, warned them on pain of death not to follow, and proceeded to set the place ablaze.[142] The event, described by Colonel Sir Hugh Gough as "the most complete burning that has taken place in this county," sent shudders of horror through gentry drawing rooms.[143] Hugh Norcott, the owner of Carker Lodge, had not been unpopular with the local peasantry, rather the reverse. But he had recently fired (without effect, it was said) on two men suspiciously lurking near his residence, thus apparently prompting the attack. Almost immediately, it was discovered that one of the incendiaries was his own gardener John Hickey, whom Norcott had reputedly treated as a friend.[144] Such treachery plunged the already dispirited gentry into even deeper gloom. As Norcott's absent daughter was told a few days later,

> The lodge was the first *gentleman's* house burnt in this country [i.e., north Cork], which has caused great agitation; your father and all the gentlemen in the country are in a state of despondence since the [burning], *rebellion* at their doors and not a *shilling* to be got. . . . Lord Combermere, the commander in chief of the troops in Ireland, is at L[or]d Doneraile's. I hope he will do something to save our poor country.[145]

Though they seethed with indignation whenever a gentry residence like Norcott's was burned, some magistrates and officials could not summon an equivalent degree of anger over the destruction of the dwellings and out-offices of ordinary farmers. This attitude of near complacency became evident in the late winter and spring of 1823 when such burnings assumed the proportions of an epidemic, much greater than that of a year earlier. Its beginnings were visible in February,[146] but a striking escalation occurred in March when, as Major General Sir John Lambert remarked, the outrages became more numerous than for many months,[147] and the violence continued at an extraordinarily high level throughout April and May before gradually subsiding. In this new wave of incendiarism the focus of attention shifted dramatically away from tithes toward the punishment of evicting landlords (especially farmer-landlords) and land grabbers (land canters in common contemporary parlance). In reporting the upsurge in outrages, Lambert had specifically observed that emissaries were going about County Cork posting notices against "all new occupiers of farms."[148] Another military officer asserted that the recent burnings in his district (north Cork) had been caused either by evictions for nonpayment of rent or by the taking of evicted farms by new tenants.[149] Prominent officials, however, found reasons to make light of the galloping incendiarism.[150] The crown solicitor Maxwell Blacker was less than distraught about a series of incendiary outrages in Tipperary: "I believe no lives were lost and little injury was done to private property, & a considerable part of the outrages, as I was informed, consisted of the burnings of cabins by ejected tenants."[151] Certainly in Cork and Limerick, where this epidemic of incendiarism was concentrated, there was no justification for such complacency, and many observers were shocked at the scale of the phenomenon and at the extent of the damages inflicted.[152]

The Rage for Burning

Even though the Rockites had already been pursuing their work furiously when Blacker's comment was penned, the incendiarism persisted for weeks afterwards. In its pages from late April through late July the *Dublin Evening Post* reported scores of cases in which dwellings or out-offices had been damaged or destroyed by fire. In County Cork, according to these accounts, no fewer than forty-three houses and fourteen out-offices were burned, and in County Limerick the corresponding figures were thirty-nine and four.[153] The statistics derived from newspaper accounts do not tell the whole story; in fact, the figures seriously understate the real totals. And in the two or three months prior to late April, scores of other houses and out-offices had already been destroyed.[154]

As when opposition to tithes had been the primary concern of the incendiaries, so too now when the focus of their antagonism was redirected toward land grabbers and evicting landlords, the two middle baronies of north Cork (Fermoy and Orrery and Kilmore) headed the list of disturbed districts in the new wave of burnings. Observers found it difficult to capture in words what was happening around them and were tired of providing specific details. "The state of the country between Mallow, Doneraile, and Fermoy," declared one overwhelmed witness in the third week of March 1823, "is beyond all description."[155] "On a recent night four fires could be seen blazing at once—near Doneraile and at Shanballymore, Ballyhooly, and Ballyduff."[156] In addition, the burnings now extended south into Barrymore barony and west into Duhallow, where they were accompanied as usual by other types of Rockite crime.[157] In northern Barrymore, declared a local magistrate in mid-April, six houses had been destroyed by fire within

This sketch of an eviction scene in 1848 effectively captures certain features that were all too common during the "clearances" associated with the Great Famine: vain pleas for mercy from the victims, the sheriff's bailiffs stripping the roof bare, other functionaries removing the ousted tenant's remaining crops and livestock, armed police protecting the eviction party, and a countryman at the far left pointing to a concealed gun under his coat. Despite the menace of reprisal implied by this last feature, resistance was rare in the late 1840s and early 1850s. The case was far different, however, where the Rockites held sway during the early 1820s. Reprisals were extremely prevalent; the targets were frequently large farmers in the position of landlords; and when tenants were dispossessed, they or their allies often simply burned down their former cabin or house before or after the holding was newly let. (*Illustrated London News*, 16 Dec. 1848)

the last fortnight,[158] and incendiary as well as other disturbances had become common near Cork city.[159] In this latter area, besides damaging farmers' property, Rockites burned the house and out-offices of a squireen-middleman named Hawkes near Ballincollig in late February and totally destroyed Riverstown House, the seat of Samuel Bennett near the city, in late April.[160]

Second only to north Cork as a center of incendiarism in the spring of 1823 were the baronies of mid-Limerick. Reported to be highly disturbed in late April were Kenry, Lower and Upper Connello, and Coshma, in all of which the so-called County Cork system of burning houses and haggards had been adopted.[161] Included within these baronies was a broad stretch of territory extending from the banks of the Shannon all the way to the northern borders of Cork. The Liberties of Limerick city and the long troubled Newcastle district were also seriously affected.[162] In part, this upsurge of incendiarism in County Limerick represented a sturdy revival of activity in old Rockite strongholds, as in Kenry, Lower Connello, and parts of Glenquin. But it also signaled a new or at least deeper penetration into districts that had not been highly disturbed previously, as in Upper Connello and Coshma. Particularly significant in these latter baronies was the influence of events in neighboring north Cork.

The bitterly sectarian burning of the Palatine village of Glenosheen was a clear instance of direct action, not just influence, from the Cork side of the border. Glenosheen lay astride the mountain road from Doneraile in Cork to Castle Oliver in Limerick.[163] The band of over one hundred Rockites who attacked the village at the end of April 1823 reportedly came over the mountains from County Cork.[164] As discussed in chapter 4, the sectarian motivation for the raid was plain from the language of the Rockites themselves,[165] but they had other ambitions as well. Glenosheen was a storehouse for arms in a double sense: the adult males in this long-established Protestant community on the estate of Oliver Gascoigne were enrolled in a body of armed volunteers (in effect, a yeomanry corps), and they were reinforced by a party of police permanently stationed there.[166] The firearms of both the police and the Palatines presented a tempting target for the attacking Rockites. The episode was reminiscent of the Rockite raid on the police barracks at Churchtown in late January 1822, and those with a longer memory of Whiteboyism in this region, such as Chidley Coote of nearby Mount Coote, could recall the day in 1793 when earlier rebels had staged an arms raid against the Palatines there.[167] Now, some thirty years later, they made a bid to wipe out the village entirely. Of the seven houses that they set on fire, they laid three in ashes.[168] Although the raiders did not avow the intention of taking over the lands farmed by the Palatines around Glenosheen and its sister village of Ballyorgan, it is not at all fanciful to conclude that they wished to drive these Protestant tenants from the district altogether, thus making the lands available for occupation by Catholics.[169]

Burnings in Other Counties

Besides portions of Limerick and Cork, parts of Kerry and Clare were also seriously affected by Rockite incendiarism early in 1823. In Kerry the burnings were confined to a limited area in the extreme north around Listowel and Ballylongford.[170] Most of the victims, like others elsewhere, appear to have been land-grabbing farmers, but at least one victim belonged to the landed gentry. Early in April a band of Rockites attacked Riversdale, the family seat of the Raymonds, located between Abbeyfeale and Listowel, while its owner was absent in England. The raiders set fire to both the house and the out-offices, and while the former was saved, the latter were destroyed.[171] An observer who reported this incident declared that on the same night no fewer than "ten fires were perceptible" in this district of north Kerry.[172] But the burnings there, though numerous, passed quickly, and the Kerry Rockites again lapsed into inactivity, this time for good. The outbreak in Clare, on the other hand, started somewhat earlier, and though it was to be interrupted in the summer and autumn of 1823, it blazed forth again in 1824 and proved extremely difficult to suppress. The initial burnings occurred north of Limerick city on the Clare side of the Shannon in January 1823, and before the end of March the eastern baronies of Bunratty and Tulla had been placed under the provisions of the Insurrection Act.[173] In a mild foretaste of what was to come, the incendiarism also touched parts of west Clare, including the districts of Killadysert and Kilrush, before abating temporarily.[174]

By mid-August 1823 incendiary crimes had been almost entirely absent for nearly three months, a situation that the authorities attributed to the shortness of the summer nights and to the higher level of employment during that season of the year.[175] The coming of autumn, however, was widely expected to bring about a fresh outbreak of burnings as the customary pressures for the collection of tithes and rents would again be applied. Official forebodings were confirmed at the change of seasons, but in most of the disturbed districts the scale of incendiarism was distinctly lower than it had been earlier. North Cork (the middle baronies) and west Limerick were again at the forefront in the burning of dwellings and out-offices during late 1823 and early 1824,[176] but other parts of these two counties were largely or entirely free of such crimes. Even in the narrowly circumscribed districts where incendiarism was still common, newspaper reports and other accounts leave the strong impression that the Rockites could not sustain their campaign at the high levels of 1822 and early 1823. Only of Clare could it be said that the frequency of the burnings in the spring of 1824 exceeded that of the previous year. The western baronies of Clonderalaw, Moyarta, and (to a lesser extent) Islands now experienced even more incendiary crimes than the eastern baronies, and the victims were either land grabbers or evicting landlords (including both farmers and gentlemen) in the great majority of cases.[177] Not until August did the house burnings in Clare finally come to an end.

In spite of the fact that hundreds of dwelling houses were reduced to smoldering ruins in these years, their occupants invariably escaped death. Indeed, with the notable exceptions of the Sheas, only two other deaths were attributable to any fire set by the Rockites.[178] Admittedly, in a few recorded cases the incendiaries certainly gave the appearance of wanting to incinerate the occupants as well as burn their houses, as had been done in the infamous outrage at Wild Goose Lodge in County Louth in 1816. Thus the band of Rockites that set fire to the dwelling of a well-to-do farmer and land grabber named Clanchy in the Kildorrery district in April 1823 first locked him and his family inside and then posted a sentry outside the front door. But the Clanchys managed to escape the flames when the Rockites suddenly decamped.[179] Much the same sequence of events, with the same fortunate outcome, was repeated a few days later in another incendiary attack in that district.[180] This suggests that rather than intending to kill anyone, the raiders wanted to heighten the terror of their victims by locking them briefly in burning houses. Frequently, in fact, the incendiaries made a point of allowing or instructing the occupants to remove their furniture and other household goods before setting fire to the dwelling.[181] Even when this scant generosity was not extended, the raiders must often have made noises calculated to rouse sleeping occupants. Despite the language of threatening notices, murder was far from being an aim of the incendiaries; their design was to destroy essential property.

Claims for Compensation

Inevitably, the successive waves of incendiarism spawned a host of claims for compensation. At the spring and summer assizes of 1823 in County Cork alone, the clerk of the peace transmitted petitions from a multitude of claimants for compensation amounting in all to nearly £13,300.[182] Although this figure covered property damaged or destroyed in other ways than by burning, incendiary offenses accounted for by far the largest portion of the total. There were eighty-six claimants (seventy-one on account of malicious burnings alone) for sums amounting to £2,750 at the Cork summer assizes of 1823,[183] so that the number of claimants at the spring assizes of that year,[184] which was not reported, must have reached several hundred, since altogether they sought to recover more than £10,500. Under the old Whiteboy Act of 1776 the liability for compensation in cases of malicious damage to property fell on the ratepayers of the parish or barony in which the crimes had occurred, unless an individual claim exceeded £100, in which case the liability fell on the county at large.[185] This system was designed to discourage the inhabitants of any particular parish or barony from countenancing the malicious destruction of property, since as ratepayers, they would be taxed for it. But the exception for claims exceeding £100 (charged to the county as a whole) weakened the effectiveness of the system: at the Cork spring assizes of 1823 about £8,600 in claims, out of the total of slightly more than £10,500, was covered by the exception and did not fall on local parishes or baronies.[186]

Like similar systems of compensation paid from public funds, this one offered opportunities for deception and fraud. Many claimants whose dwellings had been destroyed allegedly sought compensation for household property that they had actually been able to remove just before the burnings.[187] The destruction of tithe corn also gave rise to fraudulent claims. Farmers sometimes did not wait for Rockites to burn such corn but stole a march and did it themselves, intending to recoup their loss by presenting falsified petitions to the grand jury.[188] A refinement of this kind of fraud was the ploy mentioned in chapter 5 under which only the straw of wheat or oats was burned after the grain had been thrashed out, whereas the farmer's claim to the grand jury was for the value of the grain itself.[189] As the authorities discovered, even Captain Rock's own adherents were not above engaging in fraudulent practices involving both tithe corn and other kinds of property. In a celebrated case mentioned earlier, the postmaster of Abbeyfeale sought compensation for a burned building owned by him that had been used as a military barracks, even though his son and other Rockites were responsible for destroying the structure.[190]

But whether fraudulent or not, the presentation of claims for malicious injuries to property meant increased taxation, which the Rockites, like everybody else, wanted to avoid. In order to resolve this dilemma, the Rockites of some districts threatened to do worse than burn again if persons already victimized claimed compensation. Thus in Clare in March 1823 people whose property had been destroyed by fire were warned "not to attempt sending in presentments for the damages they have sustained," or else they would be exterminated.[191] The unraveling of the previous alliance between farmers and laborers in north Cork, a development noted in 1824 by the astute police magistrate Samson Carter,[192] may have resulted in part from the farmers' strong aversion to the higher taxes that fires set by laborer Rockites made necessary.

The Levying of Contributions

The much greater importance and dramatic quality of murder and incendiarism should not lead to the neglect of certain other types of Rockite activity that were considered essential to the vitality and effectiveness of the movement. Among the most common of these other activities, along with arms raids and the sending of threatening notices, was the levying of contributions to the Rockite cause. Although earlier Whiteboys had often engaged in this practice,[193] the extent to which the Rockites did so was probably unprecedented. Extorting money under an expressed or implied threat of violence was clearly a major Rockite preoccupation. In his close investigation of Rockite outrages in north Cork, Finbarr Whooley has emphatically demonstrated the great frequency with which contributions were exacted. Whooley has enumerated 331 outrages in that region in 1823, of which the forcible collection of money accounted for as many as 69, or 21 percent of the total. For 1824 the proportion is even higher: 49 of the 162 outrages, or 30 percent, took the form of levying contributions.[194] What makes these figures even more striking is that this type of crime was so often unreported to the authorities.

Both offensively and defensively, the Rockite movement required money. For offensive purposes the Rockites needed cash to buy ammunition, including gunpowder and ball cartridges, for the firearms they were so busily robbing. They also needed money to keep stolen weapons in good repair, and sometimes they must also have purchased firearms. Even with other people's money the resourceful Rockites practiced economy. The lead necessary for making bullets was frequently obtained in ways other than by purchase. It was commonly stripped from the roofs of appropriate buildings. Thus in November 1821 half a hundredweight of lead was reportedly stripped from the butter weigh house and the artillery barracks (no less) in Limerick city,[195] while in February 1822, as previously noted, a similar operation was conducted on the roof of the Protestant church of Abbeyfeale.[196] Stealing gunpowder in sufficient quantities was less easy, however, and usually it had to be bought. It was widely available for sale in the shops and stores of Limerick, Cork, and other towns, as the government soon realized. Early in December 1821, on the viceroy's orders, the authorities in the cities of Limerick and Cork seized whatever supplies they could find, including nearly three tons of gunpowder at Limerick alone.[197] But even after the government had taken this step, the surreptitious sale of gunpowder encouraged the Rockites to persist in their exactions.

During the initial phase of the movement arms raiding and the levying of contributions usually went hand in hand, though later the closeness of the connection diminished significantly. Whooley's data on outrages in north Cork appear to suggest that the Rockites there hardly attempted to levy contributions in 1822; he records only 3 cases in that year out of a total of 210 outrages.[198] This finding, however, is highly implausible. From other areas there is abundant evidence that this practice was pursued intensively in conjunction with the arms raids that were so prevalent during the winter of 1821–22. "Innumerable robberies are committed every night under the pretense of levying money for their cause," declared one County Limerick gentleman when also reporting a series of arms raids in January 1822; "no legal taxes were ever so rigorously enforced or so successfully levied."[199] When the victims of intended arms raids turned out to have no weapons to give them, the Rockites regularly demanded money instead.[200] At times they seized other possessions, such as the shoes, shirts, and greatcoats taken in November 1821 "from such opulent farmers as had not arms" in the district between Hospital and Croom in Limerick.[201] Quarts of whiskey were welcomed if money could not be obtained.[202] In one extraordinary case in January 1822 two men were arrested for having plundered a cart of forty-two gallons of whiskey on the high road near Rathcormack, Co. Cork, saying that they were "a party of Captain Rock's men moving to Newcastle and directed to levy contributions on the way for the support of his troops."[203]

Even though the number of arms raids fell steeply in 1823 and 1824, the levying of contributions persisted at a high level. The main reason was that the Rockites increasingly sought to pay the legal expenses of their comrades in jail and facing trial. To some extent they had probably done so at an earlier stage. As the police magistrate Samson Carter remarked in March 1824 as the Cork spring assizes approached, "such visitations [of country people by parties of Rockites] have been frequent prior to the assizes in this county since the commencement of the disturbances, to extort money for the benefit of the prisoners to be tried."[204] The older purposes for levying contributions certainly did not disappear after 1822, but now this new one surfaced much more often in reports of the practice, either alone or in conjunction with the others. Money was demanded "to repair arms and employ council [*sic*] for their friends against the next Cork assizes"; "to buy ammunition and assist their friends"; "to release prisoners"; or "for the relief of some of their comrades in prison."[205]

Having collected money for the defense of their friends in court, the Rockites presumably made discreet arrangements to hire counsel at the assizes, though this aspect of Rockite affairs is shrouded in mystery. As noted in chapter 4, Daniel O'Connell himself undertook the defense of accused rebels in numerous Rockite cases, including some of the most notorious in contemporary annals.[206] Curiously enough, his voluminous correspondence is completely silent on the subject of how—and how much—he was paid in such cases. But there is every reason to think that his highly prized services as a Whiteboy barrister were not rendered gratuitously or even cheaply. In a later case growing out of the famous Doneraile conspiracy, he was said to have earned a fee of 100 guineas as defense counsel.[207] O'Connell regularly used the courtroom to display his well-known hatred for agrarian violence in general and for the Rockite movement in particular. His method of doing so was capable of being interpreted in opposite ways, though this was not his intention. His vehement challenges to the credibility of approvers and other government informers provided opportunities for him to appear both as the protector of accused Rockite prisoners and as a respectable upholder of anti-Whiteboy laws.[208] This latter interpretation was of course the correct one, coinciding as it did with O'Connell's most deeply held convictions. Thus there is a special irony in the strong likelihood that the Rockite practice of levying contributions in order to fee counsel must often have put money in O'Connell's pocket and in the pockets of other Catholic barristers holding the same anti-Whiteboy views.

Besides using contributions for these traditional offensive and defensive purposes, the Rockites occasionally made financial demands for other reasons. In their posture as champions of the poor and the oppressed, they sometimes directed their victims to provide money for needy individuals or classes of people. Thus in November 1823 a County Limerick farmer was ordered "on pain of death to have three pounds ready for the widows and orphans when called upon."[209] In other cases Rockites extorted money partly as a punishment from those who had violated their laws. A band of Rockites who visited a land grabber on the Trinity College estate in south Kerry in January 1822 not only instructed him to surrender his farm but also demanded 5s. immediately and a further sum in the following week.[210] Another land grabber in north Cork was given a choice by the Rockites who called on him in March 1824: he could either quit his farm or give them £4 on their next visit.[211] In still another case the motives of punishment and championing the poor were both manifested. From the victim, a County Limerick native residing near Doneraile and locally regarded as an unwelcome stranger, the Rockites sought to collect £1, and they also directed him to coerce a neighbor to furnish money to a specified widow living in the district.[212]

The methods used by the Rockites in levying contributions displayed certain recurrent features. Parties of Rockites carried out their collections in the countryside systematically, visiting most of the houses, or at least those belonging to farmers, in a given district over a period of a week or so. Thus a report in April 1823 stating that every respectable farmer near the village of Rockmills in northeast Cork had been robbed of money also indicated that the same thing had happened to nearly all the farmers between Doneraile and Mitchelstown within the last ten days.[213] Some localities are known to have been combed more than once.[214] In order to minimize the risks of detection and perhaps to lend legitimacy to their collections, the Rockites sometimes resorted to a different tactic. They designated a small number of well-to-do farmers as their agents, and acting out of fear or perhaps sympathy, these farmers agreed to perform the service in an unobtrusive but regular fashion.[215] A resourceful magistrate in north Cork tricked a dairyman on his estate in June 1823 into revealing the names of two well-off farmers who had collected money for the Rockites of his neighborhood.[216]

In levying contributions, Rockite collectors often had to deal with the common excuse that their victims had no money at hand. To obviate this problem they announced to their victims that they would visit, or call again, on a designated night, or they directed that the sums required should be delivered to a specified place.[217] By adopting a ploy in April 1823, a magistrate in the Kilfinnane district neatly exposed this feature of Rockite methods. Taking along some troops and a few Irish speakers, he visited the houses of farmers in the area. Masquerading as Rockites, "they knocked at the doors, asked if the money was ready, and in three instances it was handed out, ready made up in paper and delivered without a question being asked. . . ."[218] In the same month Rockites levying contributions in the Kildorrery district had a standard response for those farmers who pleaded an absence of cash: the farmers were each sworn to bring or send £2 "to some public house in Doneraile in the name of 'General C. Rock,' when it would be received and accounted for."[219]

The amount of money demanded from individual victims by Rockite parties during these collections generally ranged from £2 to £5.[220] At times equality of assessment seems to have been the prevailing notion. Rockites levying contributions in the Liberties of Limerick city in November 1821, even if given higher sums, reportedly returned the surplus over a sovereign (i.e., 20s.), saying that a sovereign from each house would be sufficient for the purchase of ammunition.[221] At least some variability was evident in the assessments of another band of collectors in Doneraile parish in February 1823: they demanded 10s., 20s., and 30s. from a woman, a farmer, and a gentleman, respectively, but had to settle for less in each case. The woman gave them 2s. 6d., the farmer supplied a quart of whiskey, and the gentleman paid 7s. Yet two other farmers visited by this same band gave £1 and £1 5s., while another gentleman (Hugh Norcott of Carker Lodge) initially had to hand over his watch but was able to retrieve it in exchange for £1.[222] This series of collections suggests that there was often an element of negotiation between the assessors and the assessed, but it does not indicate a very sharp differentiation based on the varying means of the different victims. Because they were known to keep cash on hand, dairymen were targeted by Rockite collectors in some districts, but the fact that in three instances the sums extorted from them were 5s., 15s., and £1 13s. suggests that expediency and not social differentiation was paramount.[223] Yet in at least some cases the victims' wealth and social status did lead to much heavier demands. After an amicable parley in April 1823 with a band of Rockites who initially sought as much as £20, the small north Cork landlord Clifford Martin managed to reduce his contribution, but the unwelcome visit still cost him £7 or £8.[224] From another victim, apparently a wealthy farmer, Rockites near Cork city took £13 5s. in June 1823 "for the relief of some of their comrades in prison."[225]

Sowing Discord

Even though the sums exacted by the Rockites were usually small, the levying of contributions was among the least popular and least approved of their activities. To judge from the fact that the practice was condemned or forbidden in some of their own pronouncements,[226] many Rockite activists themselves must have repudiated it. One gentleman in the Newmarket district of northwest Cork, from whom a Rockite band had earlier demanded money, received a letter from "John Rock" in November 1821 denouncing the practice and promising to bring the evildoers to justice or else to "loose [sic] my life in the attempt."[227] And a Rockite notice posted on the door of Killorglin chapel in Kerry in February 1822 included among its numerous regulations one that prohibited any levying of contributions.[228] In a movement that was so loosely structured, so dependent on local initiative, and so lacking in hierarchical control, it was all too easy for the public purposes of the practice to be perverted by private peculation. How often this happened is impossible to say, but given the temptations and the absence of sanctions, the conversions of such money to private uses must have been common enough to help to discredit the practice.[229]

But it was the fairly stark class character of the practice that made it so divisive and damaging from an organizational point of view. Although the victims ranged in social status from wealthy gentlemen to poor dairymen, farmers who were described as opulent or respectable figured disproportionately among those attacked, and many less substantial farmers were also forced to contribute. Laboring households may not have been altogether spared in the general sweep of a whole neighborhood, but laborers and cottiers seem to have dominated the personnel of the levying bands. Farmers, even comfortable ones, occasionally turned up as collectors, but their service was not always voluntary. The known cases of forced service raise at least the suspicion that laborers intimidated certain farmers into becoming collectors in order to disguise the rather naked class conflict that the levying of contributions so often assumed.

The authorities naturally welcomed the sharpening of class divisions resulting from the persistent exaction of money from farmers by bands of laborers. Such outrages, while deplorable, might still have what officials considered the highly beneficial result of transforming pro-Rockite farmers into antagonists of their night-walking social inferiors.[230] This kind of victimization was not the only factor that encouraged farmers to abandon the Rockite movement or to adopt a posture of open hostility to it. Also contributing to the split between former allies by 1824 were economic improvement, effective repression, and the higher county cess arising from incendiarism. But the growing willingness of farmers to render active assistance to the forces of repression—for example, by joining police patrols at night—was largely prompted by their extreme distaste for thieving laborers.[231]

This estrangement reached a bitter climax in 1825. Like some earlier agrarian upheavals that had sputtered out in a wave of petty criminality, the Rockite movement left in its wake groups of former activists who in some districts became thieves pure and simple. This phenomenon had been observable as early as 1772 and 1773 in the aftermath of the Steelboy rising.[232] Now it was clearest in north Cork, where petty robbery under the guise of levying contributions dominated what little remained of agrarian activism by 1825. In contrast to the recent past, when farmers in that region had generally submitted tamely to the exaction of money in attacks, they now displayed a tendency toward armed opposition. In addition, in self-defense and with police encouragement they formed farmers' associations, and members of these bodies took part in police patrols.[233] The police magistrate Samson Carter, who had been assiduously promoting these developments in north Cork since late 1824, saw his tactics advanced by the spate of thefts in 1825. As he confidently observed in March of that year, after a party of thieves had visited four farmers in the Doneraile district, "this system of petty robbery . . . will raise the farmers to resistance and more zealous cooperation with the police."[234] There had, of course, always been strict limits to interclass cooperation between farmers and the rural proletariat in the Rockite movement. On important issues in dispute their interests were diametrically opposed. Yet before 1824 even those more substantial farmers who had never had any genuine liking for the Rockites had rarely sided openly with the forces of repression. Prudential reasons alone had kept them neutral. That some farmers, especially those harassed by thieving laborers, were now doing so signaled the complete collapse of even limited interclass collaboration and the total eclipse of a great agrarian movement in which it had once been a notable feature.

Nine

Repression of the Movement

Apart from the tithe war of the early 1830s, the Rockite movement confronted the authorities with their greatest challenge of the early nineteenth century. With the exception of the tithe war, no other agrarian upheaval mobilized so many rebels or produced so much violence against both persons and property. Although the army had no serious difficulty in smothering the open insurrection in northwest Cork in January 1822, the suppression of incendiarism and other major kinds of agrarian crime long proved exceedingly arduous. The burnings in particular baffled the authorities almost completely. "Such is the description of warfare carried on by the insurgents," declared a Cork gentleman in February 1823, "that the utmost exertions of the magistrates, military, and police cannot counteract their plans or prevent their depredations."[1] Many murders were also committed with impunity. This was especially true of County Kilkenny, where events tended to be overshadowed by happenings farther south. In three Kilkenny baronies from 1822 to mid-1824 a total of twenty-two murders took place, but as the king's counsel Thomas Goold pointed out reproachfully in May 1824, not a single person had yet been punished for any of these killings.[2] In countless raids Rockite bands seized the firearms of thousands of gentlemen and well-to-do farmers and then concealed them so effectively that official attempts to recover the stolen weapons were practically useless. Thus a massive military search of three to four thousand houses for unauthorized weapons early in 1823 yielded exactly one pistol and one unserviceable gun.[3] Some military leaders considered troops to be so ineffective against the Rockites that they warmly endorsed the idea of forming private armed associations headed by the resident gentry. The extension of this practice in north Cork, insisted Major General Sir John Lambert in January 1823, "would be of much more use than all the numerous military detachments for various reasons."[4] Among the reasons, presumably, were the soldiers' lack of local information, unfamiliarity with the terrain, and inadequate mobility.

Weakened Condition of the Local Magistracy

But it was unrealistic to expect a great deal of enterprise or zeal from the resident gentry and local magistrates. In general, the Anglo-Irish gentry were no longer capable of the kind of local initiatives in repression that they had sponsored in the late eighteenth century. Their numbers had been thinned by absenteeism and emigration. Their political cohesiveness had been weakened by the divisive issues of Catholic emancipation and tithe reform. And above all, their local authority had been increasingly circumscribed by the expanding power of the centralizing state. The self-confidence needed for effectiveness was demonstrably in short supply. Thrown into an unseemly panic by the initial insurrectionary phase of the Rockite movement, many gentlemen had fled with their families from their rural residences into the safety of the towns. Though most returned after the insurrection had collapsed, they often barricaded themselves inside their country houses and rarely ventured abroad if they could avoid it, especially at night. The parish priest of Doneraile blamed the wave of agrarian crime in his district in late 1822 partly on the timidity of the resident gentry, who only "muster strong on a hunting day" and who showed the common people how panic-stricken they were by seeking "security within the barricaded doors and windows of their houses."[5] Similar complaints were frequently voiced by Protestants about other disturbed districts. For example, local magistrates gave the Tralee Association little support against Rockites in that vicinity. Of the ten magistrates residing in Tralee, only two acted "promptly and with decision"; five others were incapacitated by old age or bad health, and the remaining three simply refused to stir themselves.[6]

Some magistrates (probably a small minority) were corrupt and used their official positions to reap personal gain. Known as "trading magistrates," several such men in County Cork were said to pocket from £100 to £300 a year in "illegal fees."[7] For a variety of reasons other magistrates and gentlemen were willing to turn a blind eye to serious transgressions or even to intervene on behalf of well-known agrarian offenders. This was a venerable difficulty. Writing in April 1822, the Limerick landlord Thomas Studdert claimed that "nothing was more common before the happy introduction of the Insurrection Act than gentlemen using their influence at the assizes to arrest from justice persons committed for offences of the worst description. . . ."[8]

Studdert may have exaggerated, but his complaint certainly possessed some merit, and despite a much publicized overhaul of the magistracy by Dublin Castle in 1822, the use of illegitimate influence continued to be a problem for the authorities throughout the period of the Rockite movement.[9] Jeremiah McCartie, a Cork magistrate, was accused in July 1822 of tendering the oath of allegiance and thus extending protection to Cornelius Murphy, who was "well known in Duhallow as a notorious Whiteboy leader"; McCartie could hardly have been unaware of Murphy's reputation since the local parish priest had condemned Murphy before McCartie as "the ringleader of all disturbances" in the district.[10] Timothy Cotter, another chief Rockite in Duhallow barony, who was eventually executed for the murder of the mail-coach agent William Brereton in January 1822, was long able to avoid arrest through the apparent connivance of certain magistrates.[11]

Intimidation of "Active" Magistrates

To immobilize as many magistrates and gentlemen as possible through intimidation was of course one of the Rockites' main goals. Zealous magistrates were regularly threatened with death. Thus in south Tipperary early in 1822 some Rockites amused themselves by posting a public notice offering £500 "as a reward to any man who would assassinate William Morris Meade, an active and efficient magistrate."[12] Mere threats, however, seldom discouraged truly dedicated magistrates, and the Rockites frequently resorted to sterner measures. In the case of William Allen, who kept 117 cows on a large farm in the Newmarket district of Cork, they destroyed all of his cowhouses and dairy utensils, forced his servants to quit his employment, and obliged them to swear that they would not help him to recover damages at the assizes.[13] Gentlemen in the districts of Mallow and Doneraile were repeatedly made to suffer. Early in December 1822, Rockites burned the extensive haggard of Colonel Hill, who in previous months had "uniformly and openly manifested a determination to put down insurrection," thus earning himself the popular epithet of "Bloodhound Hill." The incident marked the sixth time in the past year that members of the Hill family had been the victims of a burning or other attack.[14] Later in the same month another Rockite party set on fire and destroyed the large corn mills (worth £4,000) of John Dillon Croker near Mallow, apparently in reprisal for the anti-Whiteboy activities of his father, who was the brigade major of a yeomanry corps.[15]

But the worst was yet to come. In March 1823, Carker Lodge and its adjoining out-offices near Doneraile were reduced to ashes.[16] The owner Hugh Norcott, once "very popular with the peasantry," had forfeited some of that esteem ten days earlier near his house when he fired on two men who had refused to halt after being challenged. By this act he brought Rockite vengeance down on his head.[17] The notoriety that attached to the burning of Carker Lodge, Croker's corn mills, and Hill's haggard no doubt served to cool the zeal of numerous magistrates and gentlemen, particularly those with exposed property, and the grisly murder of the Franks family near Glanworth in September 1823 did nothing to revive it. Hideously unpopular, the Franks were killed at least partly because they "were always saying they would like to be hanging and transporting the Whiteboys."[18] By the very nature of the case it is impossible to say how much of the frequently reported timidity of local magistrates was the result of Rockite violence directed against their more zealous colleagues, but it seems fair to conclude that the Rockites were successful at least in reducing the effectiveness of the magistracy and gentry as elements in the apparatus of repression. Consequently, much more of the practical responsibility for restoring law and order had to be borne by the army and especially the police.

Severe Rockite Punishment of "Informers"

If one major weakness in the traditional apparatus of repression was a timorous magistracy, another was the popular dread of being

Among the targets of the Rockites were numerous members of the landed gentry who had antagonized these agrarian rebels in some way, perhaps by showing "excessive" zeal in their role as active magistrates. In such cases local Rockites often sought the help of outsiders who would be more difficult for witnesses of violence to identify and harder for the authorities to trace and apprehend. In this sketch the two would-be assassins waiting in ambush for their intended victim—to all appearances, a gentleman riding in a gig, accompanied by his wife and followed by four policemen—might well have been strangers to the district where the attack had been planned. (William Steuart Trench, *Realities of Irish life* [London, 1868], facing p. 245)

branded an informer. This odious name could be attached to anyone who assisted the authorities in identifying, pursuing, arresting, or convicting alleged lawbreakers. The sanctions against informing were exceptionally strong in cases arising from agrarian rebellion. This was scarcely a new problem for the authorities during the early 1820s; it was at least as old as Whiteboyism itself.[19] But the Rockites were even readier than previous agrarian rebels to deprive known, suspected, or even simply potential informers of their lives. A half dozen persons labeled as informers were murdered in Limerick, Clare, and Cork from September 1821 to March 1822 alone, a record of aggression that established at the outset of the movement an image of the Rockites as implacable enemies of treachery and betrayal. Another half dozen informers were killed subsequently, and still others only narrowly escaped violent deaths.[20] Among the latter were two men who, after being placed in the wrong part of Clonmel jail in June 1822, were beaten severely by the other prisoners (they "would have murdered them but for the vigilance and activity of the turnkey").[21] Also closely skirting death was the crown witness who in October 1823 was pursued for over two miles through the South Liberties of Limerick by a large crowd crying, "Stop the spy, stop the spy who hung all the men at the assizes." (He finally found refuge in a friendly Palatine house.)[22] Indeed, many more informers would have been killed if they had not been taken into protective government custody. Besides having to fear death, informers lived in dread of other calamities. George Smith, a Palatine believed to have privately given information leading to the conviction and transportation of eleven men under the Insurrection Act, had his house burned down in January 1824.[23] In another case a farm servant named Ellen Hassett was raped in February 1822 outside the farmhouse of her employer because her sister Catherine Hassett had agreed to prosecute several alleged Rockites at the special commission in Limerick and because Ellen was unwilling to contradict her sister's evidence.[24]

There were few who grieved over the punishment of informers, no matter how extreme the retribution. One of the least regretted victims was John Neill, murdered in March 1822 for lodging informations against a dozen persons charged with attacking a house at Shanagolden in the previous September. His family neglected to claim the mangled remains, and the local populace treated his funeral with derision and scorn.[25] The killing of an informer before the trial did not necessarily abort the prosecution. Under a law of 1810, which perpetuated a principle first enshrined in the Insurrection Act of 1796, a jury was entitled to receive and credit the written evidence of a murdered witness.[26] Those against whom John Neill had sworn were tried under this law at the Limerick spring assizes of 1822, and eight of them were convicted. The police, of course, were delighted with this verdict. The case, declared Richard Willcocks, "will teach those deluded people how dangerous it is to spirit away or murder crown prosecutors."[27]

But this was not the only lesson that might be learned. In fact, it was not much more risky to kill a key crown witness than to allow him to live and give his damning testimony in court. In either case the accused Rockites could well be hanged or transported. Obviously, it was much better for the accused and their friends if prosecutors could be persuaded through fear of reprisal or a bribe to contradict or neglect to confirm their original informations. Even though prosecutors usually entered into recognizances to appear at the assizes when they swore their original informations, many of them failed to come forward at the appointed time. This should have entailed the forfeiture of their recognizances, but the penalty was rarely imposed. As many as 189 recognizances were declared forfeit at the Cork spring assizes of 1823, but in only 7 cases were the fines actually levied.[28] Some individuals who belatedly changed their minds about prosecuting even allowed themselves to be convicted of perjury (a transportable offense) rather than assume what they considered a much higher risk. The timorous Michael Fitzgerald plainly told a court in April 1823 that he had "committed the perjury in apprehension of being murdered if he prosecuted."[29]

The case of the rape victim Ellen Hassett pointedly illustrates the effect of Rockite violence in deterring witnesses from telling what they had seen or heard. When she prosecuted three men for the rape at the Limerick assizes in April 1822, her employer, the farmer Michael Neligan, flatly contradicted her story. He denied that the accused men had committed the rape, attacked his house, or sought to have Catherine Hassett's anti-Rockite evidence challenged by Ellen Hassett and another sister. Neligan's wife fully corroborated her husband's rather implausible testimony. The jury, however, believed Ellen Hassett and convicted the three prisoners.[30] Almost certainly, the Neligans' testimony was a string of falsehoods prompted by their fear of severe reprisal if they had told the truth about the punitive rape of their servant girl.

Frustrating the Law: The Roles of Farmers

Unless farmers lived in slated houses affording some protection against incendiarism, and unless they thoroughly disapproved of the Rockites' objectives, they were not at all likely to give direct aid to the authorities. In fact, the great majority of farmers occupied thatched houses. Moreover, most of them genuinely sympathized with several of the Rockites' aims, and numerous farmers or their sons were activists in the movement. On the other hand, the widespread practice of demanding money from farmers to fee counsel for Rockites facing trial (or to buy arms and ammunition or to bribe witnesses) was generally unpopular among farmers. Even so, they usually suffered in silence, as Colonel Sir Hugh Gough observed of farmers attacked and "robbed" by bands of laborers in north Cork early in 1824: "Though the farmers evidently know them, still such is their dislike to prosecute, either through fear of personal danger, or what I consider much more likely, an apprehension [that] if they should come forward their own former insurrectionary acts would be divulged, that it is with difficulty I can ever bring them to acknowledge their having been attacked."[31] To the extent that exemplary reprisals were needed, the Rockites carried them out. A farmer in this district from whom 7s. 6d. was collected "for Captain Rock" in October 1823 recognized one of the attackers, "to whom he sent to say if the money was not forthwith returned, he would lodge informations against him." Almost at once, 5s. was left at the farmer's door, but on the very same night his house was burned down.[32] To pay quietly and to keep one's mouth shut was obviously smarter and less costly.

Not only did farmers deny essential information to the authorities but they also actively frustrated the work of repression in other ways. Screening lawbreakers from justice was extremely common. Rockites wanted for specific crimes could not be apprehended, despite the offer of large rewards, "so intrenched [sic] are they in the good wishes of the farmers," moaned a Tipperary gentleman late in 1824.[33] Even in proceedings under the Insurrection Act, which was designed to lower to a bare minimum the standards of evidence needed for convictions, the authorities again confronted the old problems of manipulation and intimidation. Speaking of a series of trials in May 1823 in County Cork (the situation elsewhere was not much different), the police magistrate George Warburton remarked:

> In the course of those trials several decent farmers were produced to give the prisoners good characters, but on their cross-examination by the Sergeant [Robert Torrens], it appeared that they were absolutely compelled to come forward by intimidation, and that they were in such thralldom that they have been almost invariably forced to give their money and substance when called for, and even to employ the very worst characters in their labour and in their houses.[34]

To reduce the "thralldom" of farmers inclined to be helpful, the government was often prepared to pay for the slating of their houses as a safeguard against incendiaries.[35] But this kind of benefit, usually extended only after farmers had demonstrated their courage and their independence of the Rockites, did little to abate the general dread of rebel reprisals.

Official Efforts to Protect Informers

What the authorities could do effectively was to provide those willing to become informers with protection and a subsistence allowance before and during the trial as well as with the means of making a living after the criminal proceedings were over. Safety usually required moving elsewhere in Ireland or even out of the country altogether. The details of each case were arranged separately, sometimes with the would-be prosecutor specifying his conditions in advance of stepping forward publicly.[36] The authorities often furnished a sum of money (rarely more than £20) beyond the subsistence allowance and occasionally a job along with it. A crown solicitor, for example, offered £20 to Patrick Shea, who had successfully prosecuted some of the alleged murderers of the mail-coach agent William Brereton. But Shea indignantly refused to accept the money unless he was also appointed to a post with the police. Though he was considered unfit for such a job, a local magistrate urged the chief secretary Henry Goulburn to do something more for him: "As it will not be safe for him to remain in this country [i.e., County Kerry], I trust you will have him removed to Dublin and get him employed as a labourer in some of the public works, which I am certain, with the twenty pounds, will satisfy him."[37]

Informers were also given assistance to emigrate if they wished to leave the country. It was reported in May 1824 that several of the "most noted witnesses" against Rockites hanged for murder and other crimes had just embarked at Limerick for Quebec. (This group included Patrick Dillane, "the celebrated Captain Rock," who had turned approver against his accomplices in the murder of Thomas Hoskins in 1821, and who had to be forced to board the vessel for Canada.)[38] There was nothing that the authorities could do if informers insisted on returning to their own neighborhoods, where they were rarely left to live in peace. John Neill, for instance, the sight of whose remains provoked derisive popular laughter, was murdered after he broke loose from protective custody.[39] But by far the greatest problem was that only a tiny fraction of witnesses to agrarian crimes were prepared to migrate or emigrate, and this was especially true of the more substantial farmers. The offer of £15 or £20 and perhaps a menial job in Dublin or some other Irish town might well tempt a struggling laborer or artisan to become an informer, but farmers were not inclined to uproot themselves for such modest considerations. As a result of all this, the authorities had little choice but to place heavy reliance in capital cases on the questionable evidence of accomplices who were ready to become approvers in order to save their own lives.

"The Majesty of the Law"

Despite major weaknesses in the apparatus of repression, the government worked with great determination to stage an impressive display of its coercive and judicial power. It began with a special commission for County Cork in February 1822. The criminal calendar was bloated with the names of no fewer than 306 prisoners, of whom nearly 200 stood charged with Whiteboy or Rockite offenses.[40] Most of the Rockite prisoners had been captured by the army or police during the abortive insurrection of the previous month. The authorities decided to bring about fifty of these prisoners before the special commission, leaving the greater number for trial at the spring assizes. Given the careful selection of cases (many of the prisoners tried had been taken in arms in the clashes at Carriganimmy, Deshure, and Kanturk), and given as well the loading of the petty juries with the respectable gentlemen customary on such occasions, the high conviction rate was thoroughly predictable. As many as thirty-eight of the forty-nine prisoners tried were found guilty, and the judges promptly condemned thirty-five to death.[41] The government blanched at the prospect of so many executions, however, and most of the death sentences were stayed on the orders of the lord lieutenant.[42] According to newspaper reports, a total of fifteen men were actually hanged in late February and early March.[43]

The executions took place at four different sites selected because each had recently been the scene of appalling crimes and because the authorities hoped that the separate spectacles as a whole would be viewed by a larger number of people than if all the hangings had been concentrated at a single site. But in almost every case the size of the crowd was disappointing: four to five hundred (mostly women) at Churchtown, about five hundred (again chiefly women) at Deshure, and a "small number" at Carriganimmy.[44] Only at Newmarket did "numerous bodies of the country people" attend the executions.[45] Obviously, the Rockites had called for a boycott of the demoralizing death marches and hangings, and for the most part their orders had been obeyed. Still, the authorities were initially confident that the fifteen executions in Cork, together with twelve in Limerick earlier,[46] would frighten Captain Rock's supporters into submission. To heighten the public impact of these hangings in Cork, the judges ordered the bodies of the executed men to be dissected and anatomized.[47] It was the customary prayer of condemned prisoners to beseech the court for "a long day" (i.e., a considerable delay in carrying out the death sentence) and for their bodies to be "given to their friends" for a traditional wake and burial. But what this court decreed was a very short day and an ignominious violation of the remains before they were released to relatives. In addition, the fate of the respited prisoners was declared to hinge on a speedy restoration of tranquility; they were to be held, one of the judges proclaimed, as hostages for the future peace of the county.[48] By such tactics of counterterror the authorities sought to conquer Captain Rock.

Poor Results under Ordinary Laws

It was correctly anticipated that simple adherence to the ordinary legal system would severely handicap the authorities as they struggled to master the Rockite movement. In particular, under Irish conditions it was almost inconceivable that the extraordinarily high conviction rate achieved at the special commission could be regularly duplicated at subsequent assizes. Data available for the county assizes of 1822 indicate that there were wide local variations. High conviction rates were attained in the three counties of Westmeath (54 percent), Limerick (45 percent), and Tipperary (41 percent), but the rates were much lower in Kerry (29 percent), Cork (28 percent), Clare (24 percent), and Kilkenny (22 percent).[49] The variability in conviction rates, however, needs to be treated cautiously. The returns from which these figures have been compiled often do not indicate how many prisoners were discharged by proclamation without trial (generally because prosecutors declined to appear), and this was a factor that could exert considerable influence over the conviction rate, increasing it if a large number of prisoners were discharged without trial and reducing it if the opposite occurred. Furthermore, the available statistics make no distinction between Whiteboy or Rockite cases and those unconnected with the agrarian rebellion, although there is good reason to believe that convictions for Rockite crimes were relatively more difficult to obtain.

In a county as highly disturbed as Cork, a general conviction rate of merely 28 percent could only have been a profound disappointment to the authorities, and the situation in Kerry was even worse. Not only was the general conviction rate low there, but at the Kerry spring assizes a large number of prisoners committed for Whiteboy offenses had to be discharged by proclamation when prosecutors failed to appear, an event that was blamed, particularly in north Kerry, on the negligence of the magistrates, who failed to follow up their committal orders with the careful preparation of cases for trial. As a result, relatively few cases involving alleged Rockites actually went to trial, and in these "there have been nothing but acquittals."[50] As a Kerry gentlemen remarked, "This is to be regretted, for [great] numbers were in confinement upon charges of that nature, and every case of impunity is converted to a triumph. . . ."[51] Even in Tipperary, where the general conviction rate in 1822 was comparatively high, the authorities expressed bitter frustration. At the summer assizes several important Whiteboy cases ended in the acquittal of all the accused, an outcome, it was said, which could only give great license to the rebels, who "now imagine they may engage in any atrocity with a tolerable certainty of escape."[52]

Extraordinary Legal Weapon

Largely because all past experience taught that the operation of the ordinary law could too often be frustrated, the government had earlier decided to arm itself with a potent new weapon against agrarian rebellion. This was the Insurrection Act of February 1822.[53] Though originally scheduled to expire at the beginning of the following August, this important coercive law was indefinitely extended in its duration in July 1822 and was not repealed until 1825.[54] By early May 1822 the act had been brought into force in Cork, Limerick, and Kerry as well as parts of Tipperary, Westmeath, and Kilkenny. Eventually, it also became operative in three additional counties (Clare, King's, and Kildare)—a total of nine counties (or parts thereof) in all.[55]

The chief provisions of the act entailed the suspension of ordinary justice for noncapital crimes (capital offenses still were triable only at the assizes) and also the suspension of ordinary civil liberties. In any county or barony proclaimed under the act by the lord lieutenant, a tight curfew from sunset to sunrise was imposed. Violation of this curfew through absence from home at night was to be punishable by transportation for seven years, and the same penalty was prescribed for certain other noncapital offenses specified by the act: the tendering or taking of illegal oaths, the posting or delivery of threatening notices, the unauthorized possession of arms or ammunition, and participation in unlawful assemblies. Persons charged with such crimes were subjected to a summary form of justice in which the normal petty jury was not impaneled. Instead, cases falling within the scope of the Insurrection Act were heard and decided by courts of special sessions, with a king's counsel presiding and with a bench of magistrates substituting for a jury. Oral testimony was taken from both the prosecution and the defense; counsel was permitted to the accused, who were usually represented; cross-examination of witnesses was allowed, and the assembled magistrates rendered the verdict by majority vote.[56]

The Rockites energetically sought to arm themselves and to disarm their opponents, including the yeomanry, zealous magistrates, other members of the landed gentry, and sometimes even the police. The police spent much of their time in fruitless efforts to recover stolen weapons. This sketch, which shows the police searching for arms in the house of an uncooperative farmer, was prompted by the Land War of the early 1880s, but apart from the style of the uniforms worn by the police, the scene depicted here epitomizes the innumerable police searches for arms seized all over the south by the Rockites. As the primary enforcers of the Insurrection Act of 1822, the police also carried out systematic domiciliary visits to detect violators of the sunset-to-sunrise curfew. (*Illustrated London News*, 9 April 1881)

From the standpoint of the authorities there were numerous advantages to this summary form of justice. First, in contrast to the assize courts, which ordinarily sat only twice a year and then exclusively in the county town, the courts of special sessions were convened much more often and at scattered sites. Rarely were prisoners left in jail without trial for more than four or five weeks, and many were tried much sooner, some within a few days of their arrest. Since the venue of the courts was movable, prisoners could usually be tried close to the scene of their alleged crimes. As one king's counsel observed, he was "perpetually on the circuit of the different [country] towns where I sit," staying in each town only a day or two at a time, or just long enough to dispose of the cases that had accumulated locally.[57]

Second, the changes of venue and the brief duration of the special sessions facilitated the attendance of local magistrates, which was a desideratum for both symbolic and practical reasons. A numerous bench was thought to be of major importance in upholding authority and in visibly demonstrating official power. It also promoted the exchange of useful information and discouraged the exercise of illegitimate magisterial influence in favor of the accused. Attendance did vary widely. Courts with as few as six magistrates and as many as twenty-five or thirty were reported. Sessions held in Cork city, for example, tended to attract a small number of magistrates, whereas courts convened at Mallow in the center of a disturbed district, with many gentry seats in the vicinity, elicited a much higher turnout.[58] Francis Blackburne, administrator of the Insurrection Act in Limerick, put "the general average attendance" of magistrates in that county at about fifteen.[59] And Maxwell Blacker, administrator of the act in Cork and Tipperary, remarked, "I would not be long sitting, probably, before there would be twelve, fifteen, twenty, and up to twenty-five magistrates in attendance."[60] This was an aspect of the operation of the Insurrection Act that government officials found especially gratifying.

Third, while unanimous verdicts were required from juries for a conviction, the votes of only a simple majority of magistrates were needed to find a prisoner guilty in a court of special sessions. In practice, divided votes were relatively infrequent. Blackburne, for example, could recall only four or five "instances of a division upon a matter of fact," as distinct from a matter of law.[61] But there was no consensus as to the consequences of a division for the ultimate fate of a convicted prisoner. Blackburne asserted that in cases where the minority dissenting from a conviction was large, the sentence of transportation was not executed.[62] But Maxwell Blacker, administrator of the Insurrection Act in Cork and Tipperary, answered "Yes" when asked if punishment actually followed conviction in many cases where the proportion of dissenting magistrates was high.[63] In any case, it seems clear that magistrates on the bench consciously sought to avoid anything less than unanimity, and that a small proportion of dissenters was not likely to disturb the execution of the original decision if the convict petitioned, as he was entitled to do, for a review by the lord lieutenant of the judgment or the sentence.

The authorities were aware that the summary justice of the Insurrection Act and its harshly penal nature (the law allowed no sentence other than transportation for seven years) robbed it of legitimacy in the eyes of the common people, and for this as well as other reasons certain steps were taken to soften its implementation. Thus requests for the postponement of trials were routinely granted on grounds less strict than those that prevailed in the assize courts, and the attorneys of prisoners were granted unusually wide latitude for the reexamination of witnesses.[64] Furthermore, one mandatory provision of the law relating to the curfew was simply ignored whenever the magistrates chose to do so. The act specifically stated that any person found in a public house after nine o'clock at night was automatically subject to transportation.[65] But popular cultural tradition was strongly resistant even to this legal threat. As Maxwell Blacker observed, "I have never in any instance carried [that clause] into effect" because "we found it would be a desperate effort to transport persons (fourteen and fifteen in number at a time) . . . where it was obvious to everyone it was nothing but the effect of rashness, and when the [intoxicated] state in which they were, precluded them from doing any mischief. . . ."[66] Another provision of the Insurrection Act that was not strictly enforced was the prohibition of tumultuous or unlawful assemblies, especially those occurring in the daytime. Admittedly, this provision was applied selectively and with what was said to be considerable success against the carrying off of crops or livestock distrained for nonpayment of rent or tithes. But while magistrates were naturally eager to discourage this common offense, they were often loathe to convict the offenders and

305

sentence them to transportation. Or if they were convicted, as in one case involving fourteen men ("many of them of most excellent character . . . and situation in life") who had been apprehended for a rescue of distrained goods, they were subsequently pardoned.[67]

Implementing the Insurrection Act of 1822

On the other hand, both the composition and the actual practice of these courts of special sessions were strikingly different from those of ordinary tribunals. Magistrates who had actively participated in preparing a prosecution were not required to disqualify themselves from adjudicating the case. A magistrate who took the criminal information on which the prosecution was based, or who was responsible for having the accused committed for trial, was left perfectly free to participate later in deciding the question of guilt or innocence. The only automatic disqualification, apparently, was that imposed on a magistrate who actually became a witness for the prosecution at the trial,[68] and even in that event the prosecuting magistrate undoubtedly had means of influencing his colleagues that went beyond his public testimony.

Moreover, adherence to the normal rules of evidence was often observed only in theory. By far the commonest type of case tried before the courts of special sessions concerned an alleged violation of the curfew, and in all such cases the prosecution did have to prove strictly that the prisoner had broken the curfew. If a violation was clearly established, however, the accused might escape conviction by offering the defense that he had a good or at least lawful reason for his absence from home during the prohibited hours. Acceptance of this defense depended on whether the sitting magistrates believed that the accused was a person of generally good character and had not previously joined in Rockite offenses. Beliefs to the contrary on the part of the magistrates were often adopted independently of the evidence furnished in court. A prisoner might present almost any number of witnesses who would swear to his good character, but such was the ease with which "characters" could be procured that the magistrates were inclined to give them little weight.[69] And if one or two magistrates could convince their colleagues that the accused was a person of bad character or was guilty of a serious agrarian crime (for which he was not currently on trial, and for which he could not be convicted in an ordinary court for want of sufficient evidence), his conviction for violating the curfew would follow.

In his important study *Protest and punishment*, George Rudé has emphasized that the conviction rate in cases heard under the Insurrection Act was quite low. According to Rudé's somewhat incomplete figures, only 507 persons out of approximately 3,940 committed to jail were found guilty (a conviction rate of about 13 percent), of whom 452 were sentenced to transportation for seven years.[70] But the low overall rate is misleading for several reasons. To begin with, two of the counties on Rudé's list—Clare and Tipperary—were responsible for a heavily disproportionate share (almost 60 percent) of total committals, and in both the conviction rates were so low (3.4 percent in Clare, 6.7 percent in Tipperary) as to reduce the overall rate greatly.[71] The exclusion of these two counties would raise the conviction rate for the rest to 22 percent. Second, and of much greater importance, those arrested and tried under the act included a very high proportion of individuals against whom the only solid evidence of wrongdoing was a violation of the curfew, such as drinking in a public house or attending a wake after sunset. "Many people can show fair and clear reasons why they are out," remarked Maxwell Blacker, "for example, that they went out to borrow a horse for the next day for the farm, or went out to see a sick relation, or went out on any of the other numerous occasions which they may have, which will explain the reason of their being . . . absent."[72]

Admittedly, not all such alleged occasions of absence were perfectly innocent, and if the district in which the prisoner lived happened to be highly disturbed, the excuse of paying a brief visit to a sick relative or of borrowing a horse, even if attested by witnesses, might not save him from conviction. The authorities were only too well aware of how quickly a man might put a few burning pieces of turf in a kettle, run across several fields, place the lighted turf in a neighbor's thatched roof or his tithe corn, and scamper home again.[73] But unless other attendant circumstances suggested that the accused had been promoting Captain Rock's cause under the cover of seemingly harmless diversions, the magistrates were strongly disposed to discharge them with merely a warning. Thus in April 1823 the authorities let go without punishment thirty-one persons taken up at a wake in the Doneraile district.[74] Similar leniency was shown to sixteen men arrested in an unlicensed public house near Feakle in Clare. No liquor was on the table in front of them (a very suspicious circumstance) when they were taken into custody. But after it was established that they had gathered to pay their rents (among the best of reasons), and that the agent of the property had shared their confinement to jail, they were quickly released.[75] There were in fact legions of similar cases.

Dispensing Rough Justice

On the other hand, magistrates refused to overlook curfew violations when the accused appeared to be guilty of more serious crimes. One court ordered to the convict ships eight men caught drinking in a County Tipperary public house where they were apparently conducting the business of a Rockite committee.[76] Another court inflicted the same penalty on two "respectable farmers," even though they had received excellent references for character, when it seemed that their absence from home had been occasioned by their participation in a scheme to defraud a landlord of rent by hiding another tenant's corn crop.[77] A third court did not hesitate to banish to New South Wales a "very comfortable" farmer caught at night near the scene of a burning out-office, for he was regarded as a Rockite leader in the disturbed North Liberties of Limerick.[78] And a fourth court enforced the penalty of the curfew law against a County Limerick schoolmaster whose reputation with the authorities, like that of most in his calling, was dishonorable; he was, according to report, "a great writer of Rockite notices and of bad character."[79] When they appeared before a bench of magistrates convened under the Insurrection Act, no other class of prisoners stood in greater need of unimpeachable testimonials to their good character than Catholic schoolmasters; the predisposition to have them transported was extremely strong.

Indeed, what made the Insurrection Act so useful to the authorities was the wide latitude that it afforded them for ridding the country of persons strongly suspected of Rockite felonies but against whom the evidence was insufficient to convince an ordinary jury of guilt. Thus William Kelly, acquitted earlier by a jury of joining in an arms robbery near Rathkeale, was convicted in May 1822 by magistrates of a lesser but transportable offense.[80] Similarly, Edmond Magner, sentenced to Botany Bay for violating the curfew, was, according to the police magistrate Samson Carter, "a most notorious Whiteboy leader, against whom I could not proceed capitally, wanting a link in the chain of evidence, for burning Hennessy's mills at Ballywalter in this barony, Fermoy, and for which offence his brother and four others were executed at last summer assizes."[81] A low conviction rate was of little concern to the authorities as long as the guilty verdicts came in the right cases. Though only four of sixty prisoners were convicted in a series of special sessions held in Clare and Limerick in August 1823, the presiding king's counsel was more than satisfied with the outcome. The short list of the guilty included a known participant in the murder of Major Richard Going, a second prisoner who "by his own confession was the hired ringleader of the party who flogged a man" near Rathkeale, a third prisoner who had the reputation of being "one of the most notorious of the Cratloe Whiteboys," and a fourth culprit "whose influence in Cratloe and its neighbourhood was most powerful" and whose house had been a regular meeting ground for local Rockites.[82]

Popular Attitudes to Transportation

There are pointed suggestions in the sources that transportation was a sentence lightly regarded by those upon whom it fell, failing to terrify them in the way that the government had hoped. The demeanor of John O'Brien, a County Limerick blacksmith thus punished in January 1824, was reportedly "marked by the most daring hardihood and betrayed a perfect indifference to his fate."[83] And John Halloran, a wealthy Limerick farmer ordered transported in March 1823, defiantly "declared when going off that he would soon be back, as he had interest sufficient to bring him home again."[84] Maxwell Blacker was asked before a parliamentary committee in May 1824 whether prisoners seemed indifferent to the sentence of banishment. He replied: "Many young men of desperate fortunes, who have not much to lose, having no families, no connexions, think very little of it and receive it sometimes with a tone of triumph."[85] A Cork gentleman provided an apt, and for him distressing, illustration of such behavior in May 1823: "The two men who were convicted here [Mallow] as well as the two sentenced at Fermoy were moved on Tuesday from the convict depot to the ship and in this progress were huzzaing and cheering as if a triumph, not a punishment, awaited them; the truth is that transportation is not considered the least punishment. . . ."[86] Or as another gentleman declared, Botany Bay "has now become no place of punishment; many commit crimes for the purpose of being sent there."[87] Certainly, to be sentenced to transportation was regarded as no disgrace. Indeed, as Maxwell Blacker remarked, some of those convicted under the act "think themselves heroes and patriots."[88] Such an attitude was clearly evident in a letter of April 1823 written by one of two brothers named Horan to their mother soon after their conviction in County Tipperary (the letter was intercepted):

> My dear mother, you [should] not be ashamed or stoop your head for our leaving our country, as our crimes are no way dishonorable, and you need not be afraid to say that we ware [sic] loyal to our country, and if they all proved so, they would not be so many in gaol. But we expect that we will hear and see satisfaction for our emigration; especially I expect that my noble countrymen will revenge the injuries done to us and the rest of my countrymen, and I hope there will not be balls wanting to place them in the hearts of Dempster and Gosnell. . . .[89]

But if some prisoners accepted their fate with sturdy courage or apparent indifference, though no doubt with burning resentment as well, many others were overcome with grief at the prospect of penal exile. When magistrates at Limerick ordered John Darcy transported for seven years, he "prostrated himself in the centre of the dock, crying out to the court for 'home confinement and whipping.'"[90] As they heard their sentence announced by a court at Cashel in December 1823, eight prisoners along with their relatives and friends in the courtroom "burst out crying"; their friends later surrounded the jail and "continued shrieking most piteously the remainder of the evening."[91] Such scenes were extremely common following convictions in Insurrection Act courts throughout the period of the Rockite movement. At Rathkeale in May 1823 "the bitterest lamentations were heard in all quarters" when the bench announced "the unexpected conviction of so many [in fact, twenty-one] of their friends."[92] And at Cork city in the following October, when eleven men were found guilty, the courtroom "resounded with the shrieks and lamentations of their friends, which continued throughout the streets on their way back to the gaol."[93]

The Authorities and Transportation

Concluding from such public outpourings of grief that their repression was effective, the authorities adopted procedures designed to

310

stimulate them. As soon as prisoners were convicted, they were taken directly from the dock and dispatched almost immediately by car to a special depot at Cork, preparatory to their embarkation for New South Wales. This tactic was deliberately pursued as a means of attracting large crowds at the point of departure for Cork and along the route as well. It was also used as another method for impressing the common people with the majesty of the law and for instilling the belief that resistance to its dictates was utterly futile. To some extent at least, the tactic worked. As seven convicts were escorted by police and troops through the streets of Rathkeale on their journey to Cork in March 1822, a crowd of some five thousand persons gathered to watch, and "the air was rent with shrieks and sobs by their relatives."[94] And in the same month the king's counsel A.C. Macartney noted with undisguised satisfaction both the immediate dispatch of the convict John Kelly from Tralee to Cork and the fact that as Kelly passed through Killarney, Millstreet, and Macroom, "the banditti had appeared in great force" to witness the procession. The lesson, Macartney was sure, could not be missed. The same practice, when adopted in County Louth in 1817, had operated "with most beneficial effect, indeed with greater [impact] than the execution of the convicts, had their crimes been capital, would have had."[95] Whether those who viewed such proceedings were immediately converted to submission may seriously be doubted, but the long-term effect was almost certainly demoralizing to the movement.

A significant proportion of those sentenced to be transported under the Insurrection Act were either pardoned or had their sentences reduced to confinement in Ireland. According to George Rudé's figures, based mostly on shipping lists preserved in Australia, about 330 persons found guilty in courts of special sessions actually reached Sydney between November 1822 and September 1825.[96] Since a minimum of 452 prisoners had been sentenced to penal banishment under the act, it appears that clemency was granted to perhaps 120 of them, or nearly 27 percent. Any convict who petitioned for the remission or commutation of his sentence was considered entitled to a review by the lord lieutenant. In such cases a list of the petitioners was sent to the presiding barristers, who were asked to report whether they knew of any circumstance that would justify the suspension or remission of the sentence.[97] In cases that raised questions of law, the presiding barrister made a recommendation based on his own knowledge or on the advice of the law officers of the crown. In the far more frequent cases involving questions of fact, he consulted with the magistrates, who often examined additional witnesses, with the result, according to Maxwell Blacker, that mercy was extended "in a great many of those cases."[98]

The Many Challenges of the Assizes

While the Insurrection Act courts handled noncapital cases, capital crimes could be tried only at the assizes, with all the pitfalls that awaited the authorities in that arena. Even in times much less disturbed than the early 1820s, officials experienced great difficulty in persuading victims and witnesses to prosecute known offenders. In periods of agrarian rebellion this problem was magnified by intimidation, bribery, and popular fidelity to oaths against prosecuting. As a result, it became essential, if convictions were to be secured, to find accomplices who were ready to betray their former comrades and to serve as approvers. In August 1823 an outlaw named Daniel Ready, who had been a secondary accomplice in the heavily publicized abduction of Miss Honora Goold, and who had escaped from Limerick jail in the previous winter, proposed in return for his pardon to betray Timothy Cotter, wanted as a principal in the murder of the mail-coach agent William Brereton in January 1822. As in earlier cases of a similar kind, the police magistrate Samson Carter strongly urged acceptance of the offer because of its effect in "destroying the confidence the disaffected have hitherto reposed generally in their accomplices."[99] In a brief note that he appended to Carter's recommendation, Lord Wellesley, the viceroy, wrote, "I see no objection to the proposal," though he advised that the chief law officers of the crown be consulted.[100] Other outlaws made similar offers,[101] and provided that important convictions were likely to result, the authorities were usually disposed to accept.

Some potential approvers were poor risks, while others improved greatly under skillful police tutelage. Timothy Sheehan confessed to complicity in the murder of the three Frankses, and at the time of his confession in October 1823, there seemed little chance of finding a more credible approver, but the police expressed some doubt as to whether he ought to be admitted in this capacity because he had given his confession in a partial manner and had sworn several separate informations; the presence of such circumstances had in the past swayed juries in County Cork against the crown.[102] (Before Sheehan could be brought forward in court, he escaped from custody and was not re-taken, but another and more effective approver was soon discovered in this case.)[103] The police had much greater success with the Rockite activist John Walsh, a participant in the attack on Churchtown bar-racks, the burning of the Palatine village at Glenosheen, and other crimes. At first reluctant to appear in court at all, Walsh soon changed his mind under the persistent prodding of the police magistrate Thomas Vokes, who expressed delight with his handiwork in January 1824: "I put him in the [Limerick] city gaol with another favourite witness and cut off all communication with his friends and so man-aged him that he has now agreed to put me in full possession and to come forward to prosecute for the attack on Glanasheen [sic]. . . ."[104]

Walsh eventually justified the initial judgment of Vokes that he was "one of the steadiest and best witnesses I ever examined."[105] Along with another approver, James Moynihan, he was instrumental in se-curing the conviction of three men charged in the Churchtown bar-racks case and executed for this crime in April 1824. At the trial Walsh stood up well under a relentless cross-examination by the redoubtable Daniel O'Connell. As defense counsel, O'Connell extracted the dam-aging admissions that Walsh, "originally a potatoe-digger," had sworn against the accused in order to save his own life, had received "blood-money" (for clothes), and had been drilled as to his evidence by the police in the jails at Limerick and Cork. For the defense O'Connell also produced John Walsh, Sr., who testified that his son the approver was not to be believed on his oath, "for he never knelt to a priest," and two other witnesses who deposed that the younger Walsh could not have been present at the scene of the crime since he was elsewhere when it occurred. In cross-examining Moynihan, the other approver, O'Connell reportedly showed him "to be even a more abandoned char-acter than Walsh." Yet in spite of his best efforts O'Connell failed to shake the testimony of the two approvers or to convince the jury of the innocence of his clients.[106]

What greatly aided the authorities in this case was that the prose-cution did not entirely depend on the approvers' testimony. One of the accused was positively identified by a policeman present at Church-town, and the circumstantial evidence against another was strong because of his wounds, which he had allegedly received during the attack on the police barracks.[107] Such additional evidence, either direct or circumstantial, was necessary before the authorities could be reasonably certain of obtaining convictions or even of holding accused Rockites in custody without the possibility of bail. In the trial at Limerick in August 1824 of Patrick Bennett, one of four men charged with the slaying of Michael Gorman in December 1822, the chief prosecution witnesses were two accomplices whose evidence against Bennett was uncorroborated. For this reason the judge advised the jurors in favor of acquittal, and yet they found Bennett guilty. This outcome, however, was almost unique. As the judge declared after the verdict, under the same circumstances "there was scarcely an instance . . . of the jury bringing in a verdict contrary to the advice of the court."[108]

But the Bennett case was not the only one in these years in which the prosecution prevailed even though there was not much evidence to corroborate the testimony of accomplices. If the defense was weak or if the approver was an especially convincing performer on the witness stand, then much might still be accomplished with little. In the trial at Cork in April 1824 of the three Creminses for the murder of the Franks family, the chief (indeed, almost the sole) prosecution witness was the approver Edmond Magner, a laborer by occupation. But partly because of his notoriety alone, Magner made a highly credible witness. His brother and four others, as mentioned previously, had already been hanged for an incendiary crime, and he himself was considered a Rockite leader, a reputation that had already earned him a conviction under the Insurrection Act. In defending the Creminses, Daniel O'Connell played on Magner's earlier conviction and subsequent lodging of an information against his clients (obviously given in hopes of securing a pardon) in an effort to discredit the approver. But the only witness for the three prisoners was their landlord, who deposed that in the twenty years he had known them, he had not heard anything bad about them. Left to choose between an insipid testimonial to their character and an entirely plausible indictment from a well-known Rockite chieftain, the jury convicted them and they were promptly executed.[109]

Another celebrated case whose outcome turned on a similar set of circumstances (weak defense and convincing approver) was that of the Minnane brothers and John Green, tried at Limerick in August 1824 for the murder of Major Thomas Hare during an attack on his house for arms in February 1822. The approver in this case, and the only significant prosecution witness, was Oliver Fitzgerald, a man with much blood on his hands. An admitted accomplice in no fewer than three murders who had also confessed to other crimes, Fitzgerald had been arrested in Canada, to which he had fled; he was brought back to County Limerick, where he proved a remorseless betrayer of his former comrades, not only in this instance but also in other major trials. The prosecution was weakened by the inability of Mrs. Hare to identify any of the three prisoners. But it was unintentionally helped by the testimony of the principal defense witness Patrick Frawley, a farmer at whose barn the Rockites had allegedly assembled before and after the attack on Major Hare's residence, and one of whose two brothers had participated in the fatal raid. Frawley simply denied that either the three accused men or Fitzgerald had appeared at his barn, and he claimed that his two brothers had left the district more than two years earlier without ever letting him know where they had gone. Left to decide between a defense witness who seemed to be engaged in concealment and an approver who, however unsavory, appeared willing now to tell all, the jury declared Green and the Minnanes guilty.[110] It could not have helped the prisoners that they had long been in hiding before their arrest, or that two of them had been discovered under the bed of a very wealthy farmer, with his wife and daughter lying on top of it.[111]

"Approvers" as Crown Witnesses: Credibility Problems

By no means were all the accomplices brought forward by the crown as approvers given full credence by juries. Of the four men tried at Cork in April 1824 for the murder of the policeman John Orpen, only one was convicted, partly because the approver who was the main crown witness "admitted on his cross-examination that he told a friend of the prisoners that he would hang them whether innocent or guilty."[112] Even that steadiest and most persuasive of approvers, John Walsh, whose damning evidence had been mainly responsible for the conviction of three men in the Churchtown barracks case and of three others for the burning of Glenosheen, failed in his assignment at the Cork summer assizes of 1824. Though he was the chief witness against eight more men charged with attacking the police at Churchtown, all these prisoners were acquitted when the jury apparently chose not to give his testimony the same credence that it had received earlier.[113]

The authorities were also unable to capitalize fully on the sweeping informations given by certain other would-be approvers in whom they had initially placed great hopes. One of these was John Dundon, a wheelwright by trade, who was "the Captain Rock" in "the vicinity of Castletownroche and barony of Fermoy generally," and whose capture by Viscount Ennismore "caused a considerable sensation" throughout much of that extensive district.[114] Dundon was very talkative in custody, not only about the identity of his former comrades but also about their ultimate goals. He told Ennismore that "their object is to get possession of all property and to *destroy the Protestants*,"[115] and to a military officer he revealed that "Pastorini had made a strong impression on their minds and that they fully expected that the prophecy would be fulfilled."[116] As many as fifty persons were apprehended on the basis of Dundon's information. Unfortunately for the government, however, thirty-six of them had to be admitted to bail "from the want of a second evidence to corroborate his testimony."[117] The same defect operated with two other approvers caught at about the same time, one of whom was David Nagle, the celebrated Rockite leader. Since the prospect of successful prosecutions on capital charges before juries at the assizes appeared bleak without corroborative evidence, the authorities decided to take their weakened cases before a court of special sessions at Mallow, where the sympathetic magistrates proceeded to convict twenty-one of the forty-three persons tried.[118]

Approvers and Rockite Demoralization

But even when approvers such as Nagle and Dundon could not be produced effectively in the assize courts, their known willingness to turn king's evidence played an important role in the work of repression. When the news of their capture and treachery was broadcast through the countryside, it spread demoralization and compelled a large number of activists to go into hiding or to seek safety abroad. The apostasy of Patrick Dillane was clearly a shocking disappointment to Rockites in the Newcastle district and indeed over a much wider area at an early stage of the movement. Dillane was an accomplice in the murder of young Thomas Hoskins, the son of Lord Courtenay's agent, and his evidence was crucial in convicting four men of the killing at the Limerick summer assizes of 1822. At the trial, which excited intense public interest, Dillane "admitted that he was the first Captain Rock in this country, and that he was 'so christened by a schoolmaster of the name of Mangan.'" The country people were reportedly astonished at Dillane's becoming a Judas. As the crown solicitor Matthew Barrington noted gleefully, some of them openly declared that "there was 'now no trusting anyone, as they thought that Paddy Dillane would sooner be hanged than become an informer.'"[119]

Fear, distrust, and demoralization also followed in the wake of turncoating by other leaders later in the movement and helped to bring about its demise. When John Dundon was captured and became an approver, the authorities reported that "many suspected of Whiteboy practices have fled," and that others "are under the greatest apprehension."[120] And when David Nagle was taken into custody soon afterward, similar responses emerged. As "the Captain Rock" in the North Liberties of Cork city,[121] Nagle was in a good position to give the police the names of those "who are committee men, fabricators of pikes, and Whiteboy leaders," and he promptly did so. With this renegade especially in mind, the police magistrate Samson Carter declared in August 1823 that the insurgents were "terrified at the apostacy of their leaders."[122] Despite his apparent cooperation with the authorities, Nagle was not admitted as an approver. Instead, at the Cork summer assizes of 1823 he was capitally convicted of leading a raid for arms in the previous May. But evidence at the trial that he had prevented plunder and any injury to the wife of the gentleman attacked by his party earned him a recommendation to mercy from the jury.[123] In the spring of 1824 the authorities were still putting Nagle, who had by then apparently been pardoned, to effective use in the continuing task of repression. The military paraded him through some of his old haunts, and Colonel Sir Hugh Gough reported the beneficial results to Dublin Castle: arms surrendered, the activists over a wide area disconcerted and disorganized by the disclosure of "all their proceedings," and even "respectable farmers" openly acknowledging their former complicity, which they claimed had been forced, and no doubt pledging adherence to the laws in future.[124]

Executions as Great Public Spectacles

Once convictions in capital cases had been secured at the assizes, the authorities had to decide which convicts, and how many of them, to execute. As in earlier periods of agrarian rebellion, so too during the years of the Rockite movement, many Whiteboy offenders, though condemned to death, soon had their sentences commuted to transportation and even in some instances to imprisonment in Ireland. But in cases of murder and other aggravated agrarian felonies the death sentence passed by the court was usually carried out with little or no delay. It had long been official practice to arrange the public execution of Whiteboy offenders in such a way as to attract and impress large crowds of onlookers. This was of course done in the prosaic hope that the "awful spectacle" would drive home the lesson that serious agrarian crime frequently brought those guilty of it to a tragic but just end on the scaffold.

In the Rockite era the largest reported attendance was that of "about twenty thousand spectators" at the combined execution of six men in August 1824 for the long-contemplated murder of the wealthy Kilkenny farmer John Marum. On this occasion a large force of some five hundred police and troops, "followed by multitudes from every quarter of the city [of Kilkenny]," escorted the condemned convicts from the county jail to the place of execution and scene of the crime about three miles from Johnstown in the highly disturbed barony of Galmoy.[125] The attendance may have been equally great for the execution in the same month of William and Darby Maher, convicted of the even more sensational burning of the Sheas in County Tipperary late in 1821. Again, the authorities fixed the scene of the crime—at Tober—as the place of execution and arranged an impressive death march from Clonmel jail, with troops, mounted police, and a masked executioner. The solemn procession reportedly struck with awe the vast multitude who had assembled round the jail from an early hour to witness "this sad spectacle," and the crowd that finally collected at Tober was described as a "numberless" throng.[126] "Immense crowds" were also attracted to the execution of Timothy Cotter in August 1824 when he was hanged at Shinnagh, the scene of the crime, for the slaying of the mail-coach agent William Brereton.[127]

Even when well-known agrarian offenders were simply executed in the county town without a long death march, the turnout was often very large. Daniel Connell, convicted of killing a resisting farmer during a Rockite attack, was hanged in August 1823 before a "vast concourse" of people gathered in front of the county jail at Limerick.[128] Another "vast concourse" of spectators witnessed the execution of Michael Mara at Clonmel in the following month.[129] And when John Green and the Minnane brothers were hanged at Limerick in August 1824 for the murder of Major Thomas Hare, the event took place, declared the police magistrate Thomas Vokes, before "the greatest assembly I ever beheld at an execution."[130]

Indeed, a small attendance at the scaffold of a noted Rockite was a great rarity after the early months of 1822. One of the few recorded instances of limited attendance occurred at the execution of the Rockite leader John Hickey, convicted of setting fire in March 1823 to Carker House near Doneraile, the residence of the gentleman Hugh Norcott, Hickey's employer.[131] The case is illuminating in several respects. Though offers of mercy were held out to Hickey if he would agree to name his confederates, he declared that he could not take it on his conscience to accuse so many—"over 300 in his own neighbourhood."[132] Hickey did disclose privately to the priests of Doneraile parish the names of those possessing arms, apparently in the hope that surrender of the weapons to the clergy might save him from the gallows.[133] But local Rockites refused to cooperate. They denounced the priests as partisans of the repressive government, they threatened reprisals against those who would give up their arms, and they issued commands that no one should attend the execution.[134] In the end hardly any weapons were surrendered and fewer than eighty of the "lower orders" turned out at Doneraile when Hickey was hanged there in April 1823. His edifying speech at the scaffold, exhorting others to avoid the acts that had brought him to this fate, was thus of little use to the authorities.[135]

What government officials wanted and sought to arrange were confessions of guilt at the gallows from the condemned, as well as anti-Rockite addresses from the attending Catholic clergy. Priests were usually obliging, although even their best efforts were sometimes misinterpreted by suspicious or uncomprehending Protestant gentlemen. After viewing several executions at Adare, Co. Limerick, in April 1822, one gentleman complained to Dublin Castle that a priest at the scaffold "made a most able and loyal address to the people which lasted above an hour, and how did he end it, by lifting up his hands and declaring, '*As sure as there is a God in heaven, these men die innocent.*'"[136] In a note appended at Dublin Castle it was explained, "This means after confession and absolution,"[137] though it is also possible that the priest was the mouthpiece for a protestation of innocence by the men about to die. Among the numerous priests who urged the onlookers at executions to shun the Rockites was Father O'Connor of Cork city, who delivered a strong anti-Whiteboy address at the hanging of two alleged agrarian rebels in August 1823 at Buttevant. At the gallows there Father O'Connor declared that during the last eighteen months he had attended to twenty persons "in this awful situation."[138] Presumably, his speech on this occasion was of a piece with others that he had given earlier. At a minimum, priests exhorted the spectators to surrender arms, and they tried to persuade the condemned convicts to make the same pleas.[139]

Mixed Results of Public Hangings

But in seeking to extract public confessions of guilt at the gallows from those on the verge of eternity, the authorities obtained very mixed results. Two reputed Rockites about to be executed at Limerick in March 1823 for the murder of Major Richard Going refused to urge the "vast concourse of people" present to give up arms, despite the entreaties of an attending priest, and they flatly declined to admit guilt. One of them, when asked by the sheriff if he had committed the crime, replied coolly, "If you'll call to me about this time tomorrow, I'll tell you."[140] All five of the men capitally convicted of setting fire to the mills and house of Charles Hennessy near Castletownroche declared through a priest as they were about to die in August 1823 that they were guiltless of the crime.[141]

This kind of contretemps at executions served, of course, to accentuate the already prevalent popular disrespect for the judicial system. Before the three Creminses (a father and his two sons) were hanged in April 1824 for the slaying of the Franks family, they vehemently protested their innocence at the place of execution,[142] and there was a loud local outcry over the perceived unfairness of their trial and sentence.[143] The police magistrate who helped to secure their conviction attributed their harmful declarations at the gallows to "the perverse disposition" of the father and to the fact that others who allegedly shared their guilt were as yet untried.[144] But in a memorial drafted prior to their execution, the Creminses asserted that they had been "put on their knees" (i.e., forced) to kill the Franks.[145] The authorities were also sorely embarrassed by the conduct of the six men executed near Johnstown, Co. Kilkenny, in August 1824 for the murder of Marum. With perhaps twenty thousand spectators gathered around the scaffold, five of the six convicts stoutly maintained their innocence to the end. At the last moment the other condemned man, Patrick Phelan, expressing a desire to speak, had his death cap raised and declared, "I acknowledge I had a hand in the killing of Marum." But he also insisted that "Dwyer, the informer, had nothing to say to it, nor did he know any of the party concerned in it." And when asked by an attending priest if those about to die with him were guilty or innocent, Phelan firmly responded, "They had nothing to say to it."[146] Naturally, this dramatic scene went far to destroy the effect that the authorities had hoped to create through the carefully arranged death march from Kilkenny city into the barony of Galmoy. Soon after the executions a provincial newspaper reported, "We have heard of very unpleasant indications of popular resentment both at Kilkenny and at Galmoy."[147]

Fortunately for the authorities, many other executions concluded much more satisfactorily. Patrick Ivers, at the age of fifty-eight one of the oldest men ever hanged for a Rockite crime (that of attacking a gentleman's house, apparently for arms), declared through a priest at the scaffold in Limerick city in August 1823 that the plan to raid the house had been hatched "by the informer, Sheehan, who swore away his life," yet Ivers admitted his guilt and forgave Sheehan, or so he said.[148] Two much younger Rockites, more typically "in the prime of life," also conceded the justice of their sentence immediately before their execution at Limerick a few days later for attacking the house of a farmer (Denis Morrissy) who was killed in the raid.[149] Especially gratifying for the authorities, given the circumstances of their arrest (in the house of a rich farmer), their own prosperous farming background, and the nature of their crime (the murder of Major Thomas Hare) was the admission of guilt by John Green and the Minnane brothers before they were hanged in front of an enormous crowd at Limerick in August 1824.[150]

The Treatment of Dead Bodies

For those convicted of agrarian murders, execution was not quite the end of their punishment, for in such cases it was a standard provision of the death sentence that their bodies be deposited in the county infirmary for dissection and anatomization.[151] This practice was clearly intended to heighten the severity of what was already the most extreme of punishments by adding to the indignity of the convict's fate. It was less barbarous than the now disused practices of beheading or of drawing and quartering to which Whiteboys and certain political offenders had been subjected in the late eighteenth century,[152] but the old objective of inflicting shame and disgrace on both the convict and his relatives as well as on his cause was still being pursued in the newer penal ritual.

Executed Rockites were also occasionally denied regular Christian burial.[153] In such cases their remains, instead of being released to their family and friends for interment, were buried with slacked lime in the jail yard or some similar place. The purpose of this procedure was both penal and practical. For the body to be dumped in an official and unhallowed hole was widely perceived at the popular level as a monstrous degradation, especially in a culture that attached so much importance to traditional funeral customs as vehicles for honoring the dead, facilitating their passage into the afterlife, and consoling the bereaved relatives and friends. Thus the withholding of the body, by making it impossible to perform the proprieties mandated by the culture, greatly intensified the punishment. But from the standpoint of the authorities, withholding the body also served the very practical purpose of preventing the traditional wakes that, in the case of agrarian and political rebels, had long been associated with the planning of further rebellious activity. Both ideologically and organizationally, the wakes of executed agrarian and political offenders were often instruments of cohesion and mobilization, not to mention occasions for plotting revenge. For this reason as well, the withholding of the body had become a frequently used weapon in combating agrarian rebellion earlier in the nineteenth century.[154]

By the 1820s, however, the authorities, armed with new weapons of repression, had largely abandoned the practice, and it was employed sparingly against the Rockites.[155] The wisdom of this restraint was not always evident, for some of the wakes given to executed Rockites were anything but reassuring to those in authority. When, for example, the bodies of the Minnane brothers were released to their family and friends in August 1824, they were carried in procession from Limerick to Rathkeale "by a vast concourse of people," and the funeral and interment on the following day were reportedly attended by "more than 7,000 persons."[156]

Increasing Effectiveness of Repression

Although this popular tribute to the Minnanes greatly annoyed officials, government repression had by this time helped to bring the Rockite movement largely under control. The condition of County Limerick was especially encouraging. In February 1824 the king's counsel Francis Blackburne remarked that serious outrages in the previous winter had been much fewer than anticipated, and that the criminal calendar for the forthcoming spring assizes was not at all heavy. Indeed, he declared that with the exception of a limited portion of the county, Limerick "had enjoyed perfect tranquillity for the last six months."[157] By July, according to report, not a single person was being held in custody under the Insurrection Act in either the county or the city jail of Limerick.[158] The authorities began to dismantle some of the apparatus of repression. In the same month the whole of Kerry, which had long been peaceful, together with all the North Liberties of Limerick and a great portion of the South Liberties as well, were relieved of the Insurrection Act.[159] Early in August the act was also withdrawn from the barony of Clonlisk in King's County.[160] In other places, admittedly, magistrates and police officers were unwilling to see the act suspended, but their desire to maintain it in force arose from the conviction that this would prevent any revival of the movement.[161]

Judged by the standards of the official response to previous agrarian rebellions, the repression of the Rockite movement had been extremely severe. The number of executions was unprecedented. The records that have been examined indicate that about one hundred persons were hanged for Rockite crimes in the five counties of Cork, Limerick, Tipperary, Kerry, and Kilkenny between 1822 and 1824. Some three-fourths of these executions were carried out in the two counties of Cork and Limerick alone, roughly half of them in 1822 and the remainder in 1823 and 1824.[162] As previously mentioned, many of those originally condemned to death had their sentences commuted to transportation, which was also the punishment that the assize courts had routinely decreed in the first instance for virtually all Rockite offenses other than murder, armed attacks on houses at night, major acts of incendiarism, and rape. Although only a rough guess is possible because of the absence of detailed criminal statistics, it is probable that two or three persons were transported as a result of assize-court convictions for every one that was hanged. To these must be added the three hundred fifty individuals (at a minimum) who were transported under the Insurrection Act. Thus the total number of those transported for Rockite offenses between 1822 and 1824 was probably about six hundred.

Because other factors, especially the improvement in economic conditions starting in 1823, also contributed to the decline and eventual disintegration of the Rockite movement, it is difficult to gauge the impact of repression alone. For their part the authorities were firmly convinced that the convictions obtained in major cases at the assizes, together with the operation of the Insurrection Act, had been of enormous importance in the gradual restoration of order in the countryside. Many officials believed that the tide had finally begun to turn in their favor during the summer of 1823, and among the reasons for this development, they attached the heaviest weight to their greatly increased ability to find prosecutors, including both victims and approvers. When ten persons were convicted under the Insurrection Act at Rathkeale early in July 1823, the *Limerick Chronicle* observed with satisfaction that "the prosecutors in each instance were country people."[163] And the results of a quarter sessions held in Rathkeale at this time also appeared to mark a decisive change: "There were forty convictions for riot, rescue, etc., a thing almost unprecedented in this county," declared an elated Limerick gentleman, who concluded from this extraordinary event that Whiteboyism was definitely on the wane.[164] But what most encouraged magistrates, police, and other officials were the multiple benefits that flowed from the arrest and turncoating of accomplices in Rockite crimes. When in August 1823 the crown solicitor Maxwell Blacker felt justified in predicting the coming of permanent tranquility, he attributed the change partly to the increased activity of magistrates and to the numerous convictions at both special sessions and assizes. But he laid particular stress on "the distrust and dismay which have been naturally excited amongst the disturbers of the publick peace" as a consequence of "discoveries" made by accomplices.[165]

In a striking number of major cases in the second half of 1823 and throughout 1824, the authorities were able to make arrests and to secure convictions at the assizes for crimes that had occurred in late 1821 or 1822. Thus the Limerick magistrate Godfrey Massey boasted in August 1823 of his success in apprehending no fewer than seventeen men ("mostly inhabitants of the mountains from Newcastle to Glanduff and very laborious and difficult to get at them") who were implicated in a whole series of offenses, including most notably the burning of Killeedy church and glebe house in February 1822. Massey's informant was a former Rockite who had since become a private in the 1st Rifle Brigade—an example of a hare shedding his old skin and deciding to run with the hounds.[166] (It was in fact fairly common for Rockites to enlist in the king's forces as a means of eluding the authorities.) Good detective work, close cooperation among scattered officials, and no doubt a fair measure of luck increasingly gave the arm of the law a long reach. As the police magistrate Thomas Vokes remarked in March 1824, "At this assizes I have one prisoner who was apprehended in Quebec, another who was taken in London, and several taken in the neighbouring counties."[167] Similarly, with the help of a carefully chosen confidential informer, three of the alleged principals in the burning of the police barracks at Churchtown were captured in London, to which they had fled soon after that well-orchestrated attack.[168]

But the case that seemed best to symbolize for the authorities what they regarded as highly effective repression was their successful prosecution in August 1824 of two men for the so-called burning of the Sheas in November 1821. For almost three years after this infamous crime, in which as many as sixteen persons perished, no convictions could be obtained because next to none of the numerous potential informants had courage enough to face the reprisals expected if they divulged what they knew. The problem was compounded by the fact that some of the potential informants were relatives of both the victims and the perpetrators of the deed. This was the unenviable situation of Mary Kelly, a country woman who "kept a [public] house of no great reputation, frequented by nightwalkers," and of her son John Kelly.[169] The latter, who may once have been a Rockite himself and who certainly had knowledge of the crime, enlisted for protection in a regiment stationed in India, from which he returned shortly before the opening of the trial at Clonmel.[170] His mother, after making initial disclosures, had fled to Mullinahone to escape the death with which she had repeatedly been threatened. While staying there, according to a police official, she was tampered with by friends of the accused: "Every effort was made to induce her to get off from us, and a large sum of money [was] offered to her not to prosecute, and when she did not consent, the general feeling was . . . that her life" would be taken.[171]

But Mary Kelly defied and beat the odds. Whether because of the urgings of a priest, as was claimed, or because she was a cousin of one of the murdered men, she pushed aside the temptations, subdued her fears, and became the chief witness for the crown at the trial of William and Darby Maher. Her testimony was damning: soon after the burning of "her people," she had accused William Maher to his face of being one of the murderers, to which he replied, "*'The devil choak 'em, we were calling 'em out, and to throw the arms out [of the Sheas' farmhouse], but they wouldn't throw them out.'*"[172] Her account seemed all the more credible because it was most unlikely that the attackers had intended to incinerate so many people. She also testified that it was from Darby Maher's house that she had seen sixteen or seventeen attackers emerge and proceed toward the Sheas' farm. Having heard the evidence of Mary and John Kelly, the jury convicted the laborer William Maher and his cousin Darby.[173] This verdict, declared a police official, gave "the greatest satisfaction to the well disposed, more particularly so as previous to the trial it was supposed impossible."[174]

Some Caveats

It should not be thought, however, that judicial repression was thoroughly effective even in 1824. Many prosecutions of major crimes still had to be abandoned or proved abortive. At the same assizes at which the Mahers were convicted, at least 100 prisoners were returned for trial, "of whom not less than thirty" faced charges of murder. But apart from the Mahers, few if any of the latter were found guilty.[175] The authorities fared somewhat better at the Limerick summer assizes of 1824, but again many prisoners managed to escape punishment. On the eve of the assizes 104 prisoners scheduled for trial lay in the Limerick county jail, and 36 of them stood charged with murder.[176] When the assizes ended, death sentences hung over 16 of the prisoners, and 8 others had been placed under rule of transportation.[177] Similarly, the county criminal calendar for the summer assizes at Cork contained the names of 190 prisoners, including 25 to 30 charged in connection with either the burning of Churchtown barracks or the killing of the Frankses.[178] But all 8 of those tried for the Churchtown attack were acquitted, along with numerous others in Rockite cases. And there were only 3 convictions for murder, all in the case of the Frankses, even though about 20 prisoners stood accused of this crime shortly before the assizes opened.[179] In short, though the authorities had regained the upper hand, they were inclined to exaggerate their judicial triumphs and to gloss over their losses.

The Peter Robinson Emigration Ventures of 1823–25

The enormity of the challenge that the Rockite movement presented to the established order helps us to understand why some landowners and government officials took a serious interest in organizing a substantial scheme of assisted emigration to Canada in the early 1820s. The scheme was aimed at precisely those areas of north Cork where agrarian violence became most intense, and the principal goal, especially in the minds of the local landowners involved, was to rid themselves of some of the worst troublemakers on their estates. Among the leading spirits in initiating this project was the third Earl of Kingston. His estates (centered on Mitchelstown) were less disturbed than those of most of the other proprietors with whom he concerted his actions: his brother-in-law Earl Mount Cashell of Moore Park at Kilworth; Arthur Hyde of Castle Hyde near Fermoy; his neighbor Sir William Wrixon-Becher of Ballygiblin House; Richard Hare, Viscount Ennismore, of Convamore at Ballyhooly; Charles Denham Jephson of Mallow Castle; and Viscount Doneraile of Doneraile Court. Associated with them was the Reverend Richard Woodward, the Anglican rector of Glanworth, another scene of serious Rockite disturbances.[180]

In June 1822, Lord Kingston, who was within months of finally acquiring control of his family's huge properties from his aged mother, sought the assistance of his well-placed friend Lord Bathurst, the secretary of state for war and the colonies in the Tory administration headed by Lord Liverpool. "I think you may help us at very small expense to the government," Kingston told Bathurst, "if you will send out a few families to Canada and provision them for one year on land given them by the government." Kingston and his landed friends understood that getting rid of "a few families" in this way would not make any immediate dent in Captain Rock's army or in local overpopulation, but the hope was that the first trickle of emigrants would soon turn into something like a flood. "I could send you thirty industrious families in a short time—well calculated for settlers," promised Kingston, "and they would have many followers."[181]

Circumstances smiled on this enterprise, which was less modest in the ambitions of its initiators than Lord Kingston admitted. Bathurst's deputy Robert Wilmot Horton was an enthusiastic proponent of state-sponsored emigration to British North America, though initially he had been thinking of its potential to drain the labor surplus in England. Horton was diverted toward Ireland partly by Kingston's request and partly by a meeting in London with John Beverly Robinson, the attorney general for Upper Canada, who wished to secure settlers for the Bathurst District (formerly the Ottawa Valley Military Settlements). It was John Robinson who persuaded his at first reluctant brother Peter, a politician, businessman, and dedicated land developer from Upper Canada, to assume the post of "superintendent of emigration from the south of Ireland to Canada." With the active support of Sir Robert Peel, the home secretary, the Colonial Office agreed to furnish £10,000 for a project that aimed to take up to five hundred settlers in 1823 from County Cork to Upper Canada, where they would receive 70 acres of land per family (the possibility of acquiring 30 more was offered), along with a dwelling, farm implements, a cow, and enough food for twelve months.[182]

Peter Robinson arrived in Ireland in May 1823 and, at the suggestion of Lord Ennismore, M.P. for County Cork (1812–27), made the garrison town of Fermoy his base of operations. He quickly found a welcome reception from the resident gentry and clergymen of all persuasions. The gentry, he later told Horton, "were unanimously of opinion that I should take as many persons as possible from the disturbed baronies in the county of Cork, which were at that time in a very distracted state."[183] Robinson worked closely with local Catholic priests in disseminating news of the provisions of the emigration scheme; he reported his surprise that in most of the parishes where he was active, the priests were on "the best terms" with their Protestant counterparts, and that all supported the project. He had copies of the gener-ous conditions of the emigration scheme distributed in the towns of Newmarket, Kanturk, Charleville, Mallow, Doneraile, Fermoy, and Mitchelstown.[184] He himself went among the common people, gave them information about the destination, answered their questions about the long journey, and sought to dispel their fears: "I explained the manner of clearing lands [in Upper Canada] and cultivating the virgin soil; I dissipated their apprehensions concerning wild beasts & the dangers of being lost in the woods."[185]

Robinson found the response more than gratifying. Within a few months he assembled a sizeable contingent of emigrants (drawn heav-ily though not exclusively from north Cork) who had impressed him as possessing the habits of industry and traits of character that would allow them to adapt successfully to the new and very different life that he was offering them in the uncleared and often densely wooded areas of the Bathhurst District, where the winters were severe indeed. In general, the emigrants of 1823 were, according to Robinson, "persons of no capital whatever" and comprised those "who might more prop-erly be called paupers."[186] Yet to judge from one of his other descrip-tions, they were not drawn from the ranks of the poorest segment of the population, and a surprising number were literate.[187] If some "troublemakers" or former agrarian militants found their way into the accepted group, it was because Robinson did not inquire into their previous conduct; he believed that such persons would in any case be transformed into useful citizens once they had the opportunity to be-come landowners in Upper Canada.[188] The historian Donald MacKay has expressed reservations about Robinson's assessment of the social background of these emigrants, whom he sees as a mixed group with numerous farmers. MacKay maintains that Robinson's "selection consisted of respectable farmers he wanted as colonists and a sprin-kling of paupers and troublemakers the landowners urged on him." The surviving data are insufficient to resolve this disputed point, though "respectable farmers" must have formed a distinct minority of the whole group, and real troublemakers must have been difficult enough to identify.[189] What we do know is that in July 1823 a total of 568 emigrants sailed from Cork (182 men, 143 women, and 243 chil-dren) on two well-provisioned ships that made the ocean crossing in about eight weeks with only modest casualties (eight children and one adult died during the voyage).[190] The total expense Horton computed at slightly less than £14,000, or about £22 per head. But this sum did not include the costs of furnishing provisions, housing, farming implements, cows, and even some clothing for the settlers for more than a year (until the end of 1824), all of which added over £8,000 more to expenditures.[191]

Robinson undertook further efforts in north Cork to recruit volunteers for assisted emigration in 1824 and 1825. His first enterprise had been sufficiently successful to stimulate widespread popular interest and to impress such members of the landed elite as Lords Doneraile, Kingston, and Mount Cashell. From his neighborhood Lord Doneraile compiled for Robinson a list containing some 400 names (73 families and 17 "single persons") who had "expressed a wish to emigrate . . . to Canada." He described them as being "principally from the labouring class and small farmers who have been reduced by the times," though he also observed that some families on his list were headed by "tradesmen reduced in circumstances."[192] Lord Mount Cashell's list of prospective emigrants was even longer, containing as many as 562 persons (83 families). He called Robinson's attention to the remarkable size of families among his group: "they nearly average seven in a family." Mount Cashell affirmed that "they are all poor and wretched, and have most of them subsisted by tilling the soil." But their destitution was a positive recommendation in Mount Cashell's eyes and (he was sure) in Robinson's: "I think they are just the sort of persons you wish to have in [British North] America, and by taking them, you will rid this country of so many paupers."[193] And Lord Kingston was eager to unload as many as 400 occupiers from his own vast estates—poor people who would "turn out bad subjects" if they stayed at home but who might be transformed into thriving Canadian settlers if they were helped to emigrate to the Bathurst District with generous government funding (but certainly not his own money).[194] Interest in assisted emigration was definitely quickening in Cork and elsewhere in Munster. "No more favourable time ever occurred for colonizing Canada," Lord Mount Cashell assured Robinson from Moore Park at Kilworth in October 1824, "for never were the people so well disposed to emigrate as at the present moment."[195] Robinson was initially sanguine about his ability to organize a major resettlement enterprise in 1824, but the failure of government ministers in London to appropriate the needed funds, the reluctance of Dublin Castle officials, and serious logistical problems thwarted his plans.[196]

His experience was starkly different in 1825. In fact, he was now besieged by an enormous throng of applicants, who eventually numbered about fifty thousand altogether—far beyond the number of fifteen hundred that the government and parliament were originally prepared to accommodate. In the spring of that year, as Donald MacKay has written with verve, "the thought of going to Canada spread through the Blackwater [region] like a beneficent contagion. People who had never dreamed of leaving their parish, let alone Ireland, were talking of crossing the ocean and were waiting impatiently for the return of Peter Robinson."[197] Even though the Rockite movement had now been dormant for nearly a year, and even though the demand for places far exceeded the available supply, many landowners were still avid to seize this renewed opportunity to put troublemakers on the vessels going to Canada at public expense. William Wrixon-Becher had pleaded with Robinson in October 1824, in anticipation of the 1825 expedition:

> The candidates for emigration are encreasing and greatly exceed the number to which you limited me. If you could give me a greater latitude, it would be desirable. You are naturally anxious to have steady men with families, and it is equally natural we should wish to get rid of idle, unmarried individuals who seem likely to keep up the present disturbances, but if possessed of a little property [in Upper Canada], would soon change their habits. If therefore we can get 5 or 6 of that description to go as one family and be satisfied with one allowance of ground amongst them all, should you have any objection?[198]

Numerous other landowners exerted similar pressure on Robinson after his return to Ireland in 1825.[199]

In the end, however, the limits of government funding (£30,000) permitted Robinson to take only about 2,000 of the 50,000 applicants.[200] As a result of previous commitments to eight proprietors in north Cork, only a small proportion of the departing emigrants came from outside that region. The third Earl of Kingston furnished the largest contingent (some 400 persons). According to MacKay, "the other seven Blackwater landowners sent slightly smaller contingents, and the remainder, one sixth of the total, were sent by other sponsors, such as the city of Cork."[201] As many as 65 percent of the emigrants of

A knot of landowners with estates centered in north Cork, an area that witnessed extremely high levels of Rockite violence, successfully promoted the idea of assisted emigration by the government as a means of detaching their poorer tenants and subtenants from allegiance to Captain Rock. The exodus to Upper Canada under Peter Robinson in 1823 and 1825 was hardly large enough or prompt enough to affect the course of the Rockite movement, but for a limited time popular enthusiasm for assisted emigration in this area reached extraordinary heights. For this enthusiasm Robinson heavily credited the Catholic priests of north Cork, who, like the big landowners of the district, were eager to find almost any means of converting the poor of their parishes to the paths of nonviolence and submission to lawful authority. As this illustration suggests, Catholic emigrants could ordinarily expect a local priest to bless them as they left home for foreign shores. (*Illustrated London News*, 10 May 1851)

1825 were reportedly children—a figure that includes 466 boys or girls over the age of fourteen as well as 851 younger children.[202] A fleet of nine ships left Cork in stages in May with 2,024 emigrants altogether—"the largest single group yet to cross the Atlantic." Proportionately, deaths during the crossing in 1825 (11 children and 4 adults) were fewer than in 1823, and as there were 15 births at sea, the vessels arriving at Quebec city coincidently held the same number of people as had embarked.[203]

Unfortunately, the newly arrived emigrants of 1825 soon confronted a whole series of adversities that their predecessors in 1823 had largely escaped. By the time that the emigrants had reached Kingston in relays during July, the intense summer heat of the Saint Lawrence Valley and the rigors of their long sea journey had taken a heavy toll on their general health. Matters then seriously worsened in their riverside tent encampment at Kingston owing to an outbreak of malaria and, in some cases, the ravages of other diseases. Over the next month thirty-three deaths occurred, and three hundred other emigrants were suffering from "fever" when Robinson rejoined the stalled horde in the second week of August. Under his direction the weakened newcomers were moved gradually (in relays again) along the banks of Lake Ontario to Coburg, a village over ninety miles away. The evacuation from Kingston took a month, and it was another three months before, in late October, all the surviving would-be settlers had been collected at Scott's Plains on the Otonabee River in the Peterborough area of southern Ontario. Sickness or disease carried off another twenty or more emigrants that autumn before they could be given allotments of land in the region. Only then did the building of rough log cabins begin; there was not enough time to do more before the snows came, blocking the trails and preventing any significant clearing of the woods. Still, by that time Robinson could claim to have settled some nineteen hundred people across six different townships, the boundaries of which extended thirty miles from north to south and fifty miles from east to west. Despite their past sufferings and current hardships, Robinson remarked of the settlers that they "seem in good spirits and much pleased with their land."[204]

The experience of these settlers in Upper Canada and the fate of assisted emigration more generally gave the lie to Robinson's optimism. Malaria arising from polluted water (commonly called "ague") was a terrible scourge in the earliest years. Before the end of 1825 almost a hundred people had perished at the Peterborough depot, and more than a thousand of the settlers in the surrounding townships had fallen sick. Though the depths of winter temporarily checked the outbreak, disease returned in the spring and killed 172 settlers during the remainder of 1826.[205] Thus the death toll was at least 15 percent among these assisted emigrants within about eighteen months of their departure from Cork. The woodlands of the Peterborough area took many years to clear. The scarcity of labor and its relatively high cost exacerbated the problems caused by the physical difficulties of the terrain. The geographer John Mannion found that the Irish of the Peterborough area "cleared on the average less than two acres of land each year," though this was a very creditable performance when compared with that of Irish settlers in certain other areas of Canada that he studied (Mirimichi and the Avalon peninsula in Newfoundland). After five years of hard toil in Ennismore township, for example, the average clearing in 1830 was still less than 5 acres per family. Time and human muscle did eventually bring results that were impressive enough. By 1851 the 75 Irish farmers of Ennismore and the 128 farmers of North Emily (the parish of Downeyville) had achieved average clearings of 30 acres—much larger than those of their counterparts in the other Canadian regions of Irish settlement examined by Mannion.[206]

Irish paupers could indeed slowly transform themselves into the sturdily independent small landowners contemplated by enlightened officials like Lord Bathurst and Sir Robert Wilmot Horton.[207] And of course, Irish proprietors, led by those of north Cork, were happy to have the British government pay all the costs of the "experiments" of 1823 and 1825. But the long-term prospects for any major expansion of these limited early ventures were fairly bleak. Emigration from Ireland without government aid was rapidly increasing in the mid and late 1820s,[208] and a belief gained hold among the British political elite that government-assisted emigration, or the prospect of more of it, was giving at least some check to this escalating voluntary outflow.[209] As if this were not discouraging enough, even more distressing was the perceived extravagance of the two experiments over which Horton and Robinson had presided. As William Forbes Adams declared long ago in his still valuable study of prefamine Irish emigration and government policy, "The combined cost of the two experiments was more than £52,000 for less than 2,600 emigrants, and Ireland needed to lose two millions," according to the most anxious of British commentators. "No government would have dared to face the cost of a general removal at this rate, and the advocates of emigration admitted that partial removal would not be effective."[210]

Nevertheless, for at least a few more years it seemed as if the advocates of government-aided emigration would succeed in gaining parliamentary approval for a major enlargement of the two previous ventures. Horton's select committees of 1826 and 1827 on emigration helped to set off a wave of discussion and debate on the subject that persisted for over two decades.[211] The acceptance at Westminster of the need in Ireland for a thoroughgoing consolidation of holdings, as indicated by the passage of the Subletting Act of 1826,[212] was now often joined to the idea that there must be assisted emigration to accommodate humanely all those Irish tenants who would be displaced from their holdings if consolidation were carried out extensively.[213] Though loud objections to Horton's forward policy were raised in certain quarters during the parliamentary inquiries into emigration in 1826 and 1827, the general outcome, according to R.D. Collison Black, "was decidedly in favour of some large-scale plan of emigration, primarily with reference to Ireland."[214]

Seeking to take advantage of this favorable climate, Horton published in June 1827 the third report of his emigration committee, which called for a series of government loans totaling £1,140,000 to facilitate the removal of 95,000 persons over three years, with the first advance of £240,000 slated for 1828–29.[215] Unfortunately for Horton, political shifts at the highest level went far to sabotage his cause. He parted company with Sir Robert Peel and Henry Goulburn, both strong supporters of the 1823 and 1825 experiments, when they resigned from office (he did not) in April 1827 rather than join the new and short-lived government of George Canning. Then, when the Duke of Wellington and Peel formed a new administration early in 1828, Wellington left Horton out in the cold, and Peel delivered a crushing blow to Horton's effort in the House of Commons in June 1829 to revive his project. Emphasizing what many had always regarded as the most glaring weakness of Horton's approach, Peel insisted that even though the proposed outlay was huge, only a small proportion of the distressed population of Ireland would be enabled to emigrate. Though Peel welcomed the idea of assisting "volunteer emigration," he refused to countenance "the policy of laying out large sums of public money to encourage emigration." This was to become an endlessly repeated formulation of the official stance of successive British governments after 1830.[216]

While the Peter Robinson ventures of 1823 and 1825 were outgrowths of the Rockite movement, it can hardly be argued that either of them significantly affected the course of this agrarian rebellion. The scale of the first scheme in 1823 was too small to do so, and the second enterprise in 1825, though much larger, came after the movement had disintegrated. It is possible that if this government-sponsored emigration had been implemented more quickly and on a greater scale, it might have contributed significantly to the repression of the movement in its north Cork heartland at least. But even that is far from certain. However much a small group of north Cork proprietors looked to these schemes to rid their estates of troublemakers, they greatly overestimated the ability of themselves and their agents to ensure that their lists of willing emigrants corresponded with the swollen ranks of Captain Rock. Their hopes that the experiments of 1823 and 1825 would stimulate a widespread popular desire to embrace the notion of assisted emigration were certainly confirmed by the astonishing number of fifty thousand applicants in 1825. But neither the British government nor Irish landowners in general were prepared to invest heavily in this type of colonization of Canada. The ventures of 1823 and 1825 were the last of their kind precisely because they cost so much in an age when those endowed with power were deeply averse to public expenditure.

Conclusion

The close examination of the Rockite movement of 1821–24 undertaken in this book has permitted us to delve deeply into the dynamics
of Irish rural society in the aftermath of the French revolutionary and
Napoleonic wars. We have seen in chapter 1 that the cradle of this
movement was the great property of some 34,000 acres owned by the
profligate Lord Courtenay and centered around Newcastle West in
County Limerick. In numerous ways the structure and experience of
this estate was typical of many others in the south and west of Ireland
in the second decade of the nineteenth century. Substantial middlemen with leases held large portions of the Courtenay property that
they mostly sublet to other, smaller tenants; those who actually occupied and tilled or grazed the ground were, in the main, people either
mired in poverty or barely rising above it.[1] Like the vast majority of
their counterparts on other Irish properties, the tenants of the Courtenay estates—regardless of their social status—found the transition
from war to peace exceedingly difficult to negotiate. What made the
payment of rents especially problematic on the Courtenay property
was the fact that new rents had mostly been established in the years
1811–13, a period that included the very height of wartime prosperity.
Even though the indulgent agent Edward Carte granted substantial
abatements beginning in 1814 in response to the first wave of price de-
clines associated with the winding down of inflated wartime demand,
he was unable (and perhaps partly unwilling) to keep arrears of rent
from soaring over the next several years. By the time that he was dismissed in the summer of 1818, there were well over £60,000 in arrears
outstanding, or more than four times the annual rental of the prop-

erty. Among the greatest beneficiaries of the Carte administration were the numerous Protestant and Catholic middlemen who were so deeply embedded on Viscount Courtenay's estate. These elements in the story of the origins of the Rockite movement—an extravagant, "playboy" proprietor who never showed his face in Ireland, the beginnings of a long-term agricultural depression, an inefficient if not corrupt agent who let the financial situation spin out of control, a body of middlemen who paid as little rent as they could get away with, a property on which rapid wartime population growth had apparently promoted fairly rampant subdivision and subletting of holdings—were no doubt common elsewhere in Munster and Connacht during these years immediately before and after Waterloo.[2] Had Edward Carte been permitted to continue presiding over this lackadaisical regime, there might have been no agrarian revolt on the scale of the Rockite movement in the region that the agitation came to dominate in the early 1820s.

Fateful Choice: Alexander Hoskins

The wild card introduced into this fairly standard deck was the appointment of the London solicitor Alexander Hoskins as Carte's successor by a group of trustees eager to place a financially ailing Irish landed property on a healthier footing. Unfortunately, they chose as the executor of their policy a man whose background, personality, and total lack of judgment soon threw the Courtenay estate and adjoining parts of west Limerick into a frenzy of disorder and violence. While living like a lord and behaving like a Mafia boss, Hoskins used exuberantly the whip hand given to landlords and magistrates by the Irish legal system and employed some of the extralegal resources offered by Irish social life of the time (for example, factions). With these weapons he tried—not very successfully—to compel tenants to pay rent, to frighten or punish his enemies, and to protect himself, his family, and his henchmen from reprisals. Perhaps the most remarkable aspect of his "tyranny" (as it was perceived) was the series of assaults waged by Hoskins against prominent middlemen of minor-gentry and Protestant background, combined with their vigorous resistance to his regime. From this clash of wills and economic interests there emerged the somewhat unusual spectacle of gentry middlemen mobilizing their own subtenants, many of whom apparently belonged to local factions, to thwart the legal stratagems and extralegal initiatives of the Hoskins administration. (Offsetting this cross-class collaboration there must have been some social conflict between nonelite agrarian groups, but the surviving sources for the extraordinary Hoskins episode have little to say about such contention.) Such bitter and widespread hostility did Hoskins arouse that at least two attempts were made to kill him, and though he luckily escaped the death intended for himself, his son Thomas was fatally wounded in July 1821. Alexander Hoskins was eventually forced to resign and to leave Newcastle West (he departed in December 1821), but not before violent and organized agrarian combinations had come to flourish on the Courtenay estate and indeed much further afield. In testimony to the firestorm of protest that he and his various bands of bailiffs, guards, and spies had set ablaze, the government had found it necessary by October 1821 to station as many as four to five thousand troops in or adjacent to County Limerick in an abortive attempt to prevent the contagion from spreading.

Murders as Spurs to Militancy

The earliest Rockites were certainly emboldened by the killing of Thomas Hoskins, by their success in driving his father out of Newcastle West, by the repeated acquittals at the Limerick assizes of persons charged with violent crimes on the Courtenay estate, and by the stoning to death of a prosecution witness aligned with Hoskins on this informer's return home from the summer assizes of 1821. In addition, the murders of Christopher Sparling and Major Richard Going in mid-October of the same year played a major role in inserting a strong element of sectarianism into the Rockite movement in its earliest phase and in propelling the movement beyond its original base in west Limerick. Sparling was a member of one of the Palatine communities settled since the early eighteenth century in the districts of Rathkeale, Askeaton, and Adare.[3] He earned hostility in a number of different ways, among which was his arrival on the Courtenay estate as a would-be pioneer in the foundation of a Protestant colony promised by Hoskins in yet another colossal misjudgment of local Catholic sensitivities. The numerous early attacks on west Limerick Palatines like Sparling were to find a kind of culmination later in the Rockite movement with the sensational burning of the Palatine village of Glenosheen on the other side of the county. Major Going's offense was partly to permit an Orange lodge to be formed in the County Limerick police establishment, of which he had previously been the head, and partly to order the burial in quicklime of two agrarian rebels who had been shot dead in a tithe affray near Askeaton. The denial of the opportunity for a decent wake and a Christian burial for these two men was bad enough, but in this case the denial of the usual ceremonies wore the appearance of a consignment of the bodies to the ignominy of a "croppy hole" and therefore for Catholics was reminiscent of the savage official and Protestant-directed repression that accompanied and followed the 1798 rebellion. Connected as he was in the popular mind with a double display of obnoxious Orange proclivities, Going came to symbolize for many Catholics and Rockites the ugly face of the Protestant church and state—kindred institutions that they would come to loathe with a special intensity in the early 1820s. The news of Going's murder was consequently the subject of widespread Catholic popular jubilation, and so too (though to a lesser extent) was that of Sparling's death.

Economic Crisis: Its Crucial Role

If discrete events arising out of local conditions were of critical significance in the initial emergence of the Rockite movement, it was driven forward into new territory and secured the allegiance of a much wider segment of the population because of the onset of one of the harshest economic downturns of the entire nineteenth century. The collapse in agricultural prices from 1819 to 1823 filled the cauldron of discontent on which the Rockite movement thrived and grew to maturity. Even if there had been no other aggravating features, the plunging prices of both grain and livestock would themselves have been sufficient to prompt a widespread agrarian rebellion like that of 1813–16 or like the tithe war of the early 1830s. But added to the onset of a punishing agricultural depression was the partial famine of 1821–22. The threat of this famine helped to ignite an insurrectionary outbreak that was almost unique among the succession of rural protest movements of the late eighteenth and early nineteenth centuries. Apart perhaps from the violent upsurge of Defenderism in 1795, the history of Irish agrarian revolt in the prefamine period can offer no other unambiguous example of the kind of open insurrection that erupted at the start of 1822. But if the dread of famine played a crucial role in spark-

ing the insurrection of early 1822, the actual onset of famine conditions over much of the south (and west) in the late spring and summer
of that year brought about the temporary demobilization of the Rockite movement. Only after famine conditions receded in the autumn of
1822 was there a renewed burst of Rockite activity, and it now occurred
over a much larger area.[4]

Among the distinctive features of the Rockite movement was its
broad social composition; it drew within its ambit a wide section of
middling and large farmers as well as multitudes from the bulging
ranks of the rural poor in the south and the southwest. This study of
the Rockites has presented abundant evidence of a major agrarian
movement that displayed both collaboration and conflict between the
landless and the land-poor on the one hand, and on the other, that
minority of tenant farmers whom contemporary commentators described as "snug," or "comfortable," or even "wealthy." Some of the
agrarian rebellions of the early nineteenth century were of an altogether different social character; they rarely embraced anyone beyond
the landless and the land-poor. In these outbreaks of rural violence
the better-off farmers almost always were only the targets of agrarian
rebels and hardly ever became willing and active participants themselves. Usually, the economic precipitants of such outbreaks were
buoyant farm prices and sharply increasing land values, with an accompanying rise in the prices of all kinds of food. In this situation
middling and large farmers were put under severe collective pressure
by those below them in the social scale to lower the price of conacre
lettings and to increase the availability of conacre land. Clear examples of such agrarian movements with a restricted social composition
are the Threshers of 1806–7, the early Caravats in 1813–14, the so-
called Ribbonmen of 1819–20, and the Terry Alts in 1829–31. By contrast, the Rockite rebellion, along with the Caravat movement of
1813–16 and the tithe war of the early 1830s, were all associated with
deep economic downturns that pushed discontent about rents and
tithes far up the social scale, reaching the wealthiest of farmers and
graziers. Indeed, such was the severity of the downturn of 1819–23
that even members of these elevated social groups were generally deprived of their customary material comforts, often to the point of insolvency. Acute economic deprivation was no certain guarantee that
previously well-to-do farmers or graziers would throw in their lot with

341

the Rockites, but this study has demonstrated that such conditions greatly enhanced the likelihood of cross-class collaboration in the attempted regulation of tithes and rents as well as in efforts to control the occupation of land.

Potent Emotions: Millenarianism and Sectarianism

As much as economic grievances fueled the Rockite movement and gave rise to sharp divisions within it, this book has repeatedly stressed that millenarian prophecies and sectarian animosities were central to the *mentalité* of the agrarian rebels of 1821–24. Millenarian beliefs had been common enough among many of the Defenders and United Irishmen in the radical and revolutionary 1790s, but none of the other great Irish agrarian movements of the late eighteenth and early nineteenth centuries were as accented by millennial hopes and sectarian hatreds as that of the Rockites. As chapter 4 has demonstrated at length, "Pastorini" (the nom de plume of Charles Walmesley, author of a notorious work on the Book of Revelation that gave rise to many popular offshoots) was a close ally, so to speak, of Captain Rock. It is worth recalling that Pastorini's book made its way into the hands of Alexander Hoskins no later than 1821, and popular variations on his tome predicting the imminent destruction of Protestantism in church and state must have been current among Hoskins's many enemies in west Limerick as the fledgling movement gathered momentum there. But just as Pastorini's prophecies seem first to have taken popular hold in the midst of the earlier famine of 1816–17 (accompanied by a typhus epidemic lasting from late 1816 to early 1819), so too the famine conditions of 1821–22, and the insurrectionary outbreak associated with this subsistence crisis, added considerably to the currency of the millennial convictions linked to his name during the early 1820s. The tithe grievance entered heavily into the equation as well. Although many earlier agrarian rebels had targeted the Irish tithe system and the Anglican clergymen who were its beneficiaries without embracing millennial beliefs, the Rockites seem to have strongly favored Pastorini's prophecies at least partly because they clashed so massively with Anglican ministers and their representatives throughout the southern and southwestern region. Unlike many of their predecessors in agrarian rebellion, the Rockites called very often for the total abolition of tithes and came much closer to realizing that objective in practice (if only temporarily) by their incendiary violence in 1822 and 1823.[5] Though the destruction of the Protestant church was in reality still far from their grasp, the blows inflicted on the Anglican church establishment in Ireland by the Rockites were heavy enough to encourage the popular belief that its demise was perhaps as imminent as Pastorini was said to have predicted. It appears reasonable to argue that there was a self-fulfilling quality to the enthusiasm with which the Rockites espoused both Pastorini's millennialism and the cause of tithe abolition.

While the intensification of the tithe grievance and the fear of famine in 1821–22 go far to explain the welling up of millennial beliefs and the deepening of anti-Protestant antagonism among the Rockites, other forces were also operating to inflame sectarian feeling at the popular level. Among the most important were the provocations offered by Protestant proselytizers. Recent research has made abundantly clear that in a variety of ways Protestant proselytism became considerably more aggressive in the decade and a half after Waterloo than it had ever been before. Besides an avalanche of anti-Catholic literature and the almost frenzied circulation of the Scriptures by an increasingly numerous and hyperactive band of Protestant missionary organizations, there was the major challenge mounted by the Protestant proselytizers in the educational arena, where sectarian combat became the order of the day. The schools established by the different Protestant evangelical societies were seen by Catholic bishops and priests as well as by leading Catholic laymen as intrinsically hostile to Catholic interests and ambitions. Through the denunciations and anathemas hurled against the proselytizers by local Catholic clergymen, these hostile feelings about Protestant schools and their local sponsors were transferred to ordinary Catholics at the grassroots. It was one sign of the sharp deterioration in Protestant-Catholic relations around 1820 that Daniel O'Connell saw fit to condemn as sectarian the policies and activities of the Kildare Place Society, whose efforts in the field of primary education had once been acceptable to Catholics generally. Though the long round of "Bible discussions" (staged theological debates between Catholic and Protestant clerical champions who attracted large crowds) only began in late 1824, and thus after the Rockite movement had gone into decline, sectarian passions had already passed into flood stage.[6]

The forces employed to contain and suppress the Rockite movement—the yeomanry, the police, and the army—helped to bring about
a massive consolidation of sectarianism. In the case of the yeomanry,
it was not their numbers but rather their reputation that mattered
more. Since there were fewer than twenty-five hundred yeomen in the
six Munster counties in 1821, they could do little by themselves to
blunt the Rockite onslaught, but their identification with Orangeism
and with bitter Catholic memories of their counterrevolutionary violence in the 1790s was enough to elicit extreme Rockite antagonism.
The police were also certain to earn the enmity of the agrarian rebels
because they were the primary enforcers of the Insurrection Act of
1822. As this book has abundantly demonstrated, police magistrates
and their constables carried the leading role in the hunting down of
Rockite offenders. Police successes represented Rockite failures, and
failures were as likely to infuriate the Rockites as to dispirit or overawe them. As if the efforts of the constabulary to transport or to execute as many Rockite activists as possible were not enough to earn
popular obloquy for the police, they too suffered in Catholic eyes from
the taint of Orange sympathies. The notoriety achieved by the conduct and fate of Major Richard Going in County Limerick went far toward fixing this taint in the popular Catholic consciousness in the
south of Ireland in the early 1820s, but there were numerous reports
of religious or sectarian partisanship on the part of the police in other
counties affected by the Rockite movement as well. Far outnumbering
either the police or the yeomanry among the forces of repression were
the regular troops who flooded into the districts disturbed by the
Rockites. (After major reinforcement the army in Ireland peaked at
about twenty-one thousand by 1823.) The newly arriving soldiers were
commonly cavalry units drawn from England and Scotland; they
marched to Protestant churches in the south and southwest, helping
to fill edifices long mostly bereft of parishioners and reminding the
Rockites that the troops had come to serve the interests of a Protestant church and state bent on the oppression (economic, political, and
religious) of Irish Catholics. That Captain Rock and his followers
viewed the army through the distorting lens of sectarianism is clear
from many threatening notices and from such extreme incidents as the
rape of the wives of men belonging to the 1st Rifle Brigade. A classic
instance of rape as punishment, so common in war and ethnic conflict,

this repulsive incident suggests the viciousness with which even less extreme sectarian feeling and conduct had taken root through the violent struggle between the Rockites and their opponents.[7]

Unprecedented Violence

Extreme violence against persons and property were defining features of the Rockite movement. The weapon of incendiarism was employed on a greater scale by the Rockites than in any other Irish agrarian rebellion of the late eighteenth and early nineteenth centuries. The burning of tithe corn, admittedly, did not menace life or limb, and neither did the burning of houses. With the exception of the catastrophic "burning of the Sheas" and one much less well-known instance, no deaths seem to have resulted from the hundreds of farmhouses and cabins destroyed by incendiary fires. Nevertheless, the firebrand was among the most feared weapons in the Rockite arsenal; no one could have sloughed off the loss of the family dwelling by fire as a matter of small consequence, and to be made to live in constant dread of this punishment, as many were, must have been an acute psychological burden. But if the Rockites did not deliver death by fire, they inflicted it on an extraordinary scale with guns, knives, pitchforks, stones, and a variety of blunt instruments. And to the scores of murders must be added the many hundreds of violent assaults. If a partial exception were made for the agrarian and political rebels of the 1790s, it would probably be no exaggeration to claim that the Rockites exceeded all previous regional agrarian protestors in the frequency and the intensity of their violence against the person (murders and woundings).[8] The Rockites employed extreme violence as a savage warning to their enemies and as an expression of their power over their opponents; at times they reveled in their violence, either because it represented a seeming reversal of traditional power relations, or because it trumpeted the avenging of wrongs they considered intolerable, or because they appreciated the paralyzing intimidation that it was often capable of inspiring.

Such an extraordinary level of sustained and extreme violence demands explanation. This book has argued that the Rockites invested enormous energy and resources in the earliest stages of their movement in seizing arms and ammunition; their attacks on the houses of the resident gentry and the better-off farmers for this purpose, and the sometimes spirited resistance to these attacks, were responsible for a remarkably high proportion of the murders and deaths associated with the rebellion. The mass attack on the police barracks at Churchtown and its burning were symptomatic of the Rockites' determination in this quest; occasionally, as in the Churchtown episode, they sought to deprive even the constabulary, regular troops, and yeomen of their weapons. Admittedly, all Irish agrarian movements of the late eighteenth and early nineteenth centuries made the acquisition of arms a primary feature of their mobilization. But part of the explanation for the extreme violence that characterized the Rockites' activities may be that they were more effective than their predecessors and thus were better supplied with the firearms used in much of their extreme violence.[9] By the early 1820s the point had long been passed when either the authorities or the agrarian rebels observed a well-understood code of mutual restraint and recognized boundaries with respect to the use of armed force. If murder and deaths in general were relatively rare in agrarian and other popular protests before 1790, and if their repression involved relatively little spilling of blood, that situation changed dramatically in the 1790s,[10] and it could be said that things were never the same again in the first three decades of the early nineteenth century. Certainly, the era of the Rockites witnessed a dramatic worsening in the lack of restraint on the popular side.

Explanations for Extreme Violence

The forces and impulses fostering an environment of extreme violence, however, were far more deep-seated than the distinct possibility that the Rockites were better armed than their predecessors. The almost unparalleled severity of the agricultural crisis of 1819–23, coupled with the impact of the famine of 1821–22, produced a general situation in which economic disaster overtook or threatened to overtake most of the rural population of the south and the west—from landless laborers all the way up to once wealthy graziers. For probably the vast majority of those trapped in this downward economic spiral, securing release from the burdens of rents and tithes (not to mention taxes) appeared to be imperative for survival. Thus the preoccupation of the Rockites with these two grievances was only to be expected.

But what made the Rockites embrace extreme violence much more readily than earlier agrarian rebels was the increased resort to eviction by a significant proportion of those whose own economic survival as "landlords"—whether they were Protestant gentlemen or large Catholic farmers—depended on their ability to get rid of tenants unable or unwilling to pay the rents demanded. Since officials then collected no statistics on the matter, it is impossible to demonstrate conclusively that the frequency of evictions rose dramatically after 1819 as compared with the rate of dispossession during the economic downturn of 1813–16, but the systematic examination of the surviving sources undertaken for this book yields the strong impression that such an increase did occur, prompting strenuous popular resistance laced with extreme violence. The Ejectment Acts of 1816–20 certainly made eviction much simpler and cheaper for landlords than it had been in the past.[11] This fact greatly increases the likelihood that the economic crisis of 1819–23 was the first major occasion when widespread insolvency among tenants triggered a dramatic upsurge in evictions. According to this reasoning, then, the extreme violence of the Rockites was to a substantial degree a response to the expanded threat of complete impoverishment and to the plunge in social status that inevitably followed dispossession. As the crisis of 1819–23 represented the second major economic downturn since the effective end of the Napoleonic wars (the first depression occurred in the years 1813–16), it is not surprising that the Rockites sought to gain for earlier evicted tenants the chance to recover their old holdings, or at least to prevent new tenants or "canters" from occupying these banned lands. (Presumably, tenants whose evictions had occurred prior to 1820 or 1821 were participants in the movement.) The Rockites' widened prohibitions on the taking of evicted farms, sometimes extending back as many as seven years, could be enforced, if at all, only through violence; the acceleration of population growth during the wartime boom up to 1813 spelled acute competition for land in Ireland long after Waterloo and the fall of Napoleon. Indeed, this aspect of the Rockite story prompts the conclusion that competition for land was now so intense that only the most extreme measures appeared capable of restraining it, and even then, only temporarily.

Conflicts rooted in class divisions no doubt contributed significantly to the profusion of extreme violence. Though it is appropriate enough to speak of the Rockite "movement" rather than to use the plural form of that word, this book has not hidden the class divisions

Evictions or even threats of dispossession were capable of stirring powerful emotions and producing extreme violence. The business of an eviction began with the posting of a notice of ejectment by a process server on the dwelling of the delinquent tenant. The process server's job and the roles of other participants were risky enough to life and limb during the Land War of the early 1880s, as portrayed in this sketch of a process server, bayonet-carrying soldiers, and an enraged tenant family at a farmhouse in County Galway. But participation in evictions was far more dangerous during the Rockite rebellion. The Ejectment Act of 1816 made eviction cheaper and quicker, but widespread use of the strong powers that this law and its sequels conferred on landlords often elicited a violent response from injured tenants. (*Illustrated London News*, 7 Feb. 1880)

that riddled the rebellion of 1821–24. On the contrary, these divisions have been portrayed as pervasive and persistent (though also as capable of being overridden under certain well-defined circumstances). Numerous examples cited in this book have underlined the hostility commonly generated among the landless and the land-poor by self-aggrandizing large farmers or grasping middlemen. From such antagonism to land monopolizers, murder, savage beatings, and incendiarism often eventuated. It is clear from still other examples that there was a substantial amount of conflict over land within classes or between people of roughly the same social status. The conclusion seems inescapable that the intensity and multiplicity of social conflicts generated by the effects of heavy population pressure and by abysmal economic conditions also help to explain the extreme violence used by the Rockites.

That many of the victims or targets of extreme Rockite violence were Protestant in religion was of course partly a function of their perceived status as economic oppressors or as agents of repression, but the anti-Protestant millenarianism and sectarianism that shaped the outlook of so many Rockites also gave warrant for the use of extreme violence against Protestants as such. On those who accepted Pastorini's notions, the bitterly anti-Protestant rhetoric and Catholic triumphalism of his prophecies had the effect of stripping members of the Anglican elite and other Protestants of their ordinary human dignity, thus making their enemies the Rockites feel morally much less reticent in wreaking extreme violence on them. Just as the currency of Pastorini's sectarian and millenarian ideas removed the taboo from the burning or desecration of some Protestant churches, so too the widespread diffusion of these notions lowered the bar as to the kind of punishment that could be doled out to Protestant landowners, agents, stewards, clergymen, and magistrates who were popularly seen as oppressive.[12]

Relative Impunity Enjoyed by Rockites

A final factor promoting the use of extreme violence was the relative impunity with which it could be perpetrated. Over the decades since the first appearance of Whiteboyism in the south of Ireland in the early 1760s, and through the long succession of popular agrarian and political movements that followed, the rebels and their sympathizers had developed a whole series of techniques that were designed to frustrate the normal operation of the legal and judicial systems. They sought through intimidation and violence to immobilize magistrates and other members of the landed gentry who might be inclined to oppose their activities, and they employed the same methods to ensure that the vast majority of the common people would be too frightened of reprisals to become informers against them. Those private prosecutors or adverse witnesses who were not readily amenable to intimidation could often be persuaded by bribes either to decline to press charges in court or even to perjure themselves at trials. Though the authorities generally took precautions to shape or "pack" petty juries in important cases with Protestants or very "safe" Catholics of means, the Whiteboys sometimes exploited corrupt influence over members of these juries in order to secure acquittals. These time-honored methods came into widespread use again (in conjunction with extreme violence against informers and zealous magistrates), and their combined effect was to make the efficient administration of the ordinary criminal law highly problematic in the early 1820s. Or as George Cornewall Lewis put the matter in relation to the Rockite upheaval, "When the Whiteboy system is in such a state of activity as has been just described, the intimidation is complete; a general terror pervades all classes; and it is less dangerous to disobey the law of the state than the law of the insurgent."[13]

The traditional system of unpaid magistrates drawn from the resident aristocracy and landed gentry displayed so many shortcomings that the constabulary act of August 1822, which created a county-based police force, took an important additional step. It provided for the appointment of an enlarged cadre of paid police magistrates who would be capable of rising above the timorous behavior of many unpaid magistrates of the Irish landed class when they came under severe Rockite pressure. In conjunction with the transformation of Peel's Peace Preservation Force of 1814 into the much larger and better-organized armed constabulary of 1822, this substantial extension of paid police magistrates to reinforce the continued work of the unpaid local magistracy (itself thoroughly reformed in 1822) represented a considerable reinvigoration of the forces of official repression.[14]

The critical need for these and other changes had been made plain by the ineffectiveness of the traditional legal and judicial systems, even when backed by military reinforcements, in halting the alarming spread of the Rockite movement. The special commission in Cork city early in 1822 saw the judges condemn thirty-five prisoners to death; these proceedings culminated in the carefully staged executions of at least fifteen of these men. But the public hanging of alleged agrarian rebels between 1822 and 1824, even though it was carried out on a scale unprecedented in the annals of Irish Whiteboyism, was far from sufficient to cripple the Rockite movement. Indeed, the great crowds that frequently turned out for these executions were meant in part to express solidarity with the condemned men and were as likely to create martyrs for the Rockite cause as to impress the assembled thousands of onlookers with the putative "majesty of the law." Though many Rockites mounted the scaffold after publicly admitting the justice of their sentence, many others loudly declared their innocence of the crimes for which they were about to suffer, or refused to make any "dying declaration" that repudiated their association with the agrarian rebels. Altogether, the authorities executed approximately one hundred persons for Rockite crimes in five of the affected counties (Cork, Limerick, Tipperary, Kerry, and Kilkenny) from 1822 through 1824, with Cork and Limerick alone accounting for about three-quarters of all the hangings for such offenses over those three years. Yet at the beginning of 1824, after most of these hangings had taken place, key parts of Cork and Limerick were still giving many signs of strong allegiance to the Rockite cause.

Transportation beyond the seas to penal colonies in Australia (mostly to "Botany Bay") constituted the other main official form of punishment designed to subdue the Rockites. This method of dealing with Irish agrarian and political offenders had been in use since the early 1790s.[15] In February 1822 the weapon of transportation as fashioned by the new Insurrection Act became the second main instrument of repression (along with the traditional courts of assize held twice annually in every county) employed by the authorities for the remainder of the Rockite movement. Not for the first time, the Insurrection Act dispensed altogether with trial by jury. This particular law substituted a summary form of justice in which the new police magistrates regularly furnished the prosecution testimony and local magistrates found the verdict by majority vote, though their power of punishment was limited to a sentence of banishment for seven years.[16]

Acting at frequently held "special sessions" in the disturbed districts, magistrates imposed this sentence on about 450 persons while the Insurrection Act was in effect. Owing to the clemency granted to perhaps 120 of them, the number who actually arrived in Sydney from November 1822 to September 1825 (according to the painstaking research of George Rudé in Australian sources) amounted to about 330.[17] But the authorities also resorted to this punishment for many Rockite offenses less serious than murder, rape, incendiarism, and armed attacks on houses at night. Recalcitrant sources reduce us to the necessity of making an educated guess as to the number transported at the assizes for these lesser Rockite crimes or transported because their death sentences had been commuted for some reason. It is thought that the combination of prisoners transported under both the normal process at the assizes and the exceptional legislation of 1822 boosted the total number of Rockite transportees to approximately 600 persons.[18]

In spite of the heavy official repression of the Rockite movement, however, the number of those who escaped punishment for murder and other serious agrarian crimes far exceeded the number who were convicted and sentenced to execution or transportation. Incendiaries among the Rockites enjoyed the greatest degree of impunity, but the authorities themselves often confessed their failures in apprehending or in securing the conviction of a host of Rockites guilty of other crimes of the deepest hue. In part the passing of the Insurrection Act in February 1822 was a confession by the authorities that if they adhered to normal legal requirements for the conviction of agrarian offenders, the Rockites would not be deterred from their violent courses. No doubt the bite of repression became considerably more effective in 1823 and 1824, but even then the relative weakness of the forces of law and order still meant that most violent agrarian crimes went unpunished. Hundreds, admittedly, were executed or transported, but there were many thousands who could be classified as active participants in the Rockite movement, and there were thousands more who cheered on these agrarian militants from the sidelines.

Economic Reasons for the Eclipse of Captain Rock

Though the authorities did enjoy increasing success in capturing and punishing leading Rockites in 1823 and 1824, it was not effective repression that finally sent the Rockite movement into eclipse as much as a distinct improvement in economic conditions. This improvement became the frequent subject of contemporary comment beginning in the early months of 1824, and observers were quick to realize its significance for inhibiting the persistence of agrarian rebellion. Some commentators insisted that nothing would or could make a greater contribution to the demobilization of the Rockites than the return of agricultural prosperity. And in the view of seasoned observers grain and livestock prices rose enough during the course of that year to justify the term "prosperity" as far as farmers were concerned. The pace of demobilization varied somewhat geographically and socially, with laborers and cottiers disengaging from rebellious activity more slowly than middling farmers and those even higher up the tenant hierarchy. Indeed, the last phase of the Rockite movement consisted of the levying of contributions and plain acts of thievery by some of the poorest elements of the population against farmers of some substance; the latter now responded in certain districts by joining local associations sponsored by police magistrates or landowners and by mounting nocturnal patrols against such depredations.[19]

The linked phenomena of rising farm prices and the eclipse of the Rockite movement meant that rents and tithes came to be paid again with reasonable punctuality. Looking back from the point of this radical alteration in economic conditions in 1824, many contemporary observers either stated explicitly or implied strongly that not much money in either rents or tithes had been collected in 1822 or 1823, and clearly the financial crisis for landlords and clergymen had set in somewhat earlier. Further research on estate records should eventually make it possible to be more precise about the magnitude of lost rents during the period of the Rockite movement, though the prevalence of the middleman system on many estates in the Rockite region means that much that we might wish to know about this matter will be forever hidden from view by the omission of subtenants from rent ledgers.[20] Yet all the qualitative evidence provided in newspapers, parliamentary papers, and letters written by officials and by members of the upper classes indicates that landowners and clergymen in the areas persistently affected by the Rockite movement from late 1821 through early or mid-1824 had an exceedingly difficult time making ends meet. In most cases in such districts probably less than half the usual rents and tithes had been paid. It would be misleading to give all the "credit" for this result to the actions of the Rockites, since the economic downturn would alone have led to much diminished payments. But both the agrarian rebels and their enemies would have been ready to attribute the general withholding of rents and tithes to Rockite mandates. And on this score the historian would also be inclined to conclude that the Rockites were highly effective in reducing the burdens of rents and tithes during the period of their ascendancy.

The Rockite Legacy

The name of Captain Rock as the avenger of agrarian wrongs long survived the great popular movement putatively led by this mythical figure.[21] The appellation commonly appeared on threatening notices during the tithe war of the 1830s and in geographical areas far outside the counties strongly affected in the early 1820s.[22] This posthumous fame hints at the considerable longer-term impact of the Rockite rebellion on Irish society and politics. The incomplete victory achieved by the Rockites on the ever-contentious tithe question made further conflict over this acute grievance almost inevitable. The Tithe Composition Act of 1823 seemed to recognize the principle that an equitable resolution of this national issue entailed the conversion of the liability to tithe payment from a set of charges on tillage crops into a combined charge affecting all landholders in a given parish, with each and every acre, regardless of its agricultural use, bearing an equal burden. Wherever a parish included a considerable portion of pasture within its boundaries, the implementation of this principle would have brought much relief to the holders of tillage land. But what the framers of the law seemingly bestowed with one hand they subtracted with the other. The legislation effectively left the power of decision in the hands of vested interests. Voting in the parish vestries was usually restricted to a relatively small number of the largest landholders, among whom livestock-rearing demesne farmers, graziers, and large farmers ruled the roost. Even where such vested interests did not block implementation, the clerical tithe owner or lay impropriator had to agree to the terms of any new settlement of the overall tithe liabilities of the body of parishioners. Taken together, these provisions constituted a recipe for revived conflict whenever the next major economic crisis came along. Tillage farmers and cottiers could not be expected to tolerate for long the blocking vote of the highly privileged in the parochial vestries, and the popular pressure on Anglican clergymen to accept commutation at a discounted overall figure would intensify under such conditions.[23] Furthermore, rather than accept any significant liability for the payment of tithes, graziers and dairy farmers would likely shift all their political and social weight behind the cause of the total abolition of tithes. In this indirect but highly significant way, the Rockites helped to set the stage for the tithe war—the greatest agrarian rebellion of the prefamine period.[24]

If the Rockite movement made it likely that the next major economic crisis would resurrect the popular demand for tithe abolition even more insistently than before, it also cast a long shadow in other ways. It showed how a combination of passive resistance and well-directed violence could be combined to make an anti-tithe campaign highly effective. In the matter of the weapons of their warfare, the anti-tithe agitators of the 1830s stood mostly on the shoulders of their predecessors, and the Rockites were among the most important of their teachers. But it was not only in the matter of tithes that the Rockites exerted a potent influence. They had displayed a ferocious antipathy to evictions. Whether their extreme violence was "copied" by later agrarian rebels in the normal sense of that word may be doubted, but it is clear that to punish land grabbers or evicting landlords with death became a fairly routine occurrence in the Ireland of the 1830s and early 1840s.[25] Nevertheless, it was relatively rare in the prefamine decades for landowners or middlemen to engage in large-scale evictions of the type that became so common during and immediately after the Great Famine of the late 1840s and early 1850s. Historians have generally attributed the relative ease with which landowners engineered the famine clearances in large part to the collapse of the capacity for collective resistance under the pulverizing effects of mass starvation and epidemic disease.[26] Along with other agrarian rebels of the prefamine period, the Rockites obviously provided a vivid and gruesome demonstration of the capacity for collective resistance to eviction, and it may well be suspected that the fear of violent reprisal gave many landlords before 1845 more than second thoughts about dispossessing any large number of tenants at once.[27]

In the short term too the Rockites played a role of particular significance in the dramatic changes that overtook Irish politics and

Anglo-Irish relations in the mid and late 1820s with the flowering of the great popular campaign for Catholic emancipation under the leadership of Daniel O'Connell. O'Connell's movement thrived on feelings of anti-Protestant hostility. O'Connell himself was not averse to playing the sectarian card when it suited his political interests, as he did most famously in key passages of his published address to the electors of County Clare in the crucial by-election of 1828.[28] The Rockite movement, with its virulent sectarianism, was hardly the only factor that made the sectarian card so attractive a political option in the 1820s, but it was certainly one of the most important. Furthermore, O'Connell had burnished his reputation as the greatest barrister of his day by his prominence as the defense counsel in a whole series of Rockite trials in the early 1820s. It was an impressive feature of his political ingenuity that in this role he managed to appear at one and the same time as a decided opponent of all agrarian violence and as the savior of many an agrarian rebel who, but for O'Connell's withering cross-examinations of prosecution witnesses, appeared destined for the hangman's noose or the convict ship. It is certainly no exaggeration to say that O'Connell considerably boosted his political popularity in rural Ireland by his remarkably skillful defense of a large number of prisoners accused of Rockite crimes.[29] Also raising his stature in the countryside was his inheritance of the special mantle created for him, as it were, by Pastorini's prophecies. Their widespread diffusion in the early 1820s, extending well beyond the counties affected by the Rockite movement, had spawned popular expectations of the imminent obliteration of the Protestant Ascendancy in Ireland. Admittedly, the predictions that Protestant domination in church and state would end in 1821 or 1825 had proved to be untrue if taken too literally, but from 1824 onward the extension of the crusade for Catholic emancipation across much of the country, and the increasing odds of ultimate success in this quest, made Daniel O'Connell appear to many among the Catholic rural masses as the divine agent who had been sent to overturn Protestant supremacy and to lead his coreligionists into the promised land, where such grievances as excessive tithes and exorbitant rents would trouble them no longer.[30]

Notes

Abbreviations

Census Ire., 1851	*The census of Ireland for the year 1851 . . .* [etc.]
DEP	*Dublin Evening Post*
LEP	*Limerick Evening Post*
NAI	National Archives of Ireland, Dublin
NLI	National Library of Ireland, Dublin
PRO, CO	Colonial Office Papers, Public Record Office, Kew, England (now the National Archives)
PRONI	Public Record Office of Northern Ireland, Belfast
SOCP 1	State of the Country Papers, Series 1
SOI evidence	*Minutes of evidence taken before the select committee appointed to inquire into the disturbances in Ireland in the last session of parliament, 13 May–18 June 1824,* H.C. 1825 (20), vii, 1.
SOI evidence	*Minutes of evidence taken before the select committee of the House of Lords appointed to examine into the nature and extent of the disturbances which have prevailed in those districts of Ireland which are now subject to the provisions of the Insurrection Act, and to report to the House, 18 May–23 June 1824,* H.C. 1825 (200), vii, 501.
SOI evidence	*Report from the select committee on the state of Ireland, ordered to be printed 30 June 1825, with the four reports of minutes of evidence,* H.C. 1825 (129), viii, 1, 173, 293, 455.
SOI evidence	*Minutes of evidence taken before the select committee of the House of Lords appointed to inquire into the state of Ireland, more particularly with reference to the circumstances which may have led to the disturbances in that part of the United Kingdom, 18 February–21 March 1825,* H.C. 1825 (181), ix, 1.
SOI evidence	*Minutes of evidence taken before the select committee of the House of Lords appointed to inquire into the state of Ireland, more particularly with reference to the circumstances which may have led to the disturbances in that part of the United Kingdom (24 March–22 June 1825), brought from the Lords, 5 July 1825,* H.C. 1825 (521), ix, 249.

Introduction

1. William Carleton, "Wildgoose Lodge," in *The works of William Carleton,* 3 vols. (New York, 1881), 3:936–44 (quotation on p. 944). See also D.J. Casey, "Wildgoose Lodge: the evidence and the lore" (parts 1 and 2), *Journal of the Louth Archaeological Society* 18, no. 2 (1974), pp. 140–64; 18, no. 3 (1975), pp. 211–31; Terence Dooley, *The murders at Wildgoose Lodge: agrarian crime and punishment in pre-famine Ireland* (Dublin and Portland, Ore., 2007), passim.

2. For the origins of the name "Captain Rock," see below, chapter 1, pp. 37–38.

3. For earlier and more restricted treatments of the Rockite movement, see Sailbheastar Ó Muireadhaigh, "Buachaillí na Carraige, 1820–25," *Galvia* 9 (1962), pp. 4–13; Finbarr Whooley, "Captain Rock's rebellion: Rockites and Whiteboys in County Cork, 1820–25" (M.A. thesis, University College Cork, 1986); Shunsuke Katsuta, "The Rockite movement in County Cork in the early 1820s," *Irish Historical Studies* 33, no. 131 (May 2003), pp. 278–96.

4. For studies of Whiteboy movements in the late eighteenth century, see Maureen Wall, "The Whiteboys," in T.D. Williams, ed., *Secret societies in Ireland* (Dublin and New York), pp. 13–25; J.S. Donnelly, Jr., "The Whiteboy movement, 1761–5," *Irish Historical Studies* 21, no. 81 (March 1978), pp. 20–54; idem, "The Rightboy movement, 1785–8," *Studia Hibernica,* nos. 17–18 (1977–78), pp. 120–202; idem, ed., "A contemporary account of the Rightboy movement: the John Barter Bennett

manuscript," *Journal of the Cork Historical and Archaeological Society* 88, no. 247 (Jan.–Dec. 1983), pp. 1–50; idem, "Irish agrarian rebellion: the Whiteboys of 1769–76," *Proceedings of the Royal Irish Academy* 83, sec. C, no. 12 (Dec. 1983), pp. 293–331; M.J. Bric, "Priests, parsons, and politics: the Rightboy protest in County Cork, 1785–1788," in C.H.E. Philpin, ed., *Nationalism and popular protest in Ireland* (Cambridge, 1987), pp. 163–90; Thomas Power, *Land, politics, and society in eighteenth-century Tipperary* (Oxford, 1993); idem, "Father Nicholas Sheehy (*c.*1728–1766)," in Gerald Moran, ed., *Radical Irish priests, 1660–1970* (Dublin, 1998), pp. 62–78. In an interesting overview of Irish violence in the eighteenth century as a whole, Sean Connolly concludes that apart from the dramatic exception of the 1790s, "Ireland does not appear to have been by contemporary standards a particularly violent society." See S.J. Connolly, "Violence and order in the eighteenth century," in Patrick O'Flanagan, Paul Ferguson, and Kevin Whelan, eds., *Rural Ireland: modernisation and change, 1600–1900* (Cork, 1987), pp. 42–61; quotation at p. 59.

5. For this early case of what would later be termed Whiteboyism, see the excellent study by S.J. Connolly, "The Houghers: agrarian protest in early eighteenth-century Connacht," in Philpin, *Nationalism and popular protest,* pp. 139–62. See also S.J. Connolly, *Religion, law, and power: the making of Protestant Ireland, 1660–1760* (Oxford, 1992), pp. 198–262.

6. For these later outbreaks of houghing, see W.E.H. Lecky, *A history*

of Ireland in the eighteenth century (new ed., 5 vols., London, 1892), 3:385–92; P.M. Hogan, "Civil unrest in the province of Connacht, 1793–1798: the role of the landed gentry in maintaining order" (M.Ed. thesis, University College Galway, 1976), pp. 38–72; Thomas Bartlett, "Defenders and Defenderism in 1795," *Irish Historical Studies* 24, no. 95 (May 1985), pp. 373–94; J.G. Patterson, "Republicanism, agrarianism, and banditry in the west of Ireland, 1798–1803," *Irish Historical Studies* 35, no. 137 (May 2006), pp. 17–39, esp. pp. 27–31.

7. M.R. Beames, *Peasants and power: the Whiteboy movements and their control in pre-famine Ireland* (Brighton, Sussex, and New York, 1983), pp. 42–53. This book has some strengths but also numerous weaknesses. Among its most serious weaknesses is its approach to the sources. Beames relied mainly on newspapers and parliamentary papers, with only occasional forays into the great mass of manuscript material once in the State Paper Office at Dublin Castle and now deposited in the National Archives of Ireland.

8. Jim Smyth, *Men of no property: Irish radicals and popular politics in the late eighteenth century* (Dublin, 1992), pp. 114–20; Hogan, "Civil unrest," chaps. 2–4; Thomas Bartlett, "An end to moral economy: the Irish militia disturbances of 1793," *Past and Present,* no. 99 (May 1983), pp. 41–64; Bartlett, "Defenders," pp. 373–94; Bartlett, *The fall and rise of the Irish nation: the Catholic question, 1690–1830* (Savage, Md., 1992), chaps. 10–11; N.J. Curtin, *The United Irishmen: popular politics in Ulster and Dublin, 1791–1798* (Oxford,

1994), pp. 145–73. But James Patterson has recently drawn attention to the extreme violence of Whiteboys in County Cork over the grievance of tithes in 1798–99, a phenomenon that he convincingly attributes to "the brutalisation of Irish society in the 1790s" under the impact of revolutionary republicanism and its severe repression. See J.G. Patterson, "'Educated Whiteboyism': the Cork tithe war, 1798–9," *History Ireland* 12, no. 4 (Winter 2004), pp. 25–29.

9. Though Whiteboyism and Ribbonism were often conflated by contemporary commentators in the early nineteenth century, Ribbonism, as historians generally understand it now, was a kind of political underground movement, or series of movements, with distinctive elements of organization, ideology, and geographical distribution that generally set it apart from Whiteboyism. See Tom Garvin, "Defenders, Ribbonmen, and others: underground political networks in pre-famine Ireland," in Philpin, *Nationalism and popular protest,* pp. 224–28; M.R. Beames, "The Ribbon societies: lower-class nationalism in pre-famine Ireland," ibid., pp. 245–63.

10. It was not, however, the only major kind of Irish rural collective action with a long pedigree. As Garvin has persuasively argued, a geographical mapping of Ribbonism as it revealed itself in the early nineteenth century strongly suggests that it was "the lineal descendant of the sectarian Defenderism of the years 1784–94," though subsequently, Defenderism spread to the southern half of the country and assumed strong agrarian

and political features while retaining the sectarianism of its beginnings. See Garvin, "Defenders, Ribbonmen, and others," pp. 225-27.

11. The most trenchant exposition of this view came from Joseph Lee in his stimulating 1973 essay entitled "The Ribbonmen," in Williams, *Secret societies in Ireland,* pp. 26-35.

12. For accounts of the Rightboys, see Donnelly, "Rightboy movement," pp. 120-202; idem, "Contemporary account," pp. 1-50; Bric, "Priests, parsons, and politics," pp. 163-90.

13. In his 1983 book Beames expressed some dissent from Lee, who had connected the three biggest eruptions of agrarian turmoil in the early nineteenth century to economic crises arising from depressions in agricultural prices, especially grain prices. Beames pointed out that in the case of one of these eruptions, agrarian unrest was evident before the downturn in prices manifested itself, and that certain other Whiteboy movements arose in times of buoyant prices and good harvests. See Beames, *Peasants and power,* pp. 111-39 and esp. p. 113. But what Beames did not provide was a coherent alternative scheme of explanation covering the specific movements of protest with which he dealt. An attempt to address this complex issue in a systematic fashion will be found in J.S. Donnelly, Jr., "The social composition of agrarian rebellions in early nineteenth-century Ireland: the case of the Carders and Caravats, 1813-16," in P.J. Corish, ed., *Radicals, rebels, and establishments* (*Historical Studies* 15), (Belfast, 1985), pp. 151-69.

14. See especially Peter Worsley, *The trumpet shall sound: a study of "cargo"*

cults in Melanesia (2nd ed., New York, 1968), pp. ix-lxix, 221-56; George Shepperson, "The comparative study of millenarian movements," in S.L. Thrupp, ed., *Millennial dreams in action: studies in revolutionary religious movements* (Schocken Books paperback ed., New York, 1970), pp. 44-52. See also Norman Cohn, "Medieval millenarianism: its bearing on the comparative study of millenarian movements" in Thrupp, *Millennial dreams,* p. 42; J.F.C. Harrison, *The second coming: popular millenarianism, 1780-1850* (London and New Brunswick, N.J., 1979), pp. 207-30.

15. Worsley, *Trumpet,* pp. 227-43; Cohn, "Medieval millenarianism," pp. 40-41; Cohn, *The pursuit of the millennium: revolutionary millenarians and mystical anarchists of the middle ages* (rev. ed., New York, 1970), pp. 281-86 and passim.

16. Cohn, "Medieval millenarianism," p. 41. See also Michael Adas's excellent study, *Prophets of rebellion: millenarian protest movements against the European colonial order* (Chapel Hill, N.C., 1979).

17. Patrick O'Farrell, "Millen[n]ialism, messianism, and utopianism in Irish history," in *Anglo-Irish Studies* 2 (1976), pp. 53-54.

18. J.S. Donnelly, Jr., "Propagating the cause of the United Irishmen" in *Studies: An Irish Quarterly Review* 64, no. 273 (Spring 1980), pp. 15-20.

19. See chapter 4 below, pp. 122-34.

20. O'Farrell, "Millen[n]ialism," p. 52.

21. Ibid., p. 47.

22. Lee, "Ribbonmen," p. 33.

23. O'Farrell, "Millen[n]ialism," p. 50; Lee, "Ribbonmen," p. 33.

24. O'Farrell, "Millen[n]ialism," pp. 49–50.

25. Lee, "Ribbonmen," p. 33.

26. Cohn, *Pursuit of the millennium,* passim. See also the essays by Cohn, Howard Kaminsky, and Donald Weinstein in Thrupp, *Millennial dreams,* pp. 31–43, 166–203.

27. S.L. Thrupp, "Millennial dreams in action: a report on the conference discussion," in Thrupp, *Millennial dreams,* p. 23.

28. Cohn, *Pursuit of the millennium,* pp. 198–222, 234–51, 284–85; idem, "Medieval millenarianism," pp. 37–39. But see also Worsley, *Trumpet,* pp. xxxix–xlii.

29. For a discussion of the integrative function of millennial cults, see Worsley, *Trumpet,* pp. 227–43. For treatment of the issues of social cohesion and class conflict within the Rockite movement, see chapter 5 below.

30. For a short but definitive account, see David Dickson, *Arctic Ireland: the extraordinary story of the great frost and forgotten famine of 1740–41* (Belfast, 1997).

31. Idem, "The other great Irish famine," in Cathal Póirtéir, ed., *The great Irish famine* (Cork and Dublin, 1995), pp. 50–59; quotation at p. 56.

32. Cormac Ó Gráda, *Ireland: a new economic history, 1780–1939* (Oxford and New York, 1994), pp. 16–17. See also Roger Wells, "The Irish famine of 1799–1801," in Adrian Randall and Andrew Charlesworth, eds., *Markets, market culture, and popular protest in eighteenth-century Britain and Ireland* (Liverpool, 1996), pp. 163–93.

33. David Large, "The wealth of the greater Irish landowners, 1750–1815," *Irish Historical Studies* 15, no. 57 (March 1966), pp. 21–47.

34. Kevin Whelan, "An underground gentry? Catholic middlemen in eighteenth-century Ireland," in J.S. Donnelly, Jr., and K.A. Miller, eds., *Irish popular culture, 1650–1850* (Dublin and Portland, Ore., 1998), pp. 118–72.

35. The figures in this paragraph, derived from data in the 1841 census and intended to indicate rough orders of magnitude, were kindly provided by Professor Cormac Ó Gráda of University College Dublin, to whom I am greatly indebted for permission to use them here. The above paragraph appeared originally in Donnelly, "Social composition of agrarian rebellions," p. 152.

36. For the "demographic adjustment," see Kevin O'Neill, *Family and farm in pre-famine Ireland: the parish of Killashandra* (Madison, Wis., and London, 1984), pp. 164–86. See also David Dickson, Cormac Ó Gráda, and Stuart Daultrey, "Hearth tax, household size, and Irish population change, 1672–1821," *Proceedings of the Royal Irish Academy* 82, sec. C, no. 6 (Dec. 1982), pp. 125–81.

37. Ó Gráda, *Ireland,* p. 94.

38. M.R. Beames, "Rural conflict in pre-famine Ireland: peasant assassinations in Tipperary, 1837–1847," *Past & Present,* no. 81 (Nov. 1978), pp. 75–91. A decade later, this article was reprinted in Philpin, *Nationalism and popular protest,* pp. 264–83.

39. Beames, *Peasants and power,* pp. 124–25.

40. Beames, "Rural conflict," pp. 88–89.

41. David Fitzpatrick, "Class, fam-

ily, and rural unrest in nineteenth-century Ireland," in P.J. Drudy, ed., *Irish studies 2: land, politics, and people* (Cambridge, 1982), pp. 37–75.

42. Ibid., p. 44 (emphasis added).

43. Ibid., p. 53.

44. Ibid., p. 47.

45. Samuel Clark, *Social origins of the Irish land war* (Princeton, 1979), pp. 66–73.

46. P.E.W. Roberts, "Caravats and Shanavests: Whiteboyism and faction fighting in east Munster, 1802–11," in Samuel Clark and J.S. Donnelly, Jr., eds., *Irish peasants: violence and political unrest, 1780–1914* (Madison, Wis., and Manchester, 1983), pp. 64–101. See also Sailbheastar Ó Muireadhaigh, "Na Carabhait agus na Sean-Bheisteanna," *Galvia* 8 (1961), pp. 4–20. For a contemporary account focused on the attempt at judicial repression, see Randall Kernan (compiler), *A report of the trials of the Caravats and Shanavests at the special commission for the several counties of Tipperary, Waterford, and Kilkenny, before the Right Hon. Lord Norbury and the Right Hon. S. O'Gready [sic], commencing at Clonmel on Monday, February 4th, 1811, taken in shorthand by Randall Kernan, esq., barrister-at-law* (Dublin, 1811).

47. For a contemporary account of the Threshers and of the judicial repression prompted by their movement, see William Ridgeway (compiler), *A report of the proceedings under a special commission of oyer and terminer and gaol delivery for the counties of Sligo, Mayo, Leitrim, Longford, and Cavan in the month of December 1806* (Dublin, 1807). For a recent scholarly analysis of the Ribbonmen of 1819–20, see David

Lenahan, "Ribbonmen of the west? The Connacht outrages of 1819–20" (M.A. thesis, University of Wisconsin–Madison, 2003). For the Terry Alts, see Peter Gorman (compiler), *A report of the proceedings under a special commission of oyer and terminer in the counties of Limerick & Clare in the months of May and June 1831, including the proceedings at the adjourned commission in Ennis, taken in short-hand* (Limerick, 1831); J.S. Donnelly, Jr., "The Terry Alt movement, 1829–31," *History Ireland* 2, no. 4 (Winter 1994), pp. 30–35.

48. Roberts, "Caravats and Shanavests," pp. 64, 66, 95–98. Roberts conceded, however, that "the worsening economic condition of the rural middle class after 1813 created a more fertile soil for cooperation across class divisions" (p. 98).

49. Lee did not claim that "the main waves of Ribbon activity" (by which he really meant Whiteboy activity) represented class conflict exclusively, but he portrayed cottiers as "the backbone of Ribbon [i.e., Whiteboy] societies" during the major agrarian upheavals of the early nineteenth century. See his "Ribbonmen," pp. 27–28.

50. Donnelly, "Social composition of agrarian rebellions," pp. 151–69. The following paragraph essentially reproduces the argument put forward on pp. 154–55 of the earlier essay.

51. Roberts, "Caravats and Shanavests," pp. 81–85.

52. All the works on the Whiteboy movements of the eighteenth century cited in note 2 above deal with the Irish tithe system, often in considerable detail. On the subject of the tithe system and its peculiarities before 1800, see

also M.R. Bric, "The tithe system in eighteenth-century Ireland," *Proceedings of the Royal Irish Academy* 86, sec. C (1986), pp. 271–88. In *Peasants and power* (1983) by Michael Beames, the section dealing with grievances surrounding the tithe issue (pp. 114–18) is seriously flawed. Besides falling into numerous factual errors, Beames seems to be saying (quite incorrectly) that the emphasis other historians have placed on tithes as a cardinal Whiteboy grievance is not to be credited.

53. The salience of the tithe issue in the major agrarian rebellions of both the late eighteenth and the early nineteenth centuries, and the capacity of this grievance to cut across the traditional lines of social division, tends to undercut any explanation of social composition that would place too much stress either on social conflict *between* nonelite social groups (the interpretation advanced by Lee) or on different kinds of conflict *within* the farming and the laboring classes (the view emphasized by David Fitzpatrick).

54. See Donnelly, "Irish agrarian rebellion," pp. 299, 302, 312; idem, "Rightboy movement," pp. 149–50; idem, "Social composition of agrarian rebellions," pp. 160–65; W.A. Maguire, ed., *Letters of a great Irish landlord: a selection from the estate correspondence of the third marquess of Downshire, 1809–45* (Belfast, 1974), pp. 91–92.

55. In arguing for a relatively late set of dates for the diffusion of a potato-dominated diet in prefamine rural Ireland, Louis Cullen has pointed to the expansion of dairying in the southwest (including much, if not most, of Munster) as the critical driving force behind its widespread adoption there in the mid and late eighteenth century. See L.M. Cullen, "Irish history without the potato," in Philpin, *Nationalism and popular protest,* pp. 131–32.

56. P.M.A. Bourke, *'The visitation of God'? The potato and the great Irish famine,* ed. Jacqueline Hill and Cormac Ó Gráda (Dublin, 1993), pp. 97–100.

57. Bric, "Tithe system," passim.

58. Donnelly, "Rightboy movement," pp. 152–54.

59. For strong evidence of this kind of collaboration during the Caravat movement of 1813–16, see Donnelly, "Social composition of agrarian rebellions," pp. 160–65.

60. For a highly illuminating example of a middleman being ground between the upper and nether millstones in this way, see F.S.L. Lyons, ed., "The vicissitudes of a middleman in County Leitrim, 1810–27," *Irish Historical Studies* 9, no. 35 (March 1955), pp. 300–18. For the systematic efforts of a landed proprietor to eradicate middlemen as part of a wider scheme to reform the administration of his Cork and Kerry estates before, during, and after the Great Famine, see J.S. Donnelly, Jr., ed., "The journals of Sir John Benn-Walsh relating to the management of his Irish estates, 1823–64 [part 1]," *Journal of the Cork Historical and Archaeological Society* 80, no. 230 (July–Dec. 1974), pp. 86–123; idem, "The journals of Sir John Benn-Walsh relating to the management of his Irish estates, 1823–64 [part 2]," *Journal of the Cork Historical and Archaeological Society* 81, no. 231 (Jan.–June 1975), pp. 15–42.

61. For evidence confirming the assertions made in this paragraph, see chapter 7 below.

62. George Cornewall Lewis, *Local disturbances in Ireland* (Tower Books ed., Cork, 1977; originally published 1836).

63. Ibid., passim. But Tom Garvin has rightly pointed out that Lewis "was careful to distinguish between local combinations of the usual 'economic' or 'Whiteboy' type and Ribbonism. He was unsure of Ribbonism's exact character, but felt it was a descendant of Defenderism, a general name for the anti-Protestant Catholic gangs and networks of south Ulster and neighbouring areas in the late eighteenth century. He believed that Ribbonism was more organized than the usual Whiteboy fraternity." See Garvin, "Defenders, Ribbonmen, and others," pp. 224–25.

64. For an older example of this tendency, see K.B. Nowlan, "Agrarian unrest in Ireland, 1800–1845," *University Review* 2, no. 6 (1959), pp. 7–16. Michael Beames remained closely wedded to the thesis of localism in his depiction of Whiteboy movements in his 1983 book *Peasants and power.*

65. Thanks to the industry of one scholar, several hundred of these threatening letters and notices have been brought together in a recently published edition. See S.R. Gibbons, ed., *Captain Rock, night errant: the threatening letters of pre-famine Ireland, 1801–1845* (Dublin and Portland, Ore., 2004). The largest proportion of the documents in this collection relates to the Rockites.

66. The recent literature on the 1798 rebellion, its background, and its immediate aftermath has swelled to very large proportions indeed. Besides the important works of Jim Smyth and Nancy J. Curtin already cited, see Bartlett, *Fall and rise of the Irish nation*, 228–67; idem, "End to moral economy"; idem, "Defenders"; Thomas Bartlett, David Dickson, Dáire Keogh, and Kevin Whelan, eds., *1798: a bicentenary perspective* (Dublin, 2003); J.D. Beatty, ed., *Protestant women's narratives of the Irish rebellion of 1798* (Dublin and Portland, Ore., 2001); Allan Blackstock, *An ascendancy army: the Irish yeomanry, 1796–1834* (Dublin and Portland, Ore., 1998); David Dickson, "Taxation and disaffection in late eighteenth-century Ireland," in Clark and Donnelly, *Irish peasants,* pp. 37–63; David Dickson, Dáire Keogh, and Kevin Whelan, eds., *The United Irishmen: republicanism, radicalism, and rebellion* (Dublin, 1993); J.S. Donnelly, Jr., "Republicanism and reaction in the 1790s," *Irish Economic and Social History* 11 (1984), pp. 94–100; J.S. Donnelly, Jr., "Sectarianism in 1798 and in Catholic nationalist memory," in L.M. Geary, ed., *Rebellion and remembrance in modern Ireland* (Dublin, 2001), pp. 15–37; Tom Dunne, *Rebellions: memoir, memory, and 1798* (Dublin, 2004); Marianne Elliott, *Partners in revolution: the United Irishmen and France* (New Haven, Conn., and London, 1982); Sean Farrell, *Rituals and riots: sectarian violence and political culture in Ulster, 1784–1886* (Lexington, Ky., 2000); Hugh Gough and David Dickson, eds., *Ireland and the French Revolution* (Blackrock, Co. Dublin, 1990); I.R. McBride, *Scripture politics: Ulster Presbyterians and Irish radicalism in the late*

eighteenth century (Oxford, 1998); D.W. Miller, "The Armagh troubles, 1784–95," in Clark and Donnelly, *Irish peasants,* pp. 155–91; D.W. Miller, ed., *Peep O'Day Boys and Defenders: selected documents on the County Armagh disturbances, 1784–96* (Belfast, 1990); Kevin Whelan, *The tree of liberty: radicalism, Catholicism, and the construction of Irish identity, 1760–1830* (Cork, 1996); Kevin Whelan, *Fellowship of freedom: the United Irishmen and 1998* (Cork, 1998).

67. Garvin, "Defenders, Ribbonmen, and others," pp. 219–44; Beames, "Ribbon societies," pp. 245–63.

68. Garvin struggled to encapsulate its ideology amid its variations and the scarcity of reliable information: "Given these difficulties of evidence, it is still possible to say that Ribbonism was nationalist, Catholic communalist if not sectarian, and vaguely radical in a populist mode, with much millennial admixture" ("Defenders, Ribbonmen, and others," p. 242). Beames was less reluctant to place Ribbonism squarely in the tradition of revolutionary republicanism. With good reason he declared that "the Ribbonmen of the 1820s expected to participate in an imminent nationalist rising." More generally, he commented that "the nationalist sentiment of the Ribbonmen was nourished not only by the memory of the revolutionary republicanism of 1798 but also by their position as Roman Catholics under a Protestant ascendancy" (Beames, "Ribbon societies," pp. 254, 256).

69. Garvin, "Defenders, Ribbonmen, and others," pp. 224–25.

70. Beames, "Ribbon societies," p. 246.

71. Garvin, "Defenders, Ribbonmen, and others," p. 225.

72. Beames, "Ribbon societies," p. 246.

73. Ibid., pp. 246–47.

74. Ibid., p. 248.

75. Supporting evidence for the arguments in this paragraph will be found in chapter 3 below.

76. For a striking reference to the widespread popular panic caused by the epidemic, see L.M. Geary, "John Milner Barry and public health in early nineteenth-century Cork," *Journal of the Cork Historical and Archaeological Society* 106 (2001), pp. 131–42.

77. For the first recorded reference to a prophecy from "Pastorini" predicting, in this instance, the wholesale destruction of Protestants in 1817, see Information of Patrick McKeen, 21 Aug. 1817 (National Archives of Ireland, Dublin, State of the Country Papers, Series 1 [hereafter cited as NAI SOCP 1], 1830/13). I owe this reference to the kindness of P.E.W. Roberts. See also chapter 4 below, pp. 124–25.

78. Gibbons, *Captain Rock,* pp. 125–26 (no. 178).

79. Ibid., pp. 116–17 (no. 160).

80. Ibid., pp. 119–20 (no. 166).

81. Ibid., p. 120 (no. 167).

82. These price data were kindly supplied to me by Professor Liam Kennedy of the Queen's University of Belfast. They have now been published in Liam Kennedy and P.M. Solar, *Irish agriculture: a price history from the mid-eighteenth century to the eve of the First World War* (Dublin, 2007), appendix tables for oats, barley, wheat, butter, and beef.

83. *Dublin Evening Post* (hereafter cited as *DEP*), 7, 10, 12 Sept., 8, 10 Oct. 1816.

84. *DEP*, 14 Nov. 1816.

85. *DEP*, 23 Aug. 1817.

86. Kennedy and Solar, *Irish agriculture,* appendix tables for oats, barley, wheat, potatoes, beef, butter, and pig meat in these years. See also *DEP,* 9 Sept., 30 Oct. 1817.

87. *DEP*, 11, 15, 20, 25, 27 Aug., 3 Sept. 1818. The dairying and grazing districts of England also suffered from serious drought, compounding the effect on prices. See *DEP*, 5 Sept. 1818.

88. *DEP*, 10 Oct. 1818. See also *DEP*, 5, 19 Sept., 3, 20 Oct. 1818. Because of the drought, the hay crop was extremely short, and the yields of oats and barley also suffered. See *DEP*, 16, 18, 30 July 1818.

89. See F. Barker and J. Cheyne, *An account of the rise, progress, and decline of the fever lately epidemical in Ireland, together with communications from physicians in the provinces and various official documents,* 2 vols. (Dublin, 1821), 1:62, 94. Laurence Geary has suggested that the figure given by Barker and Cheyne for total mortality in this epidemic may well have been exaggerated. For a full discussion of fever epidemics and fever hospitals in the eighteenth and early nineteenth centuries, see L.M. Geary, *Medicine and charity in Ireland, 1718–1851* (Dublin, 2004), chapter 4.

90. Kennedy and Solar, *Irish agriculture,* appendix tables for oats, barley, wheat, butter, beef, and pig meat in these years.

Chapter One.
Origins of the Movement

1. The acreage of the Courtenay estates given here is that recorded in a rare surviving rental for the year 1827–28. See "Rental of the Irish estates of the Right honorable William Lord Viscount Courtenay situate[d] in the baronies of Upper and Lower Connelloe and county of Limerick for one year from Lady Day 1827 to Lady Day 1828, containing Mr. Furlong's sixth general account with Lord Viscount Courtenay for that period, being [the] second rental stated in imperial measurement and currency," p. 36 (in the possession of Robert Cussen and Son, Solicitors, Bridge Street, Newcastle West, Co. Limerick). I am deeply grateful to Mr. John Cussen of this firm for allowing me to consult this valuable document (hereafter cited as Courtenay rental for 1827–28) and for many other kindnesses over the long period that this project has been in progress.

2. Anon., *Old Bailey solicitor* (n.p., [1822?]), p. 6. The only known surviving copy of this pamphlet is that in the possession of Mr. John Cussen, a solicitor with a deep interest in the history of Newcastle West and its vicinity. His copy lacks the title page and several other pages as well. For many years Mr. Cussen has tried to locate another copy, but so far without success. I am very much indebted to him for generously placing his copy at my disposal. Only someone with a thorough knowledge of the recent administration of the Courtenay estates and with the law of landlord and tenant could have written the pamphlet. The best guess as to the identity of the au-

thor is that of the late Michael Dore, a learned local historian, who informed John Cussen in July 1986: "My guess is that this pamphlet was written by Richard Creagh-Odell, who was an attorney in Newcastle during the Capt. Rock period and was involved with the defence in almost all the 'Rock' cases, invariably instructing Daniel O'Connell as counsel. The work reflects a technical knowledge of the law and a confidence in the use of terms which could only come from a practitioner." See Michael Dore to John Cussen, 22 July 1986 (in the possession of John Cussen). For decades after its publication the pamphlet was apparently well known locally. See Maurice Lenihan, *Limerick: its history and antiquities* (Dublin, 1866), p. 450.

3. *Old Bailey solicitor*, p. 60.

4. William Gregory to Charles Grant, [?] Nov. 1820 (NAI, SOCP 1, 2185/51). See also Richard Going to Grant, 19 Nov. 1820 (ibid.); *DEP*, 25 Oct. 1821.

5. T.J. Westropp, *The Desmond Castle, Newcastle West, Co. Limerick* (reprint ed., Newcastle West, 1983, originally published 1910), pp. 352–53.

6. For William Beckford the Elder and his son, see *The [concise] dictionary of national biography* (London, 1939), pp. 80–81. See also the entry for the younger Beckford at http://www .heureka.clara.net/art/beckford.htm (accessed 6 Jan. 2007); Rictor Norton, "A visit to Fonthill," in "Gay history and literature: essays by Rictor Norton," at http://www.infopt.demon.co .uk/beckfor3.htm (accessed 6 Jan. 2007); Monica Kanellis, "The Gothic Labyrinth," at http://www.omni.sytes .net/~monica/mysterious.htm (accessed 6 Jan. 2007).

7. Antipathy toward Courtenay on account of his homosexuality was strong among members of the landed elite in the south of Ireland. The prominent judge Robert Day made a scathing entry in his diary for 9 September 1811: "Visit before breakfast the Castle of the disgusting Sodomite Lord Courtenay, a very respectable mansion [at Newcastle West]. Locke, his agent, is dismissed by Lord Courtenay (who is as respectable as his cousin [Courtenay] is to the contrary), and leaves Newcastle tomorrow to make room for Mr. Carte, his successor." See Gerald O'Carroll, ed., *Robert Day (1746–1841): the diaries and the addresses to grand juries, 1793–1829* (Tralee, 2004), p. 207. For important aspects of the personal life of the third Viscount Courtenay, see the official Web site for Powderham Castle, the English seat of the family, at http://www.powderham.co.uk/history -conservation-Powderham.htm (consulted 4 Jan. 2007). Late in life the childless third viscount scored a notable success for his family when in 1831 the House of Lords granted his petition for the restoration to the Courtenays of the title of Earl of Devon. On his death in 1835 the earldom passed to his cousin, another William Courtenay. According to the Powderham Web site, the third viscount (and ninth earl) was "dearly loved by his tenants, who insisted that he be buried in stately fashion back at Powderham." This claim may distort the historical reality.

8. *Ennis Chronicle and Clare Advertiser*, 19 Jan. 1811.

9. For sales of portions of his estates, see *Leinster Journal,* 18–22 Jan. 1806; Printed advertisement for a public auction on 1 Dec. 1817 of "sundry valuable freehold estates" owned by Viscount Courtenay (in the possession of John Cussen). Lord Courtenay's estates were placed under the management of trustees as early as 1803 in response to pressure from creditors. Sales of large portions of his Irish estates between 1803 and 1808 enabled the trustees to discharge all but £30,000 of his debts, and additional sales amounting to almost 4,700 acres in the latter year probably cleared off the remaining indebtedness. But Lord Courtenay's spendthrift ways persisted, making yet further sales like those of 1817 necessary. See Printed advertisement for auction on 23 July 1808: "Sketch of the title of the Right Honorable William, Lord Viscount Courtenay, to lands in the manor of Meane and Bewly in the county of Limerick. Advertized to be sold by auction at the Royal Exchange Coffee-House on Saturday, the 23[r]d of July, 1808. [Issued by] Furlong and Chambers, solicitors for Lord Courtenay and his trustees, 60, Aungier Street, Dublin. Printed by William Porter, Grafton Street [Dublin]." Three pages. In the possession of Robert Cussen and Son, Solicitors, Bridge Street, Newcastle West, Co. Limerick.

10. Richard Gregory to Charles Grant, 19 Nov. 1820 (NAI, SOCP 1, 2185/51).

11. O'Carroll, *Robert Day,* p. 207.

12. *Limerick Evening Post* (hereafter cited as *LEP*), 16 Jan. 1813.

13. See Michael Dore, "The murder of Thomas Hoskins," in *Annual Ob-*

server (June 1983), p. 40, n. 1. The *Annual Observer* was a publication of the Newcastle West Historical Society.

14. John McAuliffe, "Richard Griffith and road-making," in *Journal of the Newcastle West Historical Society,* no. 1 (1990), pp. 20–22. See also P.J. O'Connor, "Ireland's last best west: evidence from the Courtenay/Devon estate," ibid., pp. 23–29.

15. *Old Bailey solicitor,* p. 7. See also ibid., p. 61.

16. Ibid., pp. 11–12. According to a newspaper report of September 1814, the rents on lands let before 1813 were to be lowered by 10 percent, and still larger abatements of 25 percent were ordered on lands let in 1813 or later (*DEP,* 29 Sept. 1814). An optimistic observer reported to Dublin Castle in December 1814 that the recent abatements had anticipated tenant complaints and produced widespread tranquility (NAI, SOCP 1, 1814/1560/76).

17. Deposition of Alexander Hoskins, 19 Jan. 1820 (NAI, SOCP 1, 2185/51). It appears that the selection of the Old Bailey and Lincoln's Inn solicitor Hoskins as Carte's successor was partly attributable to the influence of Viscount Courtenay's kinsman William Courtenay, another prominent London solicitor attached to Lincoln's Inn, whose name was often linked with that of Edward Carte in the administration of the property (*LEP,* 17 Jan. 1816). It is likely that William Courtenay was one of the estate trustees. No clear evidence survives that Edward Carte engaged in any abuses of power while serving as agent, though he and certain relatives were or became lease-

holders on Lord Courtenay's estate. As David Dickson has observed, "the most frequent abuse" that Irish agents committed was to arrange "the leasing of a multiplicity of farms to themselves, their families, or their nominees" at less than the full rental value of the land. Alexander Hoskins's hostile actions as the new agent against members of the Carte family do raise the possibility that Edward Carte was sacked for something more than lax management. See David Dickson, *Old world colony: Cork and south Munster, 1630–1830* (Madison, Wis., and Cork, 2005), p. 327.

18. Hoskins to Lord Manners, 19 Jan. 1820 (NAI, SOCP 1, 2185/51); Dore, "Murder of Thomas Hoskins," p. 40, n. 2.

19. *Old Bailey solicitor,* passim.

20. Richard Going to Charles Grant, 19 Nov. 1820 (NAI, SOCP 1, 2185/51).

21. *Old Bailey solicitor,* pp. 16–17.

22. Deposition of Alexander Hoskins, 19 Jan. 1820 (NAI, SOCP 1, 2185/51).

23. Col. G. Turberville to Lord Aylmer, 8 Jan. 1820 (ibid.).

24. For economic conditions, see below, pp. 52–58.

25. For the canting of lands out of lease on the Courtenay estate in 1816, see *LEP,* 17 Jan., 16 Dec. 1816.

26. *Old Bailey solicitor,* pp. 16–17.

27. Ibid., p. 61.

28. Ibid., p. 7. See also ibid., p. 14.

29. Ibid., pp. 14–15.

30. Ibid., p. 14.

31. Ibid., p. 15.

32. Ibid., pp. 15–16.

33. Ibid., p. 55.

34. Deposition of Alexander Hoskins, 19 Jan. 1820 (NAI, SOCP 1, 2185/51); Deposition of Rev. Henry Bateman, 19 Jan. 1820 (ibid.).

35. Information of Alexander Hoskins, 16 Dec. 1819 (ibid.); *Old Bailey solicitor,* p. 47.

36. Deposition of Alexander Hoskins, 19 Jan. 1820 (NAI, SOCP 1, 2185/51).

37. *Old Bailey solicitor,* pp. 45–46.

38. Ibid., p. 49.

39. Ibid., pp. 83–84. Two other explanations for the choice of the name "Captain Rock" later enjoyed some currency, though mostly among the Rockites' adversaries in the Protestant upper and middle classes. Many Protestants and some well-to-do Catholics believed or suspected that the name had sectarian or revolutionary implications. One notion was that the name betrayed an ambition on the part of lower-class Catholics to overthrow Protestantism and install Catholicism as the official religion through the agency of the Rockite movement. In support of this conjecture the biblical pronouncement, "Thou art Peter, and upon this rock I will build my church" (Matthew 16:18), was cited. The second notion was that the letters of the name "Rock" were a secret code for the initials of R[oger] O'C[onnor], K[ing of Ireland], who claimed in a work published in 1822 that he was "head of his race" and "chief of the prostrated people of this nation." Since Roger O'Connor (1762–1834) had been a United Irish leader in County Cork in the 1790s, and was the brother of the prominent United Irish national leader Arthur O'Connor (1763–1852), this notion obviously suggested that the

Rockites had revolutionary political ambitions and drew inspiration from the rebels of 1798. Thomas Moore mentioned these origin stories in his phenomenally successful 1824 novel *The memoirs of Captain Rock*. See the discussion in Luke Gibbons, "Between Captain Rock and a hard place: art and agrarian insurgency," in Tadhg Foley and Seán Ryder, eds., *Ideology and Ireland in the nineteenth century* (Dublin and Portland, Ore., 1998), pp. 26–27. See also *Oxford dictionary of national biography*, under Roger O'Connor.

40. *Old Bailey solicitor*, p. 84. In another source the surname given to the schoolmaster in question was Mangan. See below, chapter 9, p. 317.

41. For William Brown's case, see ibid., pp. 70–72.

42. Deposition of Alexander Hoskins, 19 Jan. 1820 (NAI, SOCP 1, 2185/51).

43. For Hoskins's relationship to factions, see *Old Bailey solicitor*, pp. 20–21.

44. Ibid., p. 61.

45. Ibid., p. 62.

46. Robert Carte, Sr., Robert Carte, Jr., and Robert Parker to Lord Manners, 23 Dec. 1819 (NAI, SOCP 1, 2185/51).

47. Rev. William Ashe to [?], 18 Nov. 1820 (ibid.).

48. *Old Bailey solicitor*, p. 5.

49. Richard Going to Charles Grant, 19 Nov. 1820 (NAI, SOCP 1, 2185/51).

50. "Precis of letters relating to Alexander Hoskins's representation of the disorderly state of the neighbourhood of Newcastle, etc." (ibid.).

51. George Massey to William Gre-gory, 25 June 1821 (NAI, SOCP 1, 2296/9); *DEP*, 25 Sept. 1821.

52. Deposition of Alexander Hoskins, 19 Jan. 1820 (NAI, SOCP 1, 2185/51); *DEP*, 29 Sept. 1821.

53. *Old Bailey solicitor*, p. 56.

54. Ibid., pp. 84–85.

55. Ibid., p. 85.

56. Ibid., p. 83.

57. Ibid., pp. 56–57.

58. Ibid., p. 57.

59. Maj. Gen. Sir John Elley to [?], 3 July 1821 (NAI, SOCP 1, 2296/10).

60. *DEP*, 6 Aug. 1822.

61. *Old Bailey solicitor*, pp. 103–4.

62. *DEP*, 20 Sept. 1821.

63. *Old Bailey solicitor*, p. 59.

64. Ibid.

65. Ibid., pp. 59–60.

66. Memorial of County Limerick magistrates to Lt. Col. Turner, 20 Aug. 1821 (NAI, SOCP 1, 2296/[?]).

67. *Old Bailey solicitor*, p. 70. At the end Hoskins's sworn enemies included "the very persons who boasted of having received liberal rewards from him and who were his most confidential vassals" (ibid.).

68. Ibid., pp. 77–78.

69. Ibid., p. 77.

70. *DEP*, 27 Oct. 1821.

71. *DEP*, 6 Nov. 1821.

72. *DEP*, 13, 25 Oct. 1821. The Furlongs had enjoyed a lengthy association with the third Viscount Courtenay. The Dublin solicitors' firm of Furlong and Chambers had represented Courtenay and his trustees at the time of the land sales of 1808 that helped temporarily to discharge his heavy debts. See Printed advertisement for auction on 23 July 1808, p. 1 (in the possession of Robert Cussen and Son, Solicitors,

Bridge Street, Newcastle West, Co. Limerick).

73. *Old Bailey solicitor,* p. 85.

74. Ibid., p. 86.

75. Ibid.

76. Ibid., p. 78.

77. Ibid., p. 87.

78. *DEP,* 23 Oct. 1821.

79. *Old Bailey solicitor,* p. 88.

80. *DEP,* 27 Oct., 1 Nov. 1821.

81. *DEP,* 25 Oct. 1821.

82. W.H. Wyndham-Quin to Charles Grant, 23 Oct. 1821 (NAI, SOCP 1, 2296/25).

83. *DEP,* 4 Dec. 1821.

84. *DEP,* 1, 15 Dec. 1821.

85. *DEP,* 20, 29 Sept. 1821.

86. One of Hoskins's many disputes with individual tenants led to a civil suit at the assizes in 1821 in which O'Connell, acting for the tenant, turned Hoskins into "the laughing stock of the court." During the cross-examination by O'Connell, Hoskins "became so much embarrassed, from a conviction of his guilt and deceitful equivocations . . . , that the sudden operation of nature became so predominant that he was compelled to rise suddenly from his chair and make a quick advance to a water-closet, but not until his breeches betrayed an outward and visible sign of inward _____" (*Old Bailey solicitor,* p. 69).

87. *DEP,* 25 Sept. 1821.

88. *DEP,* 29 Sept. 1821.

89. *DEP,* 10 Sept. 1822.

90. *DEP,* 20 Oct., 6, 20 Dec. 1821.

91. For the case of the Sparlings, see *Old Bailey solicitor,* pp. 80–81. The extended Sparling family suffered multiple attacks from the Rockites. George Sparling, Christopher's brother, was robbed of his arms and fatally beaten near Adare early in December 1821, and George's son-in-law Frederick Petit was waylaid and murdered at midday between Adare and Rathkeale in February 1822. See Memorial of Christopher Sparling to Lord Bathurst, 28 Jan. 1826, Colonial Office Papers, Public Record Office, Kew (hereafter cited as PRO, CO), 384/14, pp. 558/1107–8. See also Sparling's memorial to Marquis Wellesley, 27 April 1827, and Sparling's memorial to Robert Wilmot Horton, 11 May 1827 (PRO, CO 384/12). The memorialist was the eldest son of the murdered George Sparling. I owe these references to the kindness of Professor Kerby Miller.

92. *DEP,* 16, 18 Oct. 1821; Maj. George Warburton to William Gregory, 16 Oct. 1821 (NAI, SOCP 1, 2296/21); Galen Broeker, *Rural disorder and police reform in Ireland, 1812–36* (London and Toronto, 1970), pp. 122–23.

93. *Old Bailey solicitor,* p. 80.

94. *DEP,* 20 Oct. 1821.

95. Report of Richard Willcocks and George Warburton to Charles Grant, 23 Oct. 1821 (NAI, SOCP 1, 2296/24).

96. *DEP,* 23 Aug., 23 Dec. 1821.

97. *DEP,* 8 Nov. 1821.

98. *Old Bailey solicitor,* p. 80.

99. *DEP,* 20 Oct. 1821.

100. Knight of Glin to Charles Grant, 25 Sept. 1821 (NAI, SOCP 1, 2296/14); Allen McCarthy to [?], 29 Oct. 1821 (NAI, SOCP 1, 2296/27); *DEP,* 29 Sept., 20, 23, 27, 30 Oct., 3 Nov. 1821.

101. *DEP,* 13 Oct. 1821. See also *DEP,* 27 Oct., 3, 17 Nov. 1821.

102. *Old Bailey solicitor,* p. 82.

103. Ibid., p. 87.

104. Report of Richard Willcocks and George Warburton to Charles Grant, 23 Oct. 1821 (NAI, SOCP 1, 2296/24).

105. For the circulation of notions about breaking up gentry estates, see *Old Bailey solicitor*, pp. 83, 101.

106. Dublin Corn Market prices, 1785–1839 (National Library of Ireland, Dublin [hereafter cited as NLI], MS 4168).

107. *DEP*, 15 April 1819.

108. *DEP*, 16 Dec. 1820.

109. *Report from the select committee on agriculture, with the minutes of evidence taken before them, and an appendix and index*, app. no. 6, p. 631, H.C. 1833 (612), v, 1.

110. Hely Dutton, *A statistical and agricultural survey of the county of Galway, with observations on the means of improvement, drawn up for the consideration and by the direction of the Royal Dublin Society* (Dublin, 1824), p. 123.

111. William Greig, *General report on the Gosford estates in County Armagh, 1821*, with introduction by F.M.L. Thompson and D. Tierney (Belfast, 1976), pp. 113–20; *The census of Ireland for the year 1851*, pt. v: *tables of deaths*, vol. i, *containing the report, tables of pestilences, and analysis of the tables of deaths* (hereafter cited as *Census Ire., 1851*), pp. 343, 362 [2087-I], H.C. 1856, xxix, 261.

112. *Report from the select committee on agriculture*, p. 631.

113. *DEP*, 12, 19 Oct. 1820; Dutton, *Galway*, p. 124.

114. *DEP*, 17 May, 27 Sept., 9 Oct. 1821.

115. *DEP*, 6 Sept. 1821; *Report from the select committee on agriculture*, p. 631.

116. *DEP*, 9 Oct. 1821.

117. *DEP*, 28 Feb. 1822. See also *DEP*, 23 Feb., 29 Aug., 7, 12 Sept. 1822.

118. *DEP*, 3 Oct. 1822.

119. *DEP*, 12 Oct. 1822.

120. *DEP*, 10, 17 Oct. 1822.

121. *DEP*, 29 Oct. 1822. See also *DEP*, 17, 26 Oct., 2, 16 Nov., 10 Dec. 1822; Dutton, *Galway*, p. 124.

122. *DEP*, 12 Dec. 1822.

123. *DEP*, 7 Sept. 1822.

124. *DEP*, 8 April 1823. See also *DEP*, 15, 27 May, 3, 12, 28 June, 12 July 1823.

125. *DEP*, 19 July 1823. See also *DEP*, 7, 21, 28 Aug., 23 Sept., 7 Oct. 1823.

126. *DEP*, 14, 18 Oct. 1823.

127. *DEP*, 28 Oct. 1823.

128. *DEP*, 13 Nov., 2, 4, 9, 16 Dec. 1823. Despair rapidly gave way to exultation in 1824 over the brisk business done at much enhanced rates (*DEP*, 13, 27 Jan., 6, 13, 25 May, 3, 24 June, 12, 31 Aug., 7 Sept., 7, 28 Oct., 7 Dec. 1824).

129. John O'Donovan, *The economic history of livestock in Ireland* (Cork, 1940), p. 153.

130. *Census Ire., 1851*, pt. v, vol. i, p. 346. See also ibid., pp. 343–44.

131. O'Donovan, *Economic history*, p. 153; *Report from the select committee on the butter trade of Ireland*, p. 272, H.C. 1826 (406), v, 135.

132. O'Donovan, *Economic history*, p. 159.

133. *DEP*, 27, 29 Aug., 19 Oct., 2 Nov. 1822; *Census Ire.*, 1851, pt. v, vol. i, p. 360.

134. *DEP*, 7 July, 6, 8 Sept., 4 Oct., 20, 22, 24 Dec. 1821; *Census Ire., 1851*, pt. v, vol. i, pp. 344, 362.

135. *DEP,* 21 Feb., 2, 26 March, 13, 18, 20, 25 April, 4, 30 May, 11 June 1822; *Census Ire., 1851,* pt. v, vol. i, p. 362.

136. *Report from the select committee on agriculture,* p. 620.

137. Maj. Richard Willcocks to William Gregory, 19 Feb. 1822 (NAI, SOCP 1, 2350/71).

138. Thomas Studdert to Henry Goulburn, 25 Feb. 1822 (NAI, SOCP 1, 2350/81).

139. Ibid.; *DEP,* 18 Oct. 1823. See also *DEP,* 21 Aug., 20, 27 Sept., 2, 16 Oct. 1823.

140. *DEP,* 18 Oct. 1823.

141. *DEP,* 16, 18 Oct., 15 Nov. 1823; 27 May, 10, 17, 24 June, 1 July 1824.

Chapter Two.
Expansion and Retreat

1. Thomas W. Sandes to Daniel Mahony, 14 Aug. 1821, copy (NAI, SOCP 1, 2295/3). See also NAI, SOCP 1, 1821/2295/4.

2. Lord Headley to Charles Grant, 22 Oct. 1821 (NAI, SOCP 1, 2295/13). See also NAI, SOCP 1, 1821/2295/8; *DEP,* 25 Oct., 24 Nov., 27 Dec. 1821.

3. NAI, SOCP 1, 1821/2295/14; *DEP,* 1, 3 Nov. 1821.

4. *DEP,* 29 Dec. 1821.

5. NAI, SOCP 1, 1821/2295/17; *DEP,* 20, 24 Nov. 1821.

6. *DEP,* 1 Dec. 1821. See also *DEP,* 18 Dec. 1821; 8, 12, 29 Jan. 1822.

7. *DEP,* 8, 10, 22 Jan. 1822.

8. *DEP,* 8, 12 Jan. 1822.

9. M.R. O'Connell, ed., *The correspondence of Daniel O'Connell, 1815–1823,* 8 vols. (New York, 1973), 2:341–42.

10. *DEP,* 10, 22 Jan. 1822; O'Connell, *Correspondence of Daniel O'Connell,* 2:347–48.

11. *DEP,* 26 Jan. 1822; O'Connell, *Correspondence of Daniel O'Connell,* 2:351–53. For the destruction of pounds elsewhere in Kerry, see *DEP,* 15 Jan., 5 Feb. 1822.

12. James O'Connell to Daniel O'Connell, 18 Jan. 1822 (O'Connell, *Correspondence of Daniel O'Connell,* 2:354–55).

13. Ibid., pp. 356–57.

14. James O'Connell to Henry Goulburn, 26 Feb. 1822 (NAI, SOCP 1, 2348/67).

15. James O'Connell to Daniel O'Connell, 17 Feb. 1822 (O'Connell, *Correspondence of Daniel O'Connell,* 2:356–57).

16. George Warburton to William Gregory, 15 Feb. 1822 (NAI, SOCP 1, 2341/16).

17. *DEP,* 1 Nov. 1821.

18. NAI, SOCP 1, 1822/2341/25.

19. *DEP,* 24 Nov. 1821.

20. NAI, SOCP 1, 1822/2341/25.

21. Notice (copy) enclosed in Ensign Connop to Lt. Col. Creagh, 25 Nov. 1821 (NAI, SOCP 1, 2296/45).

22. Notice enclosed in B. Lamb to Lt. Col. Creagh, 25 Nov. 1821 (ibid.).

23. *DEP,* 29 Dec. 1821; 5, 26 Jan. 1822; NAI, SOCP 1, 1822/2341/31–32. Early in March 1822, Commissioner John Ormsby Vandeleur's house near Newmarket-on-Fergus was raided for arms and his steward was severely wounded in the attack. But the rarity of such violence in Clare is suggested by the revealing comment that this raid was "the commencement of a system organized by emissaries from the counties of Limerick and Kerry." See A. Fitzgerald to William Gregory, 5 March 1822 (NAI, SOCP 1, 2341/22).

24. NAI, SOCP 1, 1822/2341/27.

25. This report was apparently prompted by a general destruction of the windows in the Protestant church of Tomgraney. See J. Read to William Gregory, 28 March 1822 (NAI, SOCP 1, 2341/26).

26. The Reverend Sir William Read, the rector of Tomgraney parish, was hated both as a zealous magistrate and as an exacting tithe owner. As a magistrate, he had been conspicuous in suppressing the Ribbonmen of 1819–20 in his district (NAI, SOCP 1, 1822/ 2341/12).

27. In late October 1821 police fired upon and dispersed three to four hundred rebels who had gathered near Sixmilebridge with the plan of seizing the weapons of a local yeomanry corps. In this clash some of the insurgents were wounded, at least one of them fatally (*DEP*, 1, 3, 6 Nov. 1821). In the following December the authorities apprehended five men for administering oaths in the parish of Feakle near Scarriff and an important Rockite leader near Cratloe, a man whose capture led to the murder later the same day of the suspected informer Stack (*DEP*, 3 Jan. 1822).

28. NAI, SOCP 1, 1821/2293/8, 15; 1821/2296/16; *DEP*, 6 Oct. 1821.

29. *DEP*, 13 Nov., 13 Dec. 1821.

30. NAI, SOCP 1, 1822/2342/45.

31. *DEP*, 17 Nov. 1821. When the viceroy "proclaimed" a certain geographical district, this entailed the suspension of ordinary legal processes and the introduction of martial law or other extraordinary measures of repression.

32. *DEP*, 20 Nov. 1821.

33. *DEP*, 29 Nov., 1 Dec. 1821.

34. *DEP*, 24 Nov., 1 Dec. 1821.

35. *DEP*, 15 Dec. 1821. A "stand of arms" signified the full equipment of one soldier, including a musket, bayonet, cartridge box, and belt, though the phrase commonly meant one soldier's musket and bayonet alone.

36. *DEP*, 13, 15 Nov. 1821; NAI, SOCP 1, 1822/2342/15.

37. *DEP*, 13, 22 Dec. 1821; 5, 8, 12, 17 Jan. 1822.

38. *DEP*, 13 Dec. 1821, 5 Jan. 1822; NAI, SOCP 1, 1822/2342/13, 33.

39. *DEP*, 5, 10 Jan. 1822; NAI, SOCP 1, 1822/2342/6.

40. *DEP*, 17 Jan. 1822.

41. NAI, SOCP 1, 1822/2342/53.

42. Rev. Matthew Purcell to [?], 22 Jan. 1822 (NAI, SOCP 1, 2342/59).

43. Lt. Col. H.J. Evelegh to Maj. Gen. George Wulff, 8 Jan. 1822 (NAI, SOCP 1, 2342/20).

44. NAI, SOCP 1, 1821/2296/49.

45. Lord Ennismore to Henry Goulburn, 23 Jan. 1822 (NAI, SOCP 1, 2342/62).

46. *DEP*, 1 Jan. 1822.

47. NAI, SOCP 1, 1822/2342/11, 14, 26, 33; *DEP*, 12, 22 Jan. 1822.

48. Lord Carbery to Henry Goulburn, 23 Jan. 1822 (NAI, SOCP 1, 2342/66).

49. J.R. Elmore to [?], 21 Jan. 1822 (NAI, SOCP 1, 2342/54). See also *DEP*, 29 Jan. 1822.

50. *DEP*, 22 Jan. 1822.

51. *DEP*, 24 Jan. 1822. See also *DEP*, 31 Jan. 1822. The last burial place of this soldier lies in the Church of Ireland graveyard at Inchigeelagh, where the unpopular local landlord James Barry arranged for a slab with the following inscription to be placed at the

spot: "Here rest the remains of John Smith, late of the 39 Reg. Aged 32 years. This stone was erected to his memory by Major Logan's Comp[any], 23rd Batt[alion], Rifle Brigade, in testimony of the high esteem they hold the 39th Reg. AD 1822." The Rockites paid less respect to Smith's remains before their final burial at Inchigeelagh. According to local tradition, his body was first put under ground at Toorenduff, only to be exhumed some time later in the dead of night and "dumped into the bog" at Gortnaloughra, "at a place that became known locally as Smith's Hole." See Brian Brennan, *Máire Bhuí Ní Laoire: a poet of her people* (Cork, 2000), p. 43.

52. Brevet Maj. Charles Carthew to Maj. Gen. Sir John Lambert, 21 Jan. 1822, quoted in *Papers presented by his majesty's command relative to the disturbed state of Ireland,* pp. 753–54, H.C. 1822 (2), xiv, 741. See also Earl of Bantry to Henry Goulburn, 22 Jan. 1822, quoted ibid. Two Rockites known to have been killed were Amhlaoibh (Humphrey) Lynch of neighboring Derry townland and Barry O'Leary of Garryantornora in Inchigeelagh parish. According to Brian Brennan, "local lore suggests that as many as ten other Rockites were killed." Brennan also notes that of thirty-six local Rockites "sentenced to be hanged, only a handful were actually executed." Among them was William Ring of Keamcorravooly in Inchigeelagh parish, and another may have been John Burke, one of the six sons of the celebrated local poet Máire Bhuí Ní Laoire and her husband, the large farmer James Burke (Séamas de Búrca), who held about 150 acres in the townlands of Inchi More and Inchi Beg, located southwest of Ballingeary. Two more of their sons, long on the run, were eventually apprehended and charged with the soldier John Smith's murder, but they were released after Daniel O'Connell, who had a family connection with the O'Learys, successfully defended them, thereby earning a praise poem from their famous mother. See Brennan, *Máire Bhuí Ní Laoire,* pp. 24, 37, 42–44.

53. Brennan, *Máire Bhuí Ní Laoire,* pp. 42–44. For a different translation of this famous praise poem, see Sean McMahon and Jo O'Donoghue, eds., *Taisce duan: a treasury of Irish poems with translations in English* (Swords, Co. Dublin, 1992), pp. 130–37. I am indebted to Dr. Louis de Paor for bringing this second translation to my attention and for his comparative analysis of the two translations.

54. Maj. Gen. Sir John Lambert to Lieut. Gen. Sir David Baird, 24 Jan. 1822, quoted in *Papers presented by his majesty's command,* p. 755.

55. Capt. Arthur Bernard to Lambert, 25 Jan. 1822 (NAI, SOCP 1, 2343/31).

56. Lt. Col. Sir Charles Broke to Lt. Col. Thomas Sorell, 26 Jan. 1822 (NAI, SOCP 1, 2343/31).

57. H.E. Orpen to [?], 25 Jan. 1822 (NAI, SOCP 1, 2343/6).

58. NAI, SOCP 1, 1822/2343/22.

59. A. Hill to Henry Goulburn, 26 Jan. 1822 (NAI, SOCP 1, 2343/21).

60. Thomas Pakenham, *The year of liberty: the story of the great Irish rebellion of 1798* (London, 1969), pp. 114–15.

61. Maj. J. Darcy to Lt. Col. Sir Charles Broke, 24 Jan. 1822, copy (NAI,

SOCP 1, 2343/23). See also *DEP*, 26, 29, 31 Jan. 1822. The exact purpose of the attack on the mail coach is unclear. The suggestion that the rebels hoped to find a particular individual obnoxious to them, who was supposed to be traveling in the coach, is plausible, though it cannot be confirmed definitely. See NAI, SOCP 1, 1822/2348/43.

62. NAI, SOCP 1, 1822/2343/35; *DEP*, 29, 31 Jan., 2 March 1822.

63. NAI, SOCP 1, 1822/2343/5; *DEP*, 29 Jan. 1822; *Papers . . . relative to the disturbed state of Ireland,* p. 755.

64. *DEP*, 29, 31 Jan. 1822.

65. Col. J. Mitchel to Lt. Col. Sir Charles Broke, 25 Jan. 1822, copy (NAI, SOCP 1, 2343/23). See also *DEP*, 31 Jan. 1822; *Papers . . . relative to the disturbed state of Ireland,* p. 756.

66. NAI, SOCP 1, 1822/2343/8, 9.

67. *DEP*, 31 Jan., 2 Feb. 1822.

68. *DEP*, 31 Jan. 1822.

69. Rev. John E. Orpen to Emanuel N. Orpen, 2 Feb. 1822 (NAI, SOCP 1, 2344/8).

70. *DEP*, 19 Feb. 1822.

71. *DEP*, 2 Feb. 1822.

72. Ibid.

73. *DEP*, 22, 27 Nov. 1821.

74. *DEP*, 24 Nov. 1821.

75. *DEP*, 22 Nov. 1821.

76. *DEP*, 27 Nov. 1821.

77. *DEP*, 20 Nov. 1821.

78. *DEP*, 10 Nov. 1821.

79. *DEP*, 13, 27, 29 Nov. 1821.

80. *DEP*, 10 Jan. 1822. See also *DEP*, 27 Dec. 1821, 31 Jan. 1822.

81. *DEP*, 10 Jan. 1822.

82. Rev. J. Conolly to William Gregory, 11 Feb. 1822 (NAI, SOCP 1, 2355/26). See also SOCP 1, 1822/2355/64.

83. Capt. John Goff to Col. J. Stratton, 4 March 1822, copy (NAI, SOCP 1, 2355/45).

84. Edward Wilson to William Gregory, 12 March 1822 (NAI, SOCP 1, 2355/48).

85. For the Caravat movement, see Roberts, "Caravats and Shanavests," p. 65 (map).

86. Francis Despard to [?], 25 Jan. 1824 (NAI, SOCP 1, 2606/6).

87. NAI, SOCP 1, 1821/2292/14; 1822/2355/5.

88. Richard B. Osborne to [?], 7 Jan. 1822 (NAI, SOCP 1, 2355/5).

89. Lieut. D. McLachlan to Col. Norcott, 2 Feb. 1822, copy (NAI, SOCP 1, 2369/7).

90. Earl of Carrick, Robert Langrishe, and John Flood to William Gregory, 24 March 1822 (NAI, SOCP 1, 2369/15).

91. Richard B. Osborne to [?], 7 Jan. 1822 (NAI, SOCP 1, 2355/5).

92. NAI, SOCP 1, 1822/2369/20.

93. John Flood to [?], 12 April 1822 (NAI, SOCP 1, 2369/27).

94. Lt. Col. M. Lindsay to Assistant Quartermaster-Gen. Lord Greenock, 1 April 1822, copy (NAI, SOCP 1, 2369/20).

95. According to one report in April 1824, there had been fourteen murders in Galmoy barony alone since April 1822 (NAI, SOCP 1, 1824/2606/32). Careful recent research has shown that this number was somewhat exaggerated. In his fascinating study of agrarian violence in Galmoy in the years 1819–24, Pádraig Ó Macháin has enumerated a total of fourteen murders in that barony, but five of them occurred before April 1822. As he notes, "not all

of these killings were attributable to the core issue of agrarian agitation; some were incidental, some opportunistic." See Ó Macháin, *Six years in Galmoy: rural unrest in County Kilkenny, 1819-1824* (Dublin, 2004), pp. 128-29.

96. Memorial of Magistrates Garrett Neville, John Fitzpatrick, and Dennis Delany to Lords Justices, n.d. but ca. May 1821 (NAI, SOCP 1, 2292/13). See also SOCP 1, 1824/2606/19. Contrary to the wishes of the local magistrates, the government "proclaimed" the whole barony of Galmoy in mid-May 1821, not simply the three parishes identified as disturbed. About six weeks passed before detachments of police established three barracks in the area. See Ó Macháin, *Six years in Galmoy*, pp. 53-56.

97. Lt. Col. M. Lindsay to William Gregory, [?] March 1824 (NAI, SOCP 1, 2606/15).

98. NAI, SOCP 1, 1824/2606/19-20.

99. For the murder of Marum, see NAI, SOCP 1, 1824/2606/15-16, 19-21, 35. See also Ó Macháin, *Six years in Galmoy*, pp. 99-122 and passim. What led to the killing of Marum and what followed from it stand at the center of this fine study.

100. Lt. Col. M. Lindsay to William Gregory, [?] March 1824 (NAI, SOCP 1, 2606/15).

101. NAI, SOCP 1, 1822/2369/57.

102. NAI, SOCP 1, 1822/2369/11.

103. Rev. Peter Roe to William Gregory, 22 March 1823 (NAI, SOCP 1, 2606/32).

104. Lt. Col. M. Lindsay to Lt. Col. J. Finch, 3 April 1824 (NAI, SOCP 1, 2606/32).

105. Ibid.

106. For the Templemartin or Lyrath gang, see NAI, SOCP 1, 1823/2506/31-33; 1824/2606/7, 18, 54. See also below, chapter 5, pp. 174-77.

107. Lt. Col. M. Lindsay to William Gregory, 25 Jan. 1824 (NAI, SOCP 1, 2606/7).

108. NAI, SOCP 1, 1821/2292/12; 1823/2506/32.

109. NAI, SOCP 1, 1824/2606/7.

110. Lt. Col. M. Lindsay to William Gregory, 10 Dec. 1823 (NAI, SOCP 1, 2506/32).

111. NAI, SOCP 1, 1822/2350/45, 59; 1822/2351/21.

112. William Thornton to Lt. Col. Thomas Sorell, 12 Feb. 1822, copy (NAI, SOCP 1, 2350/61).

113. Ibid.

114. NAI, SOCP 1, 1822/2350/53, 55, 81.

115. Quoted in *DEP*, 9 March 1822.

116. Ibid.

117. Richard Willcocks to William Gregory, 21 Feb. 1822 (NAI, SOCP 1, 2350/72). See also *DEP*, 23 Feb. 1822.

118. NAI, SOCP 1, 1822/2350/42; 1822/2351/10; *DEP*, 5 Feb., 16, 30 April 1822.

119. *DEP*, 21 Feb., 2, 5, 12, 28, 30 March 1822; NAI, SOCP 1, 1822/2350/71; 1822/2351/5, 60.

120. NAI, SOCP 1, 1822/2350/87; *DEP*, 28 Feb., 14 March, 6 April 1822.

121. NAI, SOCP 1, 1822/2344/22; 1822/2350/39; *DEP*, 12 Feb. 1822.

122. *DEP*, 16 Feb. 1822; NAI, SOCP 1, 1822/2350/39.

123. NAI, SOCP 1, 1822/2350/53; *DEP*, 12 Feb. 1822.

124. NAI, SOCP 1, 1822/2350/73, 94; *DEP*, 26 Feb. 1822.

125. NAI, SOCP 1, 1822/2350/93; *DEP*, 2 March 1822.

126. NAI, SOCP 1, 1822/2350/60, 73.

127. See below, chapter 8, pp. 269–71.

128. *DEP*, Feb.–April 1822.

129. NAI, SOCP 1, 1822/2350/81.

130. See, e.g., *DEP*, 2 April 1822.

131. *DEP*, 30 March 1822.

132. Hill Clements to William Gregory, 16 March 1822 (NAI, SOCP 1, 2349/17). See also NAI, SOCP 1, 1822/2349/40.

133. M. Fitzgerald (Knight of Kerry) to Gregory, 14 April 1822 (NAI, SOCP 1, 2349/35).

134. NAI, SOCP 1, 1822/2345/65; 1822/2352/6.

135. NAI, SOCP 1, 1822/2353/1.

136. Col. William Thornton to Lt. Col. Thomas Sorell, 4 April 1822, copy (NAI, SOCP 1, 2352/6).

137. See, e.g., NAI, SOCP 1, 1822/2346/3.

138. *DEP*, April–Aug. 1822.

139. See below, chapter 9.

140. NAI, SOCP 1, 1822/2349/17, 22; 1822/2355/78.

141. T.P. O'Neill, "The state, poverty, and distress in Ireland, 1815–45" (Ph.D. thesis, University College Dublin, 1971).

142. See, e.g., NAI, SOCP 1, 1822/2345/80, 88; 1822/2346/2.

143. Maj. Samson Carter to William Gregory, 7 June 1822 (NAI, SOCP 1, 2345/78).

144. Col. William Thornton to Lt. Col. Thomas Sorell, 4 April 1822, copy (NAI, SOCP 1, 2352/6).

145. Richard Aldworth to William Gregory, 30 April 1822 (NAI, SOCP 1, 2345/69).

146. NAI, SOCP 1, 1822/2345/13, 33.

147. NAI, SOCP 1, 1822/2352/14.

148. NAI, SOCP 1, 1822/2345/7, 40; 1822/2350/83.

149. *DEP*, 9 March 1822.

150. *DEP*, 12 March 1822.

151. Rev. John E. Orpen to William Gregory, 4 March 1822 (NAI, SOCP 1, 2345/50).

152. NAI, SOCP 1, 1822/2352/14.

153. *DEP*, 9 March 1822.

154. NAI, SOCP 1, 1822/2352/28; 1822/2353/2, 35.

Chapter Three.
Ideology and Organization

1. Threatening notice, 6 Jan. 1822, copy (NAI, SOCP 1, 2342/33).

2. Threatening notice, 17 Feb. 1822 (NAI, SOCP 1, 2344/48).

3. Threatening notice, [?] March 1822 (NAI, SOCP 1, 2345/28). See also NAI, SOCP 1, 1823/2513/27.

4. Threatening notice, 15 March 1823, copy (NAI, SOCP 1, 2512/5).

5. Threatening notice, 18 Sept. 1824, copy (NAI, SOCP 1, 2616/35).

6. Threatening notice, 17 Feb. 1822 (NAI, SOCP 1, 2344/48).

7. *DEP*, 7 Feb. 1822.

8. Threatening notice, n.d. (NAI, SOCP 1, 1822/2296/45).

9. *DEP*, 27 Dec. 1821.

10. *DEP*, 12 Feb. 1822.

11. *DEP*, 16 Feb. 1822.

12. Threatening notice, 18 July 1822, copy (NAI, SOCP 1, 2346/21).

13. *DEP*, 8 May 1823.

14. Threatening notice, [?] Jan. 1822 (NAI, SOCP 1, 2344/68).

15. *DEP*, 25 March 1823.

16. *DEP,* 29 April 1823.

17. *DEP,* 2 Oct. 1823.

18. *DEP,* 2 Feb. 1822.

19. *DEP,* 4 Dec. 1823.

20. Threatening notice, 14 Feb. 1822 (NAI, SOCP 1, 2349/5).

21. *DEP,* 27 Dec. 1821.

22. *DEP,* 12 Jan. 1822.

23. Threatening notice, [?] April 1822 (NAI, SOCP 1, 2355/78).

24. NAI, SOCP 1, 1822/2349/40.

25. The writer of a notice in County Tipperary directed that from the beginning of May 1822 all farmers "must mind their own cattle and their farms, for I won't allow one herd accept [i.e., except to] gentlemen." Said this writer, "I have no call to the estated gentlemen, for the farmers has the land from them for a trifle . . ." (Threatening notice, 13 Feb. 1822, NAI, SOCP 1, 2355/27).

26. Threatening notice, 17 Feb. 1822 (NAI, SOCP 1, 2344/45).

27. Threatening notice, [?] March 1822 (NAI, SOCP 1, 2345/28).

28. Threatening notice, [?] March 1822, copy (ibid.).

29. Threatening notice, [?] March 1822 (NAI, SOCP 1, 2345/52). For other notices concerned with potato-ground and conacre rents, see NAI, SOCP 1, 1822/2342/51; 1822/2355/27; 1823/2512/13; 1823/2519/12.

30. *DEP,* 25 March 1823 (two years), 10 April 1823 (seven years).

31. Threatening notice, [?] March 1823, copy (NAI, SOCP 1, 2511/61).

32. NAI, SOCP 1, 1822/2342/33.

33. Donnelly, "Rightboy movement," pp. 180-81.

34. *DEP,* 25 March 1823.

35. *DEP,* 27 Dec. 1821.

36. *DEP,* 5 Feb. 1822; Threatening notice, 14 Feb. 1822 (NAI, SOCP 1, 2349/5).

37. *DEP,* 14 Jan. 1823.

38. *DEP,* 25 March 1823.

39. *DEP,* 15 April 1823.

40. Ibid.

41. *Leinster Journal,* 30 April 1823. This notice, posted on the Old Market House in Cork city, was prompted by a case of land canting at Killeens in the Liberties.

42. *DEP,* 22, 27 Dec. 1821, 5, 9 Feb. 1822.

43. Threatening notice, 14 March 1822 (NAI, SOCP 1, 2345/28).

44. Threatening notice, 5 April 1823 (NAI, SOCP 1, 2512/13).

45. Threatening notice, [?] March 1822 (NAI, SOCP 1, 2345/28).

46. *DEP,* 10 April 1823.

47. *DEP,* 5 Jan. 1822.

48. *DEP,* 10 Jan. 1822.

49. *DEP,* 25 March 1823.

50. Threatening notice, 15 March 1823, copy (NAI, SOCP 1, 2512/5).

51. See, e.g., *DEP,* 25 March, 10, 15 April 1823.

52. *DEP,* 22 Dec. 1821.

53. Threatening notice, [?] Jan. 1822 (NAI, SOCP 1, 2342/34).

54. *DEP,* 22 Dec. 1821.

55. See below, chapter 8, p. 414, note 33.

56. *DEP,* 27 Dec. 1821.

57. Threatening notice, 9 Jan. 1822, copy (NAI, SOCP 1, 2350/27). The murderers of Father John Mulqueen, who stopped him on the road at night, were said not to have known that he was a priest. See S.J. Connolly, *Priests and people in pre-famine Ireland, 1780–1845* (Dublin, 1982), p. 243.

58. *DEP*, 23 Sept. 1823.

59. Threatening notice, 18 Sept. 1824, copy (NAI, SOCP 1, 2616/35). This usage of the murder of the Franks family, which occurred on 9 September 1823, began almost immediately. See the notice printed in *DEP*, 14 Oct. 1823. For detailed discussion of these killings, see below, chapter 7, pp. 234-36; chapter 8, pp. 262-63.

60. Threatening notice, [?] Nov. 1821 (NAI, SOCP 1, 2296/44).

61. *DEP*, 10 Jan. 1822.

62. Threatening notice, 15 March 1823, copy (NAI, SOCP 1, 2512/5).

63. *DEP*, 14 Jan. 1823.

64. *DEP*, 25 March 1823.

65. Information of Chief Constable William Nash, 6 April 1822 (NAI, SOCP 1, 2355/74). The contents of the manuscript book were printed in *DEP*, 13 April 1822. See also NAI, SOCP 1, 1822/2355/74-75; *DEP*, 16 April 1822.

66. "The royal United Irishman's test," 26 Dec. 1822, copy (NAI, SOCP 1, 2511/41). See also NAI, SOCP 1, 1822/2511/37.

67. NAI, SOCP 1, 1823/2517/4.

68. NAI, SOCP 1, 1822/2511/2.

69. E.g., Lee, "Ribbonmen," p. 33. There is some disagreement among historians about the geographical distribution of Ribbonism, which ideologically was a form of nationalism combining popular political radicalism or republicanism with anti-Orange sectarianism. According to Tom Garvin, the heartlands of Ribbonism in the early nineteenth century were Dublin city, north Leinster, north Connacht, and south Ulster; Ribbonism was also "reported from all over north Ulster at various times during the 1815-45 period" ("Defenders, Ribbonmen, and others," pp. 224-25). Michael Beames, on the other hand, has given a somewhat different picture of its geographical distribution: "It was strongest in Dublin, the counties of the eastern seaboard, and parts of Ulster. It could also be found in various inland towns through which the Royal and Grand Canals passed, as well as in the industrial areas of Great Britain" ("Ribbon societies," p. 246). Nevertheless, both Garvin and Beames agree that the Munster counties were virtually free of organized Ribbonism throughout the early nineteenth century. Garvin stated that in his extensive survey of the surviving evidence he had "only come across one reference to Ribbonism proper in Munster—a letter from the Dublin leaders, picked up by the police in the Munster county of Tipperary in 1839" ("Defenders, Ribbonmen, and others," p. 225). Similarly, Beames almost completely dissociated Ribbonism from agrarian unrest in the southern counties: "Ribbonism was in fact strongest in those areas least affected by peasant disturbances in the pre-famine period" ("Ribbon societies," p. 246). This study of the Rockite movement, however, suggests that Ribbonism was not absent from Munster in the early 1820s and did play a role of modest significance in the agrarian rebellion of those years. It did so mainly by furnishing points of contact between radicalized urban workers or artisans and aggrieved country dwellers.

70. Thomas P. Vokes to William Gregory, 15 Nov. 1821 (NAI, SOCP 1, 2296/35).

71. Information of Charles Bastable, Jr., [?] Feb. 1822 (NAI, SOCP 1, 2344/25).

72. Rev. J. Conolly to William Gregory, 14 Feb. 1822 (NAI, SOCP 1, 2355/28).

73. NAI, SOCP 1, 1823/2511/38, 40.

74. NAI, SOCP 1, 1822/2342/23; 1822/2344/16–17; 1823/2355/63; 1823/2511/3, 61.

75. NAI, SOCP 1, 1822/2345/77.

76. NAI, SOCP 1, 1822/2349/61.

77. Information of William Connell, 21 April 1822 (NAI, SOCP 1, 2345/62).

78. Curtin, *United Irishmen,* pp. 56–57, 90–92, 245–46; Smyth, *Men of no property,* pp. 45–46, 85–88, 91–92, 96–97, 113–14.

79. "Whiteboy catechism," Jan. 1822 (NAI, SOCP 1, 2342/38).

80. Edward Newsom to Henry Goulburn, 20 Jan. 1822 (NAI, SOCP 1, 2345/19).

81. Brigade Maj. Daniel Mahony to Lt. Col. Turner, 29 Dec. 1822, copy (NAI, SOCP 1, 2511/2).

82. NAI, SOCP 1, 1823/2511/37.

83. Brigade Maj. Daniel Mahony to Lt. Col. Turner, 29 Dec. 1822, copy (NAI, SOCP 1, 2511/2).

84. See chapter 4 below.

85. W.H.W. Newenham to [?], 13 April 1823 (NAI, SOCP 1, 2512/16).

86. *DEP,* 29 Dec. 1821.

87. Francis Blackburne to William Gregory, 17 Aug. 1823 (NAI, SOCP 1, 2517/67).

88. Informer's report, enclosed in Col. R.H. Dick to Lt. Col. Sir Charles Broke, 3 Dec. 1821 (NAI, SOCP 1, 2296/50).

89. Rev. J. Conolly to William Gregory, 14 Feb. 1822 (NAI, SOCP 1, 2355/28).

90. [Anon.] to Marquis Wellesley, n.d. but 1822 (NAI, SOCP 1, 2356/77).

91. See, e.g., NAI, SOCP 1, 1823/2518/29.

92. NAI, SOCP 1, 1824/2621/1; *DEP,* 22 Jan. 1824.

93. Col. Sir Hugh Gough to Maj. J. Finch, 1 Jan. 1823, copy (NAI, SOCP 1, 2511/3).

94. NAI, SOCP 1, 1821/2296/50.

95. NAI, SOCP 1, 1824/2511/15.

96. NAI, SOCP 1, 1823/2515/19–20, 34.

97. A. Crossley to Maj. Samson Carter, 27 Sept. 1823 (NAI, SOCP 1, 2515/55). See also NAI, SOCP 1, 1823/2515/41.

98. NAI, SOCP 1, 1823/2515/35.

99. Information of William Connell, 20 April 1822 (NAI, SOCP 1, 2345/62). See also Connell's information dated 21 April 1822 (ibid.).

100. NAI, SOCP 1, 1821/2296/14; *DEP,* 29 Sept. 1821.

101. NAI, SOCP 1, 1821/2293/8,15; 1821/2296/16; *DEP,* 6 Oct. 1821.

102. *DEP,* 1, 3, 6 Nov. 1821.

103. NAI, SOCP 1, 1821/2296/40; 1822/2348/16; *DEP,* 15 Dec. 1821.

104. *DEP,* 12, 17, 29 Jan. 1822.

105. Capt. A. Warburton to Lt. Col. Mitchell, 1 Oct. 1822 (NAI, SOCP 1, 2347/6). See also NAI, SOCP 1, 1822/2347/1.

106. *DEP,* 20 Feb., 18 Sept. 1823; NAI, SOCP 1, 1823/2511/39. For similar large gatherings, see *DEP,* 25 Sept. 1823; NAI, SOCP 1, 1824/2619/36.

107. NAI, SOCP 1, 1823/2512/26; 1823/2513/76.

108. NAI, SOCP 1, 1822/2344/2, 16, 26, 33; *DEP*, 5 Feb. 1822, 17 April 1824.

109. NAI, SOCP 1, 1823/2517/25, 27, 59; *DEP*, 3 May 1823.

110. NAI, SOCP 1, 1823/2511/4; *DEP*, 11 Jan. 1823.

111. NAI, SOCP 1, 1822/2356/28.

112. NAI, SOCP 1, 1823/2514/4–5; 1823/2515/53.

113. *DEP*, 5 July 1823.

114. NAI, SOCP 1, 1823/2514/36.

115. NAI, SOCP 1, 1823/2515/55.

116. NAI, SOCP 1, 1823/2514/43; *DEP*, 19 Aug. 1823.

117. NAI, SOCP 1, 1824/2615/28.

118. *DEP*, 6 Aug. 1822.

119. *DEP*, 15 Oct. 1822.

120. NAI, SOCP 1, 1822/2350/18; 1822/2351/10.

121. *DEP*, 11 Jan. 1823.

122. *DEP*, 12 Nov. 1822.

123. NAI, SOCP 1, 1822/2350/35.

124. G. Nagle to [?], 17 Jan. 1822 (NAI, SOCP 1, 2342/45).

125. Lord Carbery to William Gregory, 13 July 1823 (NAI, SOCP 1, 2514/17).

126. For a discussion of the disguises employed by Whiteboys in the early nineteenth century, see Beames, *Peasants and power*, pp. 98–101. Beames rightly stresses the links between, on the one hand, seasonal folk customs or life-cycle ceremonies involving disguises and, on the other, the repertoire of the Whiteboys.

127. *DEP*, 20 Sept. 1821.

128. *DEP*, 25 March 1823. What the folklorist Alan Gailey has written about costume and disguise in Irish life-cycle ceremonies, with their "amazingly consistent" patterns, is highly relevant to the disguises adopted by the Rockites and other Whiteboys: "Face blackening, the wearing of long garments, usually white shirts, or of coats turned inside out, and the prevalence of coloured ribbons sewn to the costume and hanging from the head-dress, have all been noted. Masked faces and the need to preserve the anonymity of the performers [in folk drama] are perhaps best seen in the use of straw costume in almost every one of the customs described, and in some of the seasonal and life-cycle ceremonies there are traces of animal-disguise also." See Alan Gailey, *Irish folk drama* (Cork, 1969), p. 99. Since the Rockites so often visited or attacked a succession of houses on the same night, it is not without significance that all the folk customs discussed by Gailey, with the exception of wake games, had "a processional character, [with] the dwellings of a locality being visited one after the other" (ibid.).

129. *DEP*, 1 Jan. 1824.

130. *DEP*, 5 Feb. 1822.

131. NAI, SOCP 1, 1822/2511/19.

132. NAI, SOCP 1, 1822/2351/57; 1822/2345/56.

133. *DEP*, 11 April 1822.

134. A. Norcott to Col. William Thornton, 8 April 1822, copy (NAI, SOCP 1, 2352/12).

135. NAI, SOCP 1, 1822/2345/56.

136. See note 70 above. See also *DEP*, 11 April 1822.

137. *DEP*, 4, 16 Sept. 1823. See also N.Z. Davis, "Women on top," in Davis, *Society and culture in early modern France* (Stanford, Calif., 1975), p. 149; Luke Gibbons, "Identity without a centre: allegory, history, and Irish na-

tionalism," in Gibbons, *Transformations in Irish culture* (Cork, 1996), pp. 141–42.

138. Davis, "Women on top," pp. 124–51.

139. For the Lady Clares, see Lewis, *Local disturbances*, pp. 223–25; Donnelly, "Terry Alt movement," pp. 30–35. For the origins and activities of the Irish Molly Maguires, and for the connections between them and the trade-union militants of the anthracite coalfields of Pennsylvania in the 1860s and 1870s, see Kevin Kenny, *Making sense of the Molly Maguires* (New York and Oxford, 1998), pp. 13–44.

140. Rev. J. Orpen to William Gregory, 18 Oct. 1821 (NAI, SOCP 1, 2293/12).

141. Capt. Gunn to Lt. Col. Gordon, 14 Nov. 1821, copy (NAI, SOCP 1, 2296/40).

142. NAI, SOCP 1, 1823/2511/10.

143. NAI, SOCP 1, 1823/2511/55.

144. For the strong revival of faction fighting in early 1824, see *DEP*, 6, 8, 10 Jan., 19 Feb., 18, 25 March, 27 April, 20 May, 1, 15, 17 June 1824. For discussions of faction fighting, see Patrick O'Donnell, *The Irish faction fighters of the 19th century* (Dublin, 1975); J.S. Donnelly, Jr., "Factions in prefamine Ireland," in Audrey Eyler and Robert Garratt, eds., *The uses of the past: essays on Irish culture* (Newark, Del., and London, 1988), pp. 113–30; Gary Owens, "'A moral insurrection': faction fighters, public demonstrations, and the O'Connellite campaign, 1828," *Irish Historical Studies* 30, no. 120 (Nov. 1997), pp. 513–39; Hugh Dorian, *The outer edge of Ulster: a memoir of social life in nineteenth-century Donegal*, ed.

Breandán Mac Suibhne and David Dickson (Dublin, 2000), pp. 164–73.

145. Deposition of Alexander Hoskins, 19 Jan. 1820 (NAI, SOCP 1, 2185/51).

146. Lt. Col. Thomas Arbuthnot to Maj. J. Finch, 17 Jan. 1823, copy (NAI, SOCP 1, 2511/12).

147. Arbuthnot to Finch, 11 Oct. 1823, copy (NAI, SOCP 1, 2515/51).

148. NAI, SOCP 1, 1822/2356/55, 57.

149. Edward Wilson to William Gregory, 29 April 1823 (NAI, SOCP 1, 2518/24).

150. Maj. Samson Carter to Gregory, 6 Jan. 1822 (NAI, SOCP 1, 2342/7).

151. *DEP*, 16 Oct. 1823.

152. Thomas P. Vokes to Henry Goulburn, 7 Sept. 1824 (NAI, SOCP 1, 2619/24). See also NAI, SOCP 1, 1824/2619/29.

153. Information of Patrick Hayes, 7 Sept. 1824 (NAI, SOCP 1, 2619/25).

154. Matthew Barrington to William Gregory, 13 Aug. 1823 (NAI, SOCP 1, 2514/43). See also NAI, SOCP 1, 1823/2514/4, 36, 46; *DEP*, 19 Aug. 1823.

155. Maj. Gen. Sir John Lambert to Maj. J. Finch, 3 July 1823, copy (NAI, SOCP 1, 2514/5). See also *DEP*, 5 July 1823; NAI, SOCP 1, 1823/2515/53.

156. Dr. Thomas Wood to Henry Goulburn, 2 Oct. 1823 (NAI, SOCP 1, 2516/4). See also NAI, SOCP 1, 1824/2615/28.

157. *Minutes of evidence taken before the select committee of the House of Lords appointed to examine into the nature and extent of the disturbances which have prevailed in those districts of Ireland*

which are now subject to the provisions of the Insurrection Act, and to report to the House, 18 May–23 June 1824 (hereafter cited as *SOI evidence*, for *State of Ireland evidence*), p. 206, H.C. 1825 (200), vii, 501.

158. Roberts, "Caravats and Shanavests," p. 81.

159. W.E. Vaughan and A.J. Fitzpatrick, eds., *Irish historical statistics: population, 1821–1971* (Dublin, 1978), pp. 86–87.

160. NAI, SOCP 1, 1822/2355/74; 1823/2513/87.

161. *DEP*, 2 Sept. 1823.

162. *DEP*, 25 March 1824.

163. *DEP*, 27 April 1824.

164. *DEP*, 4 Dec. 1823.

165. *DEP*, 28 Aug. 1824.

166. *DEP*, 24 Dec. 1821.

167. Maj. Samson Carter to Henry Goulburn, 20 Dec. 1823 (NAI, SOCP 1, 2516/50).

168. *DEP*, 26 Aug. 1824.

169. *DEP*, 26 Aug. 1823.

170. *DEP*, 21 Aug. 1823.

Chapter Four.
Pastorini and Captain Rock

1. Wills G. Crofts to William Gregory, 26 April 1822 (NAI, SOCP 1, 2345/65).

2. Rev. John E. Orpen to Gregory, 4 March 1822 (NAI, SOCP 1, 2345/10).

3. Orpen to Gregory, 6 May 1823 (NAI, SOCP 1, 2513/7).

4. *DEP*, 2 Feb. 1822.

5. *DEP*, 11, 18, 23 April 1822.

6. See, e.g., *DEP*, 26 Jan. 1822, 21 Aug. 1823.

7. *Minutes of evidence taken before the select committee appointed to inquire into the disturbances in Ireland in the last session of parliament, 13 May–18 June 1824* (hereafter cited as *SOI evidence*, for *State of Ireland evidence*), p. 275, H.C. 1825 (20), vii, 1. For Bishop James Doyle's pastoral address of 1822, see *Report from the select committee on the state of Ireland, ordered to be printed 30 June 1825, with the four reports of minutes of evidence* (hereafter cited as *SOI evidence*, for *State of Ireland evidence*), pp. 665–72, H.C. 1825 (129), viii, 455.

8. *SOI evidence*, p. 276, H.C. 1825 (20), vii, 1.

9. *DEP*, 18 March 1824.

10. Quoted in *DEP*, 10 June 1824.

11. *DEP*, 15 June 1824.

12. *DEP*, 11 Dec. 1824.

13. *Minutes of evidence taken before the select committee of the House of Lords appointed to inquire into the state of Ireland, more particularly with reference to the circumstances which may have led to disturbances in that part of the United Kingdom, 18 February–21 March 1825* (hereafter cited as *SOI evidence*, for *State of Ireland evidence*), p. 167, H.C. 1825 (181), ix, 1.

14. Ibid., p. 168.

15. [Charles Walmesley], *The general history of the Christian church from her birth to her final triumphant state in heaven, chiefly deduced from the Apocalypse of St. John the apostle, by Sig[nor] Pastorini* (Dublin, 1790).

16. Copies of these editions are available in the National Library of Ireland, Dublin.

17. Walmesley, *General history* (6th ed., Cork, 1820), pp. 477–78.

18. [Mortimer O'Sullivan], *Captain Rock detected, or the origin and character*

of the recent disturbances, and the causes, both moral and political, of the present alarming condition of the south and west of Ireland, fully and fairly considered and exposed by a Munster farmer (London, 1824), p. 284. As noted below, the arrival of the new year in 1825 became the point of maximum Protestant anxiety, but once that point of crisis had passed, there were Rockites and others who retained the millenarian dream of retribution against Protestants. See, e.g., Threatening notice, 8 Feb. 1825 (NAI, SOCP 1, 2725/15).

19. "I have no occasion for the book [Walmesley's *General history*] I sent you; I have another copy, printed in 1810 in Dublin, being the 4th edition," the Earl of Rosse informed Lord Redesdale on 3 May 1822 (Public Record Office of Northern Ireland, Belfast [hereafter cited as PRONI], Redesdale papers, T.3030/C.34/13/3).

20. Earl of Rosse to Lord Redesdale, 19 April 1822 (ibid., T.3030/C.34/13/2).

21. O'Sullivan, *Captain Rock detected*, pp. 282–83.

22. Information of Patrick McKeen, 21 Aug. 1817 (NAI, SOCP 1, 1830/13). McKeen, it seems, was some kind of itinerant worker and (despite his name) a Protestant; Murray was apparently a Catholic and perhaps the keeper of an alehouse. Besides introducing the "prophecy book," Murray also produced a Bible, probably for the purpose of tendering an oath to McKeen. I owe this reference and others for the years 1817–20 to the kindness of Paul E.W. Roberts.

23. Information of Charles Farrell, [?] June 1819 (NAI, SOCP 1, 2079/12).

24. Earl of Rosse to Lord Redesdale, 19 April 1822 (PRONI, Redesdale papers, T.3030/C.34/13/2).

25. For the spread of the movement, see *DEP*, 26 Oct., 6, 11 Nov., 2, 11, 24 Dec. 1819; 13, 18, 22, 25 Jan., 10, 26 Feb., 2, 7, 9 March 1820.

26. *DEP*, 2 Dec. 1819; 13 Jan., 24 Feb., 2, 7, 16 March 1820.

27. *DEP*, 11 Dec. 1819.

28. *DEP*, 9 March 1820.

29. Rev. John Burke to [?], 23 Feb. 1820 (NAI, SOCP 1, 2171/74). See also NAI, SOCP 1, 1820/2173/11.

30. *SOI evidence*, pp. 279–80, 311, 323, H.C. 1825 (20) vii, 1; *SOI evidence*, p. 82, H.C. 1825 (200), vii, 501.

31. Threatening notice, 28 June 1822, enclosed in Major Samson Carter to William Gregory, 29 June 1822 (NAI, SOCP 1, 2345/86). Another notice posted in June 1824 on one of the gates at the entrance to Bowen's Court proclaimed: "Capten Rock is going to comence to kill all the Protestants the [i.e., he] will find and . . . burn them alive in their houses" (Threatening notice, 9 June 1824, NAI, SOCP 1, 2615/47).

32. Threatening notice, 29 Oct. 1822 (NAI, SOCP 1, 2347/23).

33. Col. Sir Hugh Gough to Maj. J. Finch, 11 Feb. 1823 (NAI, SOCP 1, 2511/27).

34. William Lindsay to Henry Goulburn, 23 Jan. 1822 (NAI, SOCP 1, 2345/19). Lindsay named the author of the book as "Pasterina," whom he called a Roman cardinal!

35. Rev. James P. Chute to Bishop Thomas Elrington, 13 Feb. 1822 (NAI, SOCP 1, 2348/47).

36. Memorial of the Protestant in-

habitants of Ardfert, n.d., but Feb. 1822 (NAI, SOCP 1, 2348/66).

37. *SOI evidence,* pp. 116-17, H.C. 1825 (129), viii, 1.

38. *SOI evidence,* p. 142, H.C. 1825 (20), vii, 1. See also ibid., p. 124.

39. Ibid., p. 145.

40. Ibid., p. 93; *SOI evidence,* p. 103, H.C. 1825 (200), vii, 501.

41. *SOI evidence,* p. 7, H.C. 1825 (200), vii, 501. See also ibid., p. 59.

42. F. Hacke[?] to Rev. Edward Geratty, 11 Feb. 1822 (NAI, SOCP 1, 2350/60).

43. "Indigator" to [?], 29 Jan. 1822 (NAI, SOCP 1, 2350/30).

44. *SOI evidence,* p. 26, H.C. 1825 (20), vii, 1.

45. Donnelly, "Propagating the cause of the United Irishmen," pp. 19-20.

46. Rev. J. Conolly to Henry Goulburn, [?] Jan. 1824 (NAI, SOCP 1, 2621/2).

47. *SOI evidence,* p. 101, H.C. 1825 (129), viii, 1.

48. "Indigator" to [?], 29 Jan. 1822 (NAI, SOCP 1, 2350/30).

49. *SOI evidence,* p. 311, H.C. 1825 (20), vii, 1.

50. Ibid., p. 324.

51. Ibid., p. 379.

52. Threatening notice, [?] May 1823, enclosed in Rev. John Townsend to William Gregory, 27 May 1823 (NAI, SOCP 1, 2513/43).

53. Lord Carbery to Gregory, 24 June 1823 (NAI, SOCP 1, 2513/78). See also NAI, SOCP 1, 1823/2513/87.

54. *SOI evidence,* p. 298, H.C. 1825 (20), vii, 1.

55. Ibid., pp. 143-84. See also *SOI evidence,* pp. 82-84, H.C. 1825 (200), vii, 501.

56. *SOI evidence,* p. 144, H.C. 1825 (20), vii, 1.

57. Ibid., pp. 23, 325; *SOI evidence,* p. 7, H.C. 1825 (200), vii, 501.

58. Rev. J. Conolly to Henry Goulburn, [?] Jan. 1824 (NAI, SOCP 1, 2621/2). See also *SOI evidence,* pp. 141-42, H.C. 1825 (20), vii, 1.

59. *SOI evidence,* p. 84, H.C. 1825 (200), vii, 501.

60. *The treble almanack for the year 1824* (Dublin, 1824), 1:195.

61. Ibid., 1:191-92.

62. O'Sullivan, *Captain Rock detected,* p. 283.

63. On the character and conduct of Catholic schoolmasters, see William Carleton, *Traits and stories of the Irish peasantry* (London, n.d.), pp. 240-41. For their significance among the Defenders and United Irishmen of the 1790s, see Smyth, *Men of no property,* pp. 30-31, 115-16. For a memoir by a sometime County Donegal schoolmaster whose account illuminates many aspects of that occupation in the early and middle decades of the nineteenth century, see Dorian, *Outer edge of Ulster.* Some of the foremost "local celebrities" discussed by Dorian were experts in the prophecies of Pastorini or Colum Cille (ibid., pp. 127-33).

64. See below, chapter 5, pp. 179-80.

65. Carleton, *Traits and stories,* pp. 241, 262.

66. *DEP,* 2 Nov. 1824.

67. Major Samson Carter to William Gregory, 12 June 1824 (NAI, SOCP 1, 2615/42).

68. Enclosed in the letter cited in note 67 above.

69. NAI, SOCP 1, 1824/2615/42.

70. William Gregory to Charles Grant, [?] Nov. 1820 (NAI, SOCP 1, 2185/51); Maj. Richard Going to Grant, 19 Nov. 1820 (NAI, SOCP 1, 2185/51); Report of Richard Willcocks and George Warburton to Grant, 23 Oct. 1821 (NAI, SOCP 1, 2296/24); *DEP*, 25 Oct. 1821.

71. William Carleton, "The Irish prophecy man" in *Irish Penny Journal* 1, no. 50 (12 June 1841), pp. 393–94. See also Melissa Fegan, *Literature and the Irish famine, 1845–1919* (Oxford and New York, 2002), pp. 131–63.

72. Georges-Denis Zimmermann, *Songs of Irish rebellion: political street ballads and rebel songs, 1780–1900* (Dublin, 1967), pp. 22–23.

73. Ibid., pp. 28–31, 198–99, 206–7.

74. Ibid., p. 30. The evangelical Anglican preacher Reverend Peter Roe (minister of Saint Mary's church in Kilkenny city) saw a significant link between Pastorini's prophecies and some of the ballads in circulation. Roe stressed that Catholic believers in the prophecies had songs "well calculated to work upon their passions, which they sing whenever a favourable opportunity offers." See Rev. Peter Roe to Henry Goulburn, 22 Sept. 1823 (NAI, SOCP 1, 2506/21).

75. *DEP*, 20 Jan. 1824.

76. William Smith to Henry Goulburn, 19 Nov. 1824 (NAI, SOCP 1, 2619/48).

77. William Lindsay to Goulburn, 23 Jan. 1822 (NAI, SOCP 1, 2345/19).

78. Manifesto "published by Capt[ain] Storm and Steele," 5 Jan. 1822 (NAI, SOCP 1, 2350/5).

79. Rev. John E. Orpen to Emanuel N. Orpen, 2 Feb. 1822 (NAI, SOCP 1, 2344/8).

80. Jemmett Browne to William Gregory, 3 May 1823 (NAI, SOCP 1, 2153/2). See also *SOI evidence*, p. 86, H.C. 1825 (200), vii, 501.

81. John Waller to Francis Blackburne, 5 Feb. 1824 (NAI, SOCP 1, 2618/27). See also NAI, SOCP 1, 1823/2513/62; *SOI evidence*, p. 227, H.C. 1825 (200), vii, 501.

82. *DEP*, 22 Aug., 21 Sept. 1822, 1 May 1823; *SOI evidence*, pp. 776–77, H.C. 1825 (129), viii, 455; Angus Macintyre, *The liberator: Daniel O'Connell and the Irish party, 1830–1847* (London, 1965), p. 172; D.H. Akenson, *The Church of Ireland: ecclesiastical reform and revolution, 1800–1885* (New Haven, Conn., and London, 1971), pp. 105–6.

83. *DEP*, 24 Nov., 1, 11, 15 Dec. 1821, 5 Feb. 1822; NAI, SOCP 1, 1822/2342/33; 1822/2348/15.

84. Manifesto, 5 Jan. 1822 (NAI, SOCP 1, 2350/5).

85. *DEP*, 14 March 1822.

86. Threatening notice, n.d., enclosed in Edward Wilson to Henry Goulburn, 11 Dec. 1822 (NAI, SOCP 1, 2356/79).

87. For the burning of Protestant churches, see *DEP*, 1 Dec. 1821; 12 Jan., 12, 26 Feb., 2 March 1822; 20 Sept. 1823; NAI, SOCP 1, 1822/2344/22; 1822/2348/3, 5, 12; 1822/2350/39, 53–54, 60, 73, 93–94; 1823/2515/21, 27, 30.

88. *SOI evidence*, p. 28, H.C. 1825 (200), vii, 501.

89. O'Sullivan, *Captain Rock detected*, pp. 282–83.

90. *Treble almanack for the year 1824*, 1:190–91.

91. *SOI evidence,* p. 379, H.C. 1825 (20), vii, 1.

92. D.H. Akenson, *The Irish education experiment: the national system of education in the nineteenth century* (London and Toronto, 1970), pp. 80–94.

93. *SOI evidence,* pp. 326–28, 391–93, H.C. 1825 (20), vii, 1.

94. NAI, SOCP 1, 1822/2355/55; *DEP,* 28 Feb. 1822.

95. *DEP,* 14 Dec. 1824.

96. *DEP,* 13 Nov. 1824.

97. Desmond Bowen, *The Protestant crusade in Ireland, 1800–70: a study of Protestant-Catholic relations between the act of union and disestablishment* (Dublin and Montreal, 1978), pp. 98–104. See also Irene Whelan, *The bible war in Ireland: the 'Second Reformation' and the polarization of Protestant-Catholic relations, 1800–1840* (Dublin and Madison, Wis., 2005), passim.

98. *DEP,* 13 Nov. 1824.

99. *A return of the number of troops or corps of effective yeomanry . . . , so far as relates to Ireland,* pp. 3–10, H.C. 1821 (306), xix, 177.

100. William Sandes to Brigade Maj. Daniel Mahony, 27 Oct. 1821 (NAI, SOCP 1, 2296/26).

101. Rev. Ralph Stoney to Henry Goulburn, 4 March 1822 (NAI, SOCP 1, 2355/43).

102. *DEP,* 15 Dec. 1821, 22, 29 Jan. 1822; NAI, SOCP 1, 1822/2342/54, 66.

103. *DEP,* 26 Jan. 1822.

104. *DEP,* 24, 26 June 1823.

105. *SOI evidence,* p. 379, H.C. 1825 (20), vii, 1. See also NAI, SOCP 1, 1821/2293/6.

106. *DEP,* 5 Feb. 1822, 6 March, 17 May 1823.

107. NAI, SOCP 1, 1821/2296/25; *DEP,* 20 Oct. 1821.

108. NAI, SOCP 1, 1823/2517/27.

109. *SOI evidence,* p. 576, H.C. 1825 (129), viii, 455. Catholics regarded the Tarbert corps of yeomanry as a nest of Orangemen; Palatines brought from County Limerick constituted a significant element within it. Daniel O'Connell pointed to this circumstance in explaining why Rockite oaths administered in north Kerry displayed strong anti-Protestant animus. See *SOI evidence,* pp. 116–17, H.C. 1825 (129), viii, 1; *SOI evidence,* pp. 141–42, H.C. 1825 (181), ix, 1.

110. Charles D. Oliver to William Gregory, 1 May 1823 (NAI, SOCP 1, 2517/26). See also NAI, SOCP 1, 1823/2517/25, 32–33; *SOI evidence,* pp. 7–8, H.C. 1825 (20), vii, 1.

111. *DEP,* 3 May 1823.

112. *DEP,* 3 Feb. 1824; *SOI evidence,* pp. 85, 88, H.C. 1825 (200), vii, 501; *SOI evidence,* p. 576, H.C. 1825 (129), viii, 455. As noted in chapter 1, the Rockite victims in the extended Sparling family included Christopher Sparling, his brother George, and George's son-in-law Frederic Petit, all three of whom the rebels killed in late 1821 or early 1822. The Sparlings were Palatines, though Petit was an Englishman; their shared membership in one or more Protestant yeomanry corps in the Adare district helped to damn them in Rockite eyes (PRO, CO 384/14, pp. 558/1107–8). I owe this last reference to the kindness of Professor Kerby Miller.

113. *SOI evidence,* pp. 7–8, H.C. 1825 (20), vii, 1; *SOI evidence,* pp. 88, 191, H.C. 1825 (200), vii, 501. The author of an address from "the Pallen-

tines & true Protestants" of County Limerick declared in June 1823, "We are all on a look out for a ship or two to Limerick, & government will find us proper tennants" in Canada. See Memorial of Adam Shoulds to King George IV, 27 June 1823, PRO, CO 384/9, pp. 446/891–92. Christopher Sparling, whose father and uncle had been killed by Rockites, also pleaded repeatedly in the late 1820s for government assistance in sending him, his wife, and their six children to Upper Canada. He cited "the interminable hostility to such [persons] as become obnoxious to the lawless faction termed Whiteboys and Rockites." See Memorial of Christopher Sparling to Robert Wilmot Horton, 11 May 1827 (PRO, CO 384/16, p. 584/1166). See also PRO, CO 384/11, 12, 14. I owe these references to the generosity of Professor Kerby Miller.

114. *Abstract of returns relative to magistrates, constables, and sub-constables appointed under the constables' act for Ireland,* pp. 1–4, H.C. 1824 (257), xxii, 405.

115. *DEP,* 11 July 1822; 19 Aug., 2 Sept., 2 Dec. 1823; 6, 20 Jan. 1824; *SOI evidence,* p. 177, H.C. 1825 (20), vii, 1.

116. *DEP,* 19 Aug., 9 Sept., 2, 4, 6, 9, 11, 23, 30 Dec. 1823; 6 Jan., 10 April 1824.

117. *DEP,* 12 Feb. 1824. At the spring assizes of 1824 in Leinster alone, twenty-two policemen stood trial, nine of them for murder (*DEP,* 6 May 1824). Six policemen were convicted of manslaughter in the killing of a man in Queen's County and were sentenced to be transported (*DEP,* 13 May 1824).

118. *DEP,* 6 July 1824.

119. Threatening notice, n.d., but ca. Jan. 1822 (NAI, SOCP 1, 2350/5).

120. *SOI evidence,* p. 49, H.C. 1825 (200), vii, 501; *DEP,* 1, 5 June 1824.

121. *DEP,* 2 Feb. 1822; *SOI evidence,* p. 107, H.C. 1825 (20), vii, 1.

122. Report of Richard Willcocks and George Warburton to Charles Grant, 23 Oct. 1821 (NAI, SOCP 1, 2296/24).

123. *DEP,* 23 Aug., 22 Dec. 1821.

124. *SOI* evidence, pp. 104–7, H.C. 1825 (20), vii, 1.

125. *DEP,* 8 Nov. 1821. See also *SOI evidence,* pp. 49–50, H.C. 1825 (200), vii, 501.

126. *DEP,* 16, 18 Oct. 1821; NAI, SOCP 1, 1821/2296/21; Broeker, *Rural disorder,* pp. 122–23.

127. *DEP,* 20 Oct. 1821. See also NAI, SOCP 1, 1821/2296/21.

128. *DEP,* 17 May 1823.

129. *SOI evidence,* pp. 117–18, H.C. 1825 (129), viii, 1.

130. *DEP,* 20 Dec. 1823. See also *DEP,* 30 Dec. 1823.

131. NAI, SOCP 1, 1821/2296/24; *DEP,* 29 June, 10 July 1824; *SOI evidence,* p. 107, H.C. 1825 (20), vii, 1.

132. Broeker, *Rural disorder,* pp. 121, 157–58.

133. NAI, SOCP 1, 1823/2513/7.

134. Threatening notice, 19 Nov. 1821, enclosed in Ensign Connop to Lt. Col. Creagh, 30 Nov. 1821 (NAI, SOCP 1, 2296/48).

135. *DEP,* 24 June 1823.

136. NAI, SOCP 1, 1822/2343/44; 1822/2347/28, 40; 1823/2511/41, 44.

137. *DEP,* 21, 23 Feb. 1822; NAI, SOCP 1, 1822/2345/32.

138. For the raping of a servant girl whose sister had been a crown witness,

see *DEP*, 12, 14 March, 25 April 1822. See also NAI, SOCP 1, 1823/2513/62.

139. *DEP*, 11 April 1822.

140. *DEP*, 23 Feb., 11 April 1822.

141. *SOI evidence*, p. 379, H.C. 1825 (20), vii, 1.

142. See Introduction above.

143. NAI, SOCP 1, 1824/2619/55–56; *SOI evidence*, pp. 14–15, H.C. 1825 (181), ix, 1.

144. *SOI evidence*, pp. 279, 379, 440, 501, 839–40, H.C. 1825 (129), viii, 173, 293, 455; *SOI evidence*, pp. 14, 17, 19, 25, H.C. 1825 (181), ix, 1.

145. NAI, SOCP 1, 1824/2615/49, 52; *SOI evidence*, pp. 6–7, 13–16, H.C. 1825 (181), ix, 1.

146. NAI, SOCP 1, 1824/2619/51, 56; *SOI evidence*, pp. 13–16, H.C. 1825 (181), ix, 1.

147. John Palliser to William Gregory, 2 Dec. 1824 (NAI, SOCP 1, 2620/18).

148. Rev. Frederick Blood to Henry Goulburn, 21 Dec. 1824 (NAI, SOCP 1, 2613/67).

149. "A lover of the government" to Henry Goulburn, 28 Dec. 1824 (NAI, SOCP 1, 2620/20).

150. Lt. Col. M. Lindsay to Maj. J. Finch, 20 June 1823 (NAI, SOCP 1, 2506/13). For another report to the same effect, see NAI, SOCP 1, 1823/2506/19.

151. Francis Blackburne to William Gregory, 1 July 1823 (NAI, SOCP 1, 2510/46).

152. *SOI evidence*, p. 379, H.C. 1825 (129), viii, 293. Millenarian beliefs were common enough among lower-class Protestants in County Cork in the 1820s. Writing from Ballinadee near Bandon in July 1827, James Hewston told his uncle Robert Eady in Nova Scotia, "The dominion of Babylon [the Catholic church] is near about being taken away for ever; her judgements have overtaken her; her time is at hand." At the end of 1828, when news reports created an impression that the Muslim-dominated Turkish empire might become extinct, Hewston informed his uncle that this anticipated development "accords in a most remarkable manner with the predictions of Daniel and St. John that a whole people should be expecting its removal" (Hewston to Eady, 31 Dec. 1828). These letters are part of the Anstis, Eady, and Hewston letters deposited in the Cork Archives Institute. Professor Kerby Miller kindly drew my attention to this collection.

153. *SOI evidence*, p. 379, H.C. 1825 (129), viii, 293.

154. The bishop remarked that great apprehension existed even in Dublin city (ibid., p. 279). The police magistrate George Warburton told the same parliamentary committee in 1825 that Protestant fears of Catholic insurrection were prevalent around Christmas of 1824 in the districts of Ballinasloe and Gort (Galway) as well as near Boyle (Roscommon) and in parts of Sligo (ibid., pp. 839–40).

155. Ibid., p. 440. Bishop James Doyle of Kildare and Leighlin labored strenuously against Pastorini's prophecies. He not only issued two pastoral letters denouncing the prophecies (in 1822 and 1825) but also gave sermons condemning them all around his diocese. In addition, he ordered the Catholic priests of the diocese to preach against them. See ibid., p. 441, and especially Thomas McGrath, *Reli-*

gious renewal and reform in the pastoral ministry of Bishop James Doyle of Kildare and Leighlin, 1786–1834 (Dublin and Portland, Ore., 1999), pp. 142, 187–206. McGrath may somewhat exaggerate the effectiveness of Doyle's campaign.

156. *SOI evidence,* p. 501, H.C. 1825 (129), viii, 293.

157. One very curious event appears to show the sense of expectation among some Catholics in Kilfinnane parish in Limerick about the fulfillment of Pastorini's prophecies around the date of Christmas 1824. The Protestant landowner Charles D. Oliver reported to the police on 26 December that a few days earlier (on a Friday night) "several pieces of raw pork weighing from one to four pounds, perfectly fresh and good, were placed on the gravestones in the churchyard, and the sexton [of the local Protestant church] has informed me that he never saw such numbers of papists as hovered about the churchyard before, during, and after morning service on Christmas day." See Charles D. Oliver to Thomas P. Vokes, 26 Dec. 1824 (NAI, SOCP 1, 2619/59).

158. NAI, SOCP 1, 1824/2621/64. "I beg to mention one of the many reports in circulation amongst the lower classes, which is that the [Catholic] rent *is to pay a police which Mr. O'Connell is going to raise."* See Maj. George M. Drought to Henry Goulburn, 11 Nov. 1824, quoted in *SOI evidence,* pp. 13–14, H.C. 1825 (181), ix, 1.

159. *DEP,* 11 April 1822. O'Connell was also a counsel for the defense in trials arising from the murder of Thomas Hoskins, the burning of

Churchtown barracks (three policemen were killed in this incident), the abduction and rape of Miss Honora Goold, and the slaying of the Frankses—all notorious crimes connected with the Rockite movement (*DEP,* 20 Sept. 1821; 6, 8, 27 Aug. 1822; 15, 17 April, 28 Aug. 1824; NAI, SOCP 1, 1822/2354/8).

160. This ballad, probably written in 1828, was sung to the air of "St. Patrick's Day." See Douglas Hyde, *Abhráin atá leagtha ar an Reachtúire,* [or] *Songs ascribed to Raftery, being the fifth chapter of "The songs of Connacht"* (Irish University Press ed., Shannon, 1973; originally published 1903), pp. 113–23. See also Gearóid Ó Tuathaigh, *Ireland before the famine, 1798–1848* (Dublin, 1972), pp. 66–68.

Chapter Five.
Social Composition and Leadership

1. Donnelly, "Whiteboy movement, 1761–5," pp. 20–24; idem, "Whiteboys of 1769–76," pp. 293–331; idem, "Rightboy movement," pp. 120–202; idem, "Social composition of agrarian rebellions," pp. 151–69. See also Smyth, *Men of no property,* pp. 26, 114–16.

2. Roberts, "Caravats and Shanavests," pp. 64–101.

3. See above, chapter 1, pp. 52–58.

4. Rev. Peter Roe to William Gregory, 6 April 1822 (NAI, SOCP 1, 2369/24).

5. Lt. Col. M. Lindsay to Assistant Quartermaster-Gen. Lord Greenock, 1 April 1822 (NAI, SOCP 1, 2369/20).

6. John Flood to [Dublin Castle], 12 April 1822 (NAI, SOCP 1, 2369/27).

7. Col. Sir Hugh Gough to Lt. Col. J. Finch, 10 Jan. 1824 (NAI, SOCP 1, 2614/11).

8. After his house had been destroyed, McDuharty swore informations against those whom he held responsible, and arrests followed. There might have been more to this case than met the eye. The military officer who reported the incident suspected as much: "I think it proper to state that the McDuhartys and some of the men who are in custody have been at variance for years, and that the [rival] factions not very long ago encountered each other in this neighbourhood." See Capt. Edward Priestly to Finch, 17 Oct. 1824 (NAI, SOCP 1, 2613/58).

9. Thomas Gooney to William Gregory, 3 June 1824 (NAI, SOCP 1, 2606/49).

10. Col. Sir Hugh Gough to Maj. J. Finch, 22 Nov. 1823 (NAI, SOCP 1, 2516/25).

11. Maj. George Warburton to William Gregory, 15 May 1823 (NAI, SOCP 1, 2513/28). See also NAI, SOCP 1, 1822/2344/42, 61.

12. Col. Sir Hugh Gough to Lt. Col. J. French, 6 Feb. 1824 (NAI, SOCP 1, 2614/37).

13. Allen McCarthy to [?], 29 Oct. 1821 (NAI, SOCP 1, 2296/27).

14. Maj. Samson Carter to Henry Goulburn, 29 Jan. 1822 (NAI, SOCP 1, 2343/34).

15. Lord Ennismore to William Gregory, 17 Feb. 1822 (NAI, SOCP 1, 2344/44).

16. See, e.g., Robert Leslie, Jr., to Gregory, 22 March 1822 (NAI, SOCP 1, 2349/22).

17. *DEP*, 16 Feb. 1822. See also NAI, SOCP 1, 1822/2295/1.

18. Robert Leslie, Jr., to William Gregory, 1 July 1822 (NAI, SOCP 1, 2349/43).

19. M. Fitzmaurice to Gregory, 8 July 1822 (NAI, SOCP 1, 2349/47).

20. *DEP*, 6 Dec. 1821, 5 Feb., 2 March 1822.

21. NAI, SOCP 1, 1822/2349/22.

22. Hill Clements to William Gregory, 16 March 1822 (NAI, SOCP 1, 2349/17).

23. *DEP*, 2 Feb. 1822. See also *DEP*, 21 Feb. 1822; NAI, SOCP 1, 1822/2342/33.

24. *DEP*, 21 Feb. 1822.

25. *DEP*, 2 Feb. 1822.

26. Maj. Samson Carter to [Henry Goulburn], 16 Nov. 1822 (NAI, SOCP 1, 2347/34).

27. For this offense they were sentenced to be transported for seven years (*DEP*, 8, 20 Nov. 1823).

28. Information of William Nash, 6 April 1822 (NAI, SOCP 1, 2355/74).

29. Robert Curtis to William Gregory, 8 April 1822 (NAI, SOCP 1, 2355/75).

30. NAI, SOCP 1, 1822/2355/74.

31. Information of Daniel Sullivan, 1 Feb. 1822 (NAI, SOCP 1, 2348/69). Significantly, Sullivan named a socially different cast of characters as participants in the assault on the process server's house in late December 1821: a pensioner, a shoemaker, seven laborers, and himself (under compulsion).

32. *DEP*, 31 Aug. 1824. There was no corroboration for the identification made by one approver (an accomplice turned informer).

33. *DEP,* 12 June 1823. These examples are not meant to contradict the widely acknowledged fact that most farmers saw themselves as the victims of extortion in the face of the repetitive levying of contributions by Rockite laborers and cottiers.

34. *DEP,* 1 Aug. 1822. See also *DEP,* 6 Aug. 1822; NAI, SOCP 1, 1822/2345/62; 1822/2354/8.

35. NAI, SOCP 1, 1824/2618/7, 10; *DEP,* 10 Aug. 1824.

36. *DEP,* 14 Aug. 1824. See also *DEP,* 22 Jan., 12 Aug. 1824. The Flynns were the uncles of the murdered Hartnett.

37. *DEP,* 15 June 1824.

38. *DEP,* 31 July 1824. See also *DEP,* 21 Aug. 1824; NAI, SOCP 1, 1824/2621/46, 49.

39. At the time of the Limerick spring assizes in 1822, there were over two hundred prisoners in jail. At these assizes nineteen prisoners were capitally convicted, of whom fourteen were sentenced to death. The latter included a prisoner named Robert Cussen, described as "far above the rank of the preceding culprits" (*DEP,* 23 April 1822). Cussen received a reprieve (*DEP,* 25 April, 2 May 1822).

40. *DEP,* 15 April 1824.

41. *DEP,* 20 April 1824. His trial was postponed owing to the absence of a key witness, and subsequently it was apparently abandoned altogether (*DEP,* 28 Aug. 1824).

42. *DEP,* 5 Feb. 1822.

43. *DEP,* 20 March 1823. Edward Ring, Sr., was the quartermaster-general of the North Cork Militia; Rockites may have regarded his house as an arms depot, and his official status alone would have made him and his family much disliked. One informer claimed that the Rockite plan called for as many as two to three hundred men to attack Ring's house (NAI, SOCP 1, 1823/2511/58; 1823/2512/5).

44. Matthew Barrington to Henry Goulburn, 7 Aug. 1824 (NAI, SOCP 1, 1824/2619/12).

45. John Keily to William Gregory, 22 April 1823 (NAI, SOCP 1, 2517/21).

46. *DEP,* 29 April 1823.

47. NAI, SOCP 1, 1823/2517/28.

48. NAI, SOCP 1, 1824/2619/12.

49. Thomas P. Vokes to Henry Goulburn, 10 Aug. 1824 (NAI, SOCP 1, 2619/13).

50. *DEP,* 21 Aug. 1824. See also *DEP,* 17 Aug. 1824.

51. Edward Wilson to William Gregory, 29 April 1823 (NAI, SOCP 1, 2518/24).

52. Rockites burned the house of the farmer George Ellard at Ballydaniel near Doneraile in March 1824. Ellard "had ejected some tenants and had possessed himself of their lands some short time back." See Col. Sir Hugh Gough to Maj. J. Finch, 13 March 1824 (NAI, SOCP 1, 2614/63). See also NAI, SOCP 1, 1824/2614/55. Many farmers behaved as George Ellard did.

53. Lt. Col. Thomas Arbuthnot to Finch, 11 April 1823 (NAI, SOCP 1, 2512/16).

54. NAI, SOCP 1, 1823/2518/19.

55. Col. Sir Hugh Gough to Maj. J. Finch, 1 April 1823 (NAI, SOCP 1, 2512/5).

56. NAI, SOCP 1, 1823/2510/16.

57. Threatening notice, 28 Feb. 1823 (NAI, SOCP 1, 2510/17).

58. Maj. George Warburton to William Gregory, 27 March 1823 (NAI, SOCP 1, 2510/28).

59. Francis Blackburne to Gregory, 16 Aug. 1824 (NAI, SOCP 1, 2613/50). For incidents of incendiarism in this area, see NAI, SOCP 1, 1824/2613/20, 29, 43.

60. Philip Reade to Blackburne, 17 Aug. 1824 (NAI, SOCP 1, 2613/50).

61. NAI, SOCP 1, 1824/2618/53.

62. NAI, SOCP 1, 1824/2619/53, 56; *DEP*, 14 Dec. 1824.

63. Maj. Samson Carter to [Henry Goulburn], 25 Oct. 1822 (NAI, SOCP 1, 2347/19).

64. A. Crossley to Carter, 22 Oct. 1822 (ibid.).

65. NAI, SOCP 1, 1823/2514/9.

66. NAI, SOCP 1, 1823/2516/25.

67. Information of Rev. Sir William Read, 10 Nov. 1823 (NAI, SOCP 1, 1824/2613/59). See also NAI, SOCP 1, 1823/2510/71.

68. John Read to Henry Goulburn, 7 Nov. 1823 (NAI, SOCP 1, 2510/72). The persons charged with firing at the rector were later acquitted (NAI, SOCP 1, 1824/2613/15; *DEP*, 20 March 1824).

69. Lt. B. Lamb to Col. Turner, 21 Sept. 1822 (NAI, SOCP 1, 2341/35).

70. Maj. George M. Drought to Henry Goulburn, 23 Nov. 1823 (NAI, SOCP 1, 2510/78).

71. NAI, SOCP 1, 1822/2341/26.

72. NAI, SOCP 1, 1823/2515/35.

73. The source for this information was an informer named Patrick Donovan, who claimed that he had been forced at gunpoint to become a Rockite and to follow Mansfield, who had allegedly administered the oath to him. See Information of Patrick Donovan, 25 Feb. 1823 (NAI, SOCP 1, 2511/38).

74. NAI, SOCP 1, 1823/2514/36. When first apprehended in the summer of 1823, Nagle was described as "the Captain Rock" of the North Liberties of Cork city (*DEP*, 5 July 1823; NAI, SOCP 1, 1823/2514/4, 5, 43).

75. Matthew Barrington to William Gregory, 13 Aug. 1823 (NAI, SOCP 1, 2514/43). See also *DEP*, 19 Aug. 1823.

76. Col. Sir Hugh Gough to William Gregory, 22 May 1824 (NAI, SOCP 1, 2615/28).

77. Lidwill made this statement while arguing against a resolution moved by Daniel O'Connell for the total repeal of the Tithe Composition Act of 1823. Lidwill asserted with some reason that O'Connell was opposed to the effect of the new law in repealing the exemption for grassland initiated in 1735 (*DEP*, 13 Jan. 1824).

78. Col. Sir Hugh Gough to Maj. J. Finch, 8 Nov. 1823 (NAI, SOCP 1, 2516/8).

79. *DEP*, 26 Aug., 18 Dec. 1823.

80. Threatening notice, 13 Feb. 1822 (NAI, SOCP 1, 2355/27). See also NAI, SOCP 1, 1822/2344/45; 1822/2345/52.

81. Threatening notice, enclosed in Rev. Francis A. Chute to William Gregory, 19 April 1823 (NAI, SOCP 1, 2519/12).

82. Though Sheehan complied, he may have informed against at least two of his assailants. The herdsman Leary and a second prisoner were sentenced to transportation under the Insurrection Act at Mallow in June 1823 (*DEP*, 7 June 1823).

83. Maj. Samson Carter to William Gregory, 7 March 1824 (NAI, SOCP 1, 2614/52).

84. Col. Sir Hugh Gough to Maj. J. Finch, 26 April 1823 (NAI, SOCP 1, 2512/38).

85. William Thornton to Lt. Col. Sorell, 9 April 1822 (NAI, SOCP 1, 2352/15). See also NAI, SOCP 1, 1820/2182/9.

86. NAI, SOCP 1, 1823/2512/26. See also NAI, SOCP 1, 1822/2347/28; 1823/2513/27, 76; *DEP,* 18 March 1823.

87. Maj. Samson Carter to William Gregory, 23 Oct. 1823 (NAI, SOCP 1, 2515/58).

88. *DEP,* 20 Nov., 9 Dec. 1823, 4 Sept. 1824; NAI, SOCP 1, 1823/ 2515/58, 63; 1823/2516/2, 21; 1824/ 2621/62.

89. Robert Longan to Henry Goulburn, 24 Dec. 1822 (NAI, SOCP 1, 2357/34). See also NAI, SOCP 1, 1822/2369/19; 1823/2519/39.

90. *DEP,* 29 Jan. 1822. This particular attack underlined the severity of the "fuel famine" prevailing in the winter of 1821–22 as a result of the wholesale destruction of turf by heavy rains and great flooding. The police fully expected that the poor would be driven to cut down orchards and timber of all kinds without the owners' leave in order to obtain firing, if only to cook their potatoes. See NAI, SOCP 1, 1821/2296/51; 1822/2350/7.

91. Serj. Arthur Jeay to [Military Secretary], 2 Feb. 1824 (NAI, SOCP 1, 2621/14).

92. *DEP,* 20 April 1824.

93. [Stuart McDonald?] to Maj. George Warburton, 11 July 1823 (NAI, SOCP 1, 2510/51). Ryan was beaten to death with stones and the laths of a gate (NAI, SOCP 1, 1823/2510/50). To the indignation of the authorities, the persons charged with Ryan's murder were acquitted at the Clare summer assizes of 1823 (NAI, SOCP 1, 1823/ 2510/54).

94. *DEP,* 23 Aug. 1823.

95. Col. Sir Hugh Gough to Col. J. Finch, 12 June 1824 (NAI, SOCP 1, 2615/45). One Rockite notice posted in County Kerry early in 1822 required all "strangers" resident in any place for less than fourteen years to return to their "respective countrys." See Threatening notice, 14 Feb. 1822 (NAI, SOCP 1, 2349/5).

96. The Earls of Ormonde and Ossory were the owners of Kilkenny Castle.

97. NAI, SOCP 1, 1822/2355/70.

98. Ensign Robert Price to Lt. Col. M. Lindsay, 12 April 1822 (NAI, SOCP 1, 2369/32).

99. William Morris Reade and Benjamin Morris to William Gregory, 13 April 1822 (NAI, SOCP 1, 2369/28).

100. Lt. Col. M. Lindsay to Col. Lord Greenock, 13 April 1822 (NAI, SOCP 1, 2369/32).

101. As early as December 1821 there was a report that three land grabbers on the Earl of Ormonde's estate in the Killamery district had recently been beaten, with one having been "badly wounded." Earlier still, a barn holding corn distrained for arrears of rent on Ormonde's property was burned down (*DEP,* 1, 22 Dec. 1821).

102. Lt. Col. M. Lindsay to Col. Lord Greenock, 3 May 1822 (NAI, SOCP 1, 2369/41). See also NAI, SOCP 1, 1822/2369/28.

103. NAI, SOCP 1, 1822/2369/19, 28.

104. NAI, SOCP 1, 1822/2356/6, 9.

105. Francis Despard to [Henry Goulburn], 25 Jan. 1824 (NAI, SOCP 1, 2606/6).

106. Despard to William Gregory, 11 May 1822 (NAI, SOCP 1, 2356/14).

107. Despard to Gregory, 12 Aug. 1822 (NAI, SOCP 1, 2356/52). Going observed that he had been threatened many times with the death of the Sheas unless he surrendered the farm "which cost him near five hundred pounds." See Memorial of John Going, 20 Nov. 1822 (NAI, SOCP 1, 2356/78).

108. Despard to [Goulburn], 25 Jan. 1824 (NAI, SOCP 1, 2606/6).

109. Despard to Goulburn, 11 Oct. 1823 (NAI, SOCP 1, 2506/27). For the expression of similar views by another local magistrate, see NAI, SOCP 1, 1823/2506/16.

110. John Barwis to Lord Chief Justice Charles Kendal Bushe, 1 April 1824 (NAI, SOCP 1, 2606/29). Barwis was Lord Ormonde's land agent (NAI, SOCP 1, 1823/2506/17).

111. *DEP,* 8 April 1824.

112. Threatening notice, n.d., but April 1822 (NAI, SOCP 1, 2355/78).

113. A.C. Macartney to William Gregory, 19 May 1822 (NAI, SOCP 1, 2349/40). See also NAI, SOCP 1, 1822/2355/78.

114. *DEP,* 28 Aug. 1823.

115. Threatening notice, 27 Jan. 1822, enclosed in Lt. Arthur Morris to Henry Goulburn, 5 Feb. 1822 (NAI, SOCP 1, 2348/37).

116. *DEP,* 20 May 1824. See also *DEP,* 10 June, 27 July 1824.

117. John Bushy to William Gregory, 23 May 1824 (NAI, SOCP 1, 2613/36).

118. Lt. C. Barker to [Military Secretary], 18 July 1824 (NAI, SOCP 1, 2619/6).

119. *DEP,* 7 March 1822.

120. Gibbons, *Captain Rock,* pp. 116–17.

121. Threatening notice, Nov. 1821 (NAI, SOCP 1, 2296/46).

122. "Copy of a threatening notice adverted to in the detail of duty," 20 Jan. 1822 (NAI, SOCP 1, 2342/51).

123. NAI, SOCP 1, 1823/2512/13.

124. Col. Sir Hugh Gough to Maj. J. Finch, 24 May 1823 (NAI, SOCP 1, 2513/42).

125. NAI, SOCP 1, 1823/2512/36.

126. *DEP,* 27 Dec. 1821.

127. William Preston White to William Gregory, 27 Feb. 1823 (NAI, SOCP 1, 2511/35).

128. Lt. Col. M. Lindsay to Lt. Col. J. Finch, 3 April 1824 (NAI, SOCP 1, 2606/32).

129. NAI, SOCP 1, 1822/2369/11.

130. Rev. Peter Roe to William Gregory, 22 March 1823 (NAI, SOCP 1, 2506/7).

131. See below, pp. 175–77.

132. Information of John Lothian (copy), 7 Oct. 1823 (NAI, SOCP 1, 2515/45).

133. Maj. Samson Carter to [Henry Goulburn], 7 Oct. 1823 (ibid.).

134. The three witnesses against those convicted were "all poor labourers with large families"; they were said to be "the intimates of the criminals." See Carter to [Goulburn], 20 Dec. 1823 (NAI, SOCP 1, 2516/50).

135. *DEP,* 4 Dec. 1823.

136. NAI, SOCP 1, 1824/2606/54. The Rockites had conducted an arms raid against Castle Blunden as early

as February 1822. The report of the raid indicated that the steward actually occupied the castle owing to the nonresidence of the owner and that the arms seized by the raiders came in fact from the Scottish gardener, who occupied a small, detached house nearby. See NAI, SOCP 1, 1821/2292 /12. The report is clearly dated 14 Feb. 1822. But there is no evidence connecting the fatal beating of Gabriel Holmes to the "Templemartin gang." Though Lyrath House and Castle Blunden were both located close to Kilkenny city, the two residences were not near each other. I wish to thank Pádraig Ó Macháin for clarifying this point.

137. NAI, SOCP 1, 1824/2606/7. These figures do not include the murder of Gabriel Holmes in June 1824.

138. *DEP*, 1 Dec. 1821.

139. Lt. Col. M. Lindsay to William Gregory, 10 Dec. 1823 (NAI, SOCP 1, 2506/32).

140. Lindsay to [Gregory?], 17 Dec. 1823 (ibid.).

141. Lindsay to Gregory, 20 Feb. 1824 (NAI, SOCP 1, 2606/7).

142. John Barwis to Gregory, 14 March 1824 (NAI, SOCP 1, 2606/18). Though others may have been involved in the deed, John Wall and John Larrissy were accused of shooting John Phelan (or Whelan), who died of his wounds on 11 December 1823 (NAI, SOCP 1, 1824/2506/33).

143. Lt. Col. M. Lindsay to Gregory, 14 March 1824 (NAI, SOCP 1, 2606/18).

144. Lindsay to Gregory, 20 Feb. 1824 (NAI, SOCP 1, 2606/7). See also NAI, SOCP 1, 1823/2506/32.

145. Lindsay to Gregory, 25 Jan. 1824 (NAI, SOCP 1, 2606/7).

146. Lindsay to Gregory, 1 Feb. 1824 (ibid.).

147. Ibid. A spy named Michael Phelan elicited this and other important information about the Templemartin gang. Phelan was an Irish speaker and frequented public houses to which members of the gang were known to resort. He was reportedly "received with open arms by the whole party" in one such pub "on giving the signs & announcing himself as hiding from Galmoy, where he [supposedly] had wounded a proctor." See Lindsay to Gregory, 10 Dec. 1823 (NAI, SOCP 1, 2506/32). For the management by Lindsay of the spies and informers Michael Phelan and John Kennedy, see Ó Macháin, *Six years in Galmoy,* pp. 93–96.

148. Lindsay to Gregory, 10 Dec. 1823 (NAI, SOCP 1, 2506/32).

149. Lindsay to Gregory, 20 Feb. 1824 (NAI, SOCP 1, 2606/7).

150. NAI, SOCP 1, 1823/2506/32.

151. *DEP,* 21 Feb. 1822.

152. Lindsay to Gregory, 25 Jan. 1824 (NAI, SOCP 1, 2606/7).

153. Rev. J. Orpen to Gregory, 18 Oct. 1821 (NAI, SOCP 1, 2293/12).

154. [August Fitzgerald?] to Gregory, 5 March 1822 (NAI, SOCP 1, 2341/22). In their attack on Vandeleur's steward, a man named Malcolm, the Rockites used a scythe and an iron bar; Malcolm's head was badly injured and some of his fingers were cut off. His arm was later amputated, but he apparently survived this brutal assault (*Clare Journal and Ennis Advertiser,* 4, 7 March 1822). Vandeleur was a

commissioner of the revenue department in Dublin as well as the owner of one of the largest estates in Clare. I am indebted to Dr. Ciarán Ó Murchadha for the references to the *Clare Journal* and for other information related to this incident.

155. Lt. Col. Sir Thomas Arbuthnot to Maj. J. Finch, 22 Feb. 1823 (NAI, SOCP 1, 2511/32).

156. Col. Sir Hugh Gough to Finch, 17 March 1823 (NAI, SOCP 1, 2511/55).

157. *DEP*, 11 March 1823; NAI, SOCP 1, 1823/2511/43, 50; 1823/2512/3, 12.

158. Maj. Gen. John Lambert to Lieut. Gen. Lord Combermere, 1 April 1823 (NAI, SOCP 1, 2512/5).

159. *DEP*, 4 Dec. 1823.

160. Arthur B. Bernard to Maj. Gen. Sir John Lambert, 10 Jan. 1822 (NAI, SOCP 1, 2342/26).

161. Rev. J.T. Newman to St. George Daly, 9 Jan. 1822 (NAI, SOCP 1, 2342/11). Three men tried to waylay Bernard a day or two later; one of them demanded to know why Bernard had killed the carpenter, and tried to fire a pistol at him, but Bernard shot the assailant dead. See Newman to Daly, 10 Jan. 1822 (NAI, SOCP 1, 2342/14).

162. *DEP*, 6 Dec. 1821.

163. Matthew Barrington to William Goulburn, 14 Aug. 1822 (NAI, SOCP 1, 2349/54).

164. See above, chapter 4, pp. 131–33. For the case of the County Cork schoolmaster Charles McCarthy Considine, see also *DEP*, 2 Nov. 1824.

165. Samson Carter to William Gregory, 12 June 1824 (NAI, SOCP 1, 2615/42).

166. Carleton, "Irish prophecy man," pp. 393–94.

167. Rev. J. Conolly to Henry Goulburn, 12 Nov. 1824 (NAI, SOCP 1, 2621/63).

168. Carleton, "Irish prophecy man," pp. 393–94.

169. See the "Whiteboy catechism" enclosed in Maj. Samson Carter to Henry Goulburn, 15 Jan. 1822 (NAI, SOCP 1, 2342/38). See also Information of William Nash, 6 April 1822 (NAI, SOCP 1, 2355/74).

170. For an impressive collection of threatening letters from the period of the Rockite movement, see Gibbons, *Captain Rock,* pp. 111–203.

171. Maj. Richard Willcocks to William Gregory, 19 March 1822 (NAI, SOCP 1, 2351/44).

172. *DEP*, 16 Dec. 1823.

173. Matthew Barrington to William Gregory, 8 Aug. 1822 (NAI, SOCP 1, 2349/51).

174. Maj. Richard Willcocks to Gregory, 2 March 1822 (NAI, SOCP 1, 2351/3).

175. *DEP*, 7 March 1822.

176. *DEP*, 13, 16 Dec. 1823.

177. Maj. Samson Carter to [Henry Goulburn], 26 Sept. 1823 (NAI, SOCP 1, 2515/34). The Rockite leader Edmond Magner gave a "voluntary confession" in which he stated that he had been invited to participate in the murder of the Frankses about three weeks before the event in the public house of Patrick Power at Shanballymore after Sunday Mass. He named six conspirators, all but one residents of Meadstown in Farahy parish: three brothers named Cremins, along with James Magrath, Patrick Meade, and John Car-

ney. See "Voluntary confession" of Edmond Magner, 28 Jan. 1824 (NAI, SOCP 1, 2511/15). See note 182 below.

178. The three Creminses were executed for the murders of the Frankses in the spring of 1824; rumors were spread that they were innocent, but in a memorial that they helped to prepare as a plea for mercy, they admitted that "they were put on their knees to shoot them [the Frankses]." See Memorial of John, Patrick, and Maurice Cremins, n.d. (NAI, SOCP 1, 1824/2615/9). See also Maj. Samson Carter to William Gregory, 14 April 1824 (ibid.).

179. Chief Constable [?] to [Maj. Samson Carter], 18 Sept. 1823 (NAI, SOCP 1, 2515/20). See also Confession and examination of Timothy Sheehan, [?] Sept. 1823 (NAI, SOCP 1, 2515/25). Cornelius Sheehan had been sentenced to transportation for life at the Cork assizes in the autumn of 1822 at the prosecution of Henry Maunsell Franks; he had accused Sheehan of assaulting him with the intent to steal arms and of administering to him a Whiteboy oath (DEP, 13 Sept. 1823; NAI, SOCP 1, 1823/2515/9). Despite his confession, Timothy Sheehan was not an unproblematic potential witness against his former comrades. He had given his confession in a "partial manner," and he had sworn more than one information, a fact that had swayed juries against prosecutors for the crown in the past in County Cork. See Maj. Samson Carter to Henry Goulburn, 28 Oct. 1823 (NAI, SOCP 1, 2515/62). But in addition to Sheehan's confession, the authorities also possessed the vivid information provided by Mary Meyers, a maidservant in the house of the

Frankses at the time of the murders and an eyewitness to them (NAI, SOCP 1, 1823/2515/9, 34).

180. The Sheehans worked for a gentleman named Freeman of Ballymague, and the Creminses were tenants of another gentleman named James Goold near Rockmills (NAI, SOCP 1, 1823/2515/40; 1823/2615/9; DEP, 15 April 1824).

181. Convicted along with Maher was his cousin Darby Maher. The chief witnesses against them at their trial were Mary Kelly and her son John Kelly. Mary Kelly "kept a [public] house of no great reputation, frequented by night-walkers." Her husband was a carpenter, and her son had enlisted in the army out of the fear that his knowledge about the burning of the Sheas might make him an object of attack (DEP, 21 Aug. 1824).

182. For Hickey, see NAI, SOCP 1, 1823/2511/55. For Edmond Magner, see NAI, SOCP 1, 1824/2511/15; DEP, 15 April 1824.

183. Maj. Samson Carter to [Henry Goulburn], 18 Jan. 1824 (NAI, SOCP 1, 2614/17).

184. Rev. J. Conolly to William Gregory, 14 Feb. 1822 (NAI, SOCP 1, 2355/28).

185. NAI, SOCP 1, 1823/2513/79, 80, 83, 86; 1823/2514/8, 15, 49.

186. DEP, 7 Feb. 1822.

187. John Busteed to Charles Grant, 19 Nov. 1821 (NAI, SOCP 1, 2295/24).

188. [Anonymous] to Marquis Wellesley, n.d., but 1822 (NAI, SOCP 1, 2356/77).

189. R.W. Gason to William Gregory, 24 April 1822 (NAI, SOCP 1, 2355/86).

190. Jemmett Browne to Gregory, 12 May 1823 (NAI, SOCP 1, 2513/21).

191. Maj. Richard Willcocks to Gregory, 18 April 1822 (NAI, SOCP 1, 2352/25). See also *DEP,* 30 April 1822.

192. John C. Kelly to [?], 27 Feb. 1822 (NAI, SOCP 1, 2350/90).

193. Matthew Barrington to William Gregory, 23 March 1824 (NAI, SOCP 1, 2618/46).

194. Ibid.

195. *DEP,* 27 March 1824.

196. Information of William Connell, 21 April 1822 (NAI, SOCP 1, 2345/62).

197. Maj. Samson Carter to William Gregory, 16 July 1822 (NAI, SOCP 1, 2346/11).

198. Cotter was allegedly "concerned in all the outrages committed between this [the town of Mallow] and Millstreet and was the principal in stopping the Killarney coach near that town and murdering Mr. Brereton." See John Longfield to Henry Goulburn, 20 Sept. 1823 (NAI, SOCP 1, 2515/23). See also NAI, SOCP 1, 1823/2519/30; 1824/2615/38.

199. Cornelius O'Leary to [Goulburn], 8 June 1824 (NAI, SOCP 1, 2615/37).

200. Maj. George M. Drought to Goulburn, 28 March 1823 (NAI, SOCP 1, 2517/16).

201. Whooley, "Captain Rock's rebellion," pp. 154–55.

202. NAI, SOCP 1, 1822/2345/78; 1822/2346/38. See also *DEP,* 27 Aug. 1822.

203. John Deane Freeman to William Gregory, 31 March 1822 (NAI, SOCP 1, 2345/50).

204. Maj. Samson Carter to Gregory, 3 Aug. 1822 (NAI, SOCP 1, 2346/22).

205. NAI, SOCP 1, 1822/2345/50; 1822/2346/22; 1822/2354/3.

206. *DEP,* 27 Aug. 1822; NAI, SOCP 1, 1822/2346/38.

207. *DEP,* 8 July 1823.

208. NAI, SOCP 1, 1822/2345/38, 39.

209. *DEP,* 27 Aug. 1822.

210. NAI, SOCP 1, 1822/2345/62; 1822/2351/60.

211. NAI, SOCP 1, 1822/2354/8.

212. *DEP,* 1 Aug. 1822. See also *DEP,* 6, 8 Aug. 1822. The informer William Connell alleged that John Browne and James David Leahy brought the abducted heiress Honora Goold to the house of the rich farmer Maurice Leahy and eventually to the house of David Leahy before she was left elsewhere and rescued. See Information of William Connell, 21 April 1822 (NAI, SOCP 1, 2345/62).

213. See note 89 above.

214. Maj. Samson Carter to William Gregory, 7 June 1822 (NAI, SOCP 1, 2345/78). See also NAI, SOCP 1, 1823/2519/30.

215. *DEP,* 29 July 1823.

216. *DEP,* 6, 8 Aug. 1822; NAI, SOCP 1, 1822/2354/8.

Chapter Six.
The Issue of Tithes

1. Viscount Clifden to Pierce Butler, 10 Dec. 1822, quoted in *DEP,* 6 March 1823.

2. NAI, SOCP 1, 1822/2355/21; *DEP,* 17 Jan. 1822. See also Threatening notice, 18 July 1822, enclosed in

Rev. John Lombard to [?], 29 July 1822 (NAI, SOCP 1, 2346/21).

3. Donnelly, "Rightboy movement," pp. 149–52.

4. Rev. John Chester to [?], 19 Jan. 1822 (NAI, SOCP 1, 2342/50).

5. *DEP*, 1 Dec. 1821. See also *DEP*, 15 Dec. 1821.

6. Threatening notice, [?] Jan. 1822, enclosed in Lord Carbery to Henry Goulburn, 15 Jan. 1822 (NAI, SOCP 1, 2342/33). See also *DEP*, 24 Nov. 1821; NAI, SOCP 1, 1822/ 2348/15.

7. *DEP*, 11 Dec. 1821.

8. *DEP*, 5 Feb. 1822.

9. *DEP*, 9 April 1822. Generally, there was no second valuation of tithes after a harvest. See *SOI evidence*, p. 316, H.C. 1825 (20), vii, 1; *SOI evidence*, p. 192, H.C. 1825 (200), vii, 501.

10. *SOI evidence*, pp. 317–18, H.C. 1825 (20), vii, 1; *SOI evidence*, pp. 194, 208, H.C. 1825 (200), vii, 501.

11. *DEP*, 13 Dec. 1821.

12. The tithe income of the rector of Whitechurch parish near Cork city was said to have risen from £350 or £400 a year before the war to £1,500 or £1,800 during its course. He did not raise the rates after 1799, when he became the incumbent, but he may have increased them at that time (*SOI evidence*, pp. 317–18, H.C. 1825 [200], vii, 501).

13. *DEP*, 31 Jan. 1822. See also *DEP*, 14 March 1822.

14. For abatements of tithes, see *DEP*, 24 Nov., 11 Dec. 1821, 9 Nov. 1822; NAI, SOCP 1, 1822/2342/19.

15. *DEP*, 20 Jan. 1824.

16. *SOI evidence*, p. 136, H.C. 1825 (200), vii, 501; *SOI evidence*, p. 736, H.C. 1825 (129), viii, 455.

17. *SOI evidence*, p. 120, H.C. 1825 (20), vii, 1. See also *SOI evidence*, p. 53, H.C. 1825 (200), vii, 501.

18. *DEP*, 3 Dec. 1822.

19. Under 7 Geo. 3, c. 21, made perpetual by 11 & 12 Geo. 3., c. 19, no defense could be taken in the civil-bill court; a copy of the monition under the seal of the consistorial court and proof of due service at least fifteen days before the hearing of the civil-bill suit were to be considered conclusive evidence as to the sum due (John Finlay, *A treatise on the law of tithes in Ireland and ecclesiastical law connected therewith* [Dublin, 1828], pp. 377–78).

20. Under 1 Geo 2, c. 12, s. 2, 4, two magistrates were authorized to adjudicate claims for tithes amounting to 40s. or less.

21. 36 Geo. 3, c. 25.

22. 39 Geo. 3, c. 16, s. 3–4; Finlay, *Law of tithes*, pp. 374, 376–77.

23. Patterson, "'Educated Whiteboyism,'" pp. 25–29.

24. 54 Geo. 3, c. 68, s. 2, 4; Finlay, *Law of tithes*, pp. 449, 454, 460–68.

25. Instances occurred in which tithe owners selected magistrates who resided far from the parish concerned, thus forcing tithe payers to travel considerable distances in answer to the summons. In this fashion the harassed tithe payers were sometimes persuaded to satisfy claims that they considered exorbitant or to abandon setting out their tithes in kind. See *SOI evidence*, pp. 759–60, H.C. 1825 (129), viii, 455; *SOI evidence*, pp. 130–31, H.C. 1825 (181), ix, 1.

26. *DEP*, 2, 5 Feb. 1822; *SOI evidence*, pp. 371–72, H.C. 1825 (20), vii, 1.

27. *DEP*, 5, 10, 15 July 1823.

28. *DEP*, 16 Sept. 1823; *SOI evidence,* pp. 233–34, H.C. 1825 (200), vii, 501; *SOI evidence,* pp. 602–4, H.C. 1825 (129), viii, 455.

29. *DEP*, 7 Oct. 1823.

30. *DEP*, 9 April, 17, 22 Oct. 1822.

31. *SOI evidence,* p. 188, H.C. 1825 (20), vii, 1.

32. *DEP*, 5 Nov. 1822. See also *DEP*, 7 Feb. 1822.

33. 1 Geo. 2, c. 12, s. 2–4.

34. *DEP*, 9 April 1822; *SOI evidence,* pp. 759–60, H.C. 1825 (129), viii, 455.

35. 54 Geo. 3, c. 39, s. 7; *DEP*, 5 Feb. 1822; *SOI evidence,* p. 193, H.C. 1825 (200), vii, 501.

36. *DEP*, 5 Feb. 1822; *SOI evidence,* pp. 134–35, H.C. 1825 (181), ix, 1.

37. *DEP*, 30 March 1822.

38. *DEP*, 30 Oct., 20, 24 Nov. 1821; 11 Jan. 1823; NAI, SOCP 1, 1822/ 2346/3; 1822/2348/69; 1823/ 2511/4.

39. NAI, SOCP 1, 1822/2348/36. See also *DEP*, 10 Jan. 1822.

40. *DEP*, 5 Aug. 1823.

41. Threatening notice, 2 Feb. 1822 (NAI, SOCP 1, 2355/21).

42. *DEP*, 29 Dec. 1821; 22, 31 Jan. 1822; 8 May 1823; NAI, SOCP 1, 1822/2355/13.

43. NAI, SOCP 1, 1822/2347/20; *DEP*, 10, 22 Jan. 1822; 15 July 1823; O'Sullivan, *Captain Rock detected,* pp. 164–67.

44. *DEP*, 8, 10, 26 Jan. 1822; 21 Aug., 4 Sept. 1823; NAI, SOCP 1, 1823/2518/42a.

45. NAI, SOCP 1, 1822/2344/51; 1822/2346/3546; 1822/2511/2; 1823/ 2511/4, 27; 1823/2514/33; 1824/2615/

62; *DEP*, 19 Sept. 1822, 11 Jan., 20 Feb. 1823.

46. NAI, SOCP 1, 1822/2342/13, 33; *DEP*, 5 Jan. 1822.

47. NAI, SOCP 1, 1822/2347/20.

48. NAI, SOCP 1, 1822/2348/28; *DEP*, 10, 22 Jan. 1822.

49. *DEP*, 5, 15 July 1823.

50. *DEP*, 29 Sept. 1821.

51. NAI, SOCP 1, 1822/2348/6, 18, 51; *DEP*, 15 Jan. 1822.

52. NAI, SOCP 1, 1822/2355/11; 1823/2518/40.

53. Rev. Thomas Stoughton to Henry Goulburn, 15 Feb. 1822 (NAI, SOCP 1, 2348/51).

54. Rev. Thomas Russell to Sir W.C. Smith, [?] Feb. 1822 (NAI, SOCP 1, 2348/51).

55. *DEP*, 16 Sept. 1823.

56. Ibid.

57. NAI, SOCP 1, 1822/2355/13.

58. Daniel Mahony to Henry Goulburn, 22 Oct. 1822 (NAI, SOCP 1, 2349/65).

59. A.C. Macartney to Goulburn, 26 Sept. 1822 (NAI, SOCP 1, 2349/59).

60. *DEP*, 29 April 1823.

61. Threatening notice, n.d., enclosed in Robert Eames to [?], 4 Aug. 1822 (NAI, SOCP 1, 2346/25). See also NAI, SOCP 1, 1822/2346/23–24, 26, 28–29.

62. Donnelly, "Rightboy movement," pp. 154–56.

63. 29 Geo. 2, c. 12; 1 Geo. 3, c. 17.

64. *DEP*, 21 Sept. 1822.

65. *DEP*, 14 Sept. 1822.

66. Ibid.

67. *DEP*, 22 May 1823; O'Sullivan, *Captain Rock detected,* pp. 160–62. In spite of the publicity given to the question of setting out tithes in kind,

there is a wealth of evidence to the effect that parishioners either did not understand the proper legal forms to be observed or still believed that for more than three or four persons to serve notices to draw tithes constituted an illegal combination. See *SOI evidence,* pp. 37, 60, 223, 316–17, 374–75, 416, 451–52, H.C. 1825 (20), vii, 1; *SOI evidence,* pp. 136–37, H.C. 1825 (200), vii, 501; *SOI evidence,* p. 337, H.C. 1825 (129), viii, 293. Tithe owners and their proctors used a variety of devices to frustrate or to prevent notices to draw. They neglected to provide notification of the sums owed until after the crops had been taken from the fields; they refused to attend and receive in kind, and instead initiated proceedings in a bishop's court or before friendly magistrates; and they made "captious objections" to the manner in which the tithes had been set out (*DEP,* 20 Jan. 1824). In the case of potatoes one great obstacle to the proper setting out of a tenth of this crop was the common necessity of the poor to dig some of their potatoes early, simply to feed themselves. The removal of even one basket of potatoes amounted in law to a severance of the crop, after which the service of a notice to draw was invalid. See *SOI evidence,* pp. 317, 375, 416, H.C. 1825 (20), vii, 1; *SOI evidence,* pp. 136–37, 192–93, H.C. 1825 (200), vii, 501; *SOI evidence,* p. 337, H.C. 1825 (129), viii, 293.

68. NAI, SOCP 1, 1822/2346/58, 72.

69. Quoted in *DEP,* 1 Oct. 1822.

70. Rev. Richard Woodward to [?], 27 Sept. 1822 (NAI, SOCP 1, 2346/68).

71. Col. Sir Hugh Gough to [?], 20 Sept. 1822 (NAI, SOCP 1, 2346/52). See also *DEP,* 1 Oct. 1822.

72. Robert Eames to Sir Edward Lees, 24 Sept. 1822 (NAI, SOCP 1, 2346/63). See also NAI, SOCP 1, 1822/2346/53.

73. *DEP,* 19 Sept. 1822. See also NAI, SOCP 1, 1822/2346/46. Hill was awarded £1,000 in compensation for this loss by the grand jury at the spring assizes of 1823. See *DEP,* 3 April 1823; *SOI evidence,* p. 587, H.C. 1825 (129), viii, 455.

74. NAI, SOCP 1, 1822/2346/58; *DEP,* 1 Oct. 1822.

75. *DEP,* 10 Oct. 1822.

76. *DEP,* 22, 24 Oct. 1822. See also NAI, SOCP 1, 1822/2347/33.

77. NAI, SOCP 1, 1822/2347/50; 1822/2511/2. Col. Hill was also hated as a zealous magistrate; he was known among local Rockites as "Bloodhound Hill" (NAI, SOCP 1, 1822/2347/63). The family as a whole (James Hill of Graig was a relative) had become the object of intense hostility. When reporting the destruction of Colonel Hill's haggard and corn, a military officer noted that this was the sixth burning or attack suffered by members of the family in the past year (NAI, SOCP 1, 1822/2347/53).

78. NAI, SOCP 1, 1822/2347/5.

79. Rev. E.M. Kenny to Daniel Mahony, 30 Oct. 1822 (NAI, SOCP 1, 2347/24).

80. *DEP,* 3 Oct. 1822.

81. "A statement of the outrages that have occurred in the baronies of Orrery & Kilmore, Fermoy, Duhallow, [and] Condons and Clongibbons [*sic*] from 6th January to 7th day of March

1823," by Maj. Samson Carter (NAI, SOCP 1, 2511/45).

82. NAI, SOCP 1, 1823/2511/4, 27, 38; *DEP*, 11, 20 Feb., 20 March 1823.

83. NAI, SOCP 1, 1823/2512/3; *DEP*, 3 April 1823.

84. Memorial of the grand jury of County Cork to the lord lieutenant, [?] April 1823 (NAI, SOCP 1, 2512/9); *DEP*, 29 March, 3 April 1823.

85. *DEP*, 24 July 1823.

86. Ibid.

87. *DEP*, 6 March 1823.

88. *DEP*, 16 Sept. 1823.

89. *DEP*, 16 Sept. 1823.

90. Col. Sir Hugh Gough to Maj. J. Finch, 4 Oct. 1823, copy (NAI, SOCP 1, 2515/44).

91. Gough to Finch, 8 Nov. 1823, copy (NAI, SOCP 1, 2516/8).

92. NAI, SOCP 1, 1823/2516/39.

93. NAI, SOCP 1, 1823/2516/22.

94. NAI, SOCP 1, 1823/2515/58, 60–61; 1823/2516/5, 25.

95. NAI, SOCP 1, 1823/2514/9; 1823/2516/25.

96. *DEP*, 29 March 1823.

97. NAI, SOCP 1, 1823/2511/15, 43, 55; 1823/2512/3, 5; 1823/2514/4–5, 9, 36, 43; 1824/2614/17; *DEP*, 12 June, 3, 5 July, 19 Aug. 1823.

98. NAI, SOCP 1, 1823/2516/8.

99. See below, pp. 209–16.

100. *Minutes of evidence taken before the select committee of the House of Lords appointed to inquire into the state of Ireland, more particularly with reference to the circumstances which may have led to disturbances in that part of the United Kingdom (24 March–22 June 1825), brought from the Lords, 5 July 1825* (hereafter cited as *SOI evidence,* for State of Ireland evidence), p. 608, H.C. 1825 (521), ix, 249.

101. *DEP*, 26 July 1823.

102. *DEP*, 21 Aug. 1823.

103. *DEP*, 16 Sept. 1823.

104. *DEP*, 28 Oct. 1823.

105. *DEP*, 18 Sept. 1823.

106. Rev. James Butler to William Gregory, 4 April 1822 (NAI, SOCP 1, 2356/13).

107. NAI, SOCP 1, 1823/2518/42a.

108. NAI, SOCP 1, 1823/2518/51.

109. *DEP*, 10 Feb. 1824. See also *DEP*, 14 Feb. 1824.

110. Maxwell Blacker to William Gregory, 27 Feb. 1824 (NAI, SOCP 1, 2621/22). See also *SOI evidence,* pp. 19–20, H.C. 1825 (200), vii, 501.

111. *SOI evidence,* p. 53, H.C. 1825 (200), vii, 501. See also *SOI evidence,* p. 120, H.C. 1825 (20), vii, 1.

112. *SOI evidence,* p. 317, H.C. 1825 (20), vii, 1.

113. Ibid., p. 38.

114. *Hansard,* new ser., ix, 807.

115. *SOI evidence,* p. 603, H.C. 1825 (129), viii, 455.

116. Akenson, *Church of Ireland*, pp. 87–103; Donnelly, "Rightboy movement," pp. 143–50, 152–54, 191–93.

117. Broeker, *Rural disorder,* pp. 128–59.

118. Rev. Richard Woodward to [?], 27 Sept. 1822 (NAI, SOCP 1, 2346/68). The Reverend Mortimer O'Sullivan elaborated this argument in his *Captain Rock detected,* pp. 120–98.

119. *DEP*, 22 Aug., 21 Sept. 1822.

120. See the sources cited in note 119.

121. *DEP*, 6 March 1823.

122. Somerset R. Butler to David Burtchaell, 6 Dec. 1822, quoted in *DEP*, 19 Dec. 1822.

123. *DEP*, 1 May 1823.

124. *SOI evidence,* p. 62, H.C. 1825 (181), ix, 1.

125. *Hansard,* new ser., viii, 494-98.

126. Quoted in Macintyre, *Liberator,* p. 172.

127. In May 1823 the Marquis of Lansdowne presented to the House of Lords a petition seeking a commutation of tithes signed by about three-quarters of the beneficed clergymen and lay impropriators of the united diocese of Limerick, Ardfert, and Aghadoe, including all of Kerry and portions of Cork and Limerick (*Hansard,* new ser., ix, 538-40).

128. *DEP,* 23 Sept. 1823.

129. *DEP,* 1 May 1823. See also *SOI evidence,* pp. 776-77, H.C. 1825 (129), viii, 455.

130. 4 Geo. 4, c. 99.

131. *Hansard,* new ser., ix, 1491.

132. 4 Geo. 4, c. 99, s. 35.

133. In the first eighty-four parishes that agreed to compound for their tithes under the act of 1823, the average acreable rate of assessment was about 11d. Even in parishes where the land devoted to pasture was relatively restricted and the scope for shifting the burden was therefore not so extensive, the acreable rates rarely exceeded 2s. or 3s. See *SOI evidence,* p. 63, H.C. 1825 (181), ix, 1.

134. *Hansard,* new ser., ix, 370.

135. *Hansard,* new ser., x, 852. See also *DEP,* 26 Aug. 1823, 20 Jan. 1824.

136. *DEP,* 23 Sept. 1823.

137. *SOI evidence,* pp. 369, 438, H.C. 1825 (521), ix, 249.

138. 4 Geo. 4, c. 99, s. 43.

139. *Hansard,* new ser., x, 859.

140. *SOI evidence,* p. 397, H.C. 1825 (521), ix, 249.

141. 4 Geo. 4, c. 99, s. 16; Finlay, *Law of tithes,* pp. 510-13, 523-24.

142. *Hansard,* new ser., x, 856.

143. Finlay, *Law of tithes,* pp. 518-19.

144. *DEP,* 30 Sept. 1823.

145. *DEP,* 21 Oct. 1823.

146. *SOI evidence,* p. 216, H.C. 1825 (200), iv, 501.

147. *DEP,* 21 Oct. 1823.

148. See above, p. 404, n. 100.

149. Ibid.

150. 4 Geo. 4, c. 99, s. 3-6. See also George Lidwill to Rev. Henry Meggs Graves, 21 Oct. 1823, quoted in *DEP,* 18 Dec. 1823.

151. 4 Geo. 4, c. 99, s. 9.

152. *Hansard,* new ser., x, 855-56.

153. Presumably, similar sentiments persuaded Daniel O'Connell to oppose the Tithe Composition Act of 1823 altogether. Whereas some other members of the Catholic Association, such as George Lidwill, approved of it in principle and wanted to see the new law amended in such a way as to benefit tillage farmers and cottiers, O'Connell introduced a motion at a meeting of the association calling for the total repeal of the act (*DEP,* 13 Jan. 1824).

154. 4 Geo. 4, c. 99, s. 41.

155. *Hansard,* new ser., x, 855.

156. *DEP,* 4 Dec. 1823.

157. *DEP,* 21 Oct. 1823.

158. *DEP,* 15 Nov. 1823. Besides magnifying the political power of graziers and dairy farmers in the special vestries, the act of 1823 served their interests in another way. It was common for large landholders to enter into long-term agreements or "incumbency bargains" with the tithe owner; they took a lease of their own tithes at relatively

low rates for a stated term of years or for the duration of the incumbency. Such contracts were safeguarded by the act of 1823. Unless such contracts were dissolved by mutual consent, the lands to which they applied had to be excluded from any parochial composition (4 Geo. 4, c. 99, s. 31). In some places this situation prevented fruitful negotiations (*DEP*, 21 Oct. 1823).

159. Col. Sir Hugh Gough to Maj. J. Finch, 8 Nov. 1823 (NAI, SOCP 1, 2516/8).

160. *Hansard,* new ser., x, 855–56.

161. 5 Geo. 4, c. 63, s. 10.

162. 5 Geo. 4, c. 63, s. 3.

163. *Hansard,* new ser., viii, 1132–33; ix, 376, 808–10, 989–92, 1434, 1456, 1489–93.

164. 5 Geo. 4, c. 63, s. 12.

165. 4 Geo. 4, c. 99, s. 27; 5 Geo. 4, c. 63, s. 13.

166. The vestry might also agree at the outset to permit the composition to remain in effect for a full term of twenty-one years, at the expiration of which a completely new composition was to be negotiated (5 Geo. 4, c. 63, s. 23–25).

167. *Hansard,* new ser., x, 856–57, 859; *SOI evidence,* p. 397, H.C. 1825 (521), ix, 249.

168. SOI *evidence,* p. 611, H.C. 1825 (521), ix, 249; Akenson, *Church of Ireland,* p. 110.

169. *Hansard,* new ser., x, 852–53.

170. *SOI evidence,* p. 273, H.C. 1825 (521), ix, 249.

171. Ibid., pp. 605–7; Akenson, *Church of Ireland,* p. 110.

172. Macintyre, *Liberator,* pp. 177–81.

Chapter Seven.
The Issue of Rents

1. For the middleman system and its decline, see Dickson, *Old world colony,* pp. 329–40. Dickson also draws attention to another complicating feature of the land system—the prevalence of partnership tenancy. This phenomenon came under attack from landowners after 1800, but it was still common in the 1820s. Dickson cites the cases of several estates where partnership or multiple tenancy accounted for more than two-fifths of groups of holdings that were newly let in the late eighteenth and early nineteenth centuries (ibid., pp. 342–43). The widespread existence of partnership tenancy helps to explain why in numerous instances there was an in-built mechanism of resistance by tenants or subtenants when certain landlords resorted to forcible methods of collecting rent.

2. *DEP,* 5 Oct. 1820.

3. One "class renegade"—Anthony Hutchins of Ardnagashel near Bantry in west Cork—asserted early in 1822 that "the country gentlemen & the [Anglican] clergy have been blindly employed for years back in bringing about this unfortunate state of things, & their folly ought to be exposed to them promptly & the evil got rid of." By giving abatements of rents and tithes appropriate to the depressed times, claimed Hutchins, "the gentry & clergy could, if they wished, put an end in one week to all the disturbances in the counties of Cork & Kerry, & save the government from the enormous expences [*sic*] of supplying military & yeomanry corps." See Anthony Hutchins to Henry Goulburn, 23 Jan. 1822 (NAI, SOCP 1, 2342/65).

4. *DEP,* 16 Nov. 1822.

5. Gibbons, *Captain Rock,* pp. 116, 128. See also *DEP,* 1 Dec. 1821.

6. *DEP,* 9 Feb. 1822.

7. *DEP,* 27 Nov. 1821.

8. Gibbons, *Captain Rock,* p. 121.

9. Colonel James Crosbie to Henry Goulburn, 14 Jan. 1822 (NAI, SOCP 1, 2348/11).

10. Daniel Mahony to William Gregory, 19 Feb. 1822 (NAI, SOCP 1, 2348/59).

11. Earl of Clare to Gregory, 24 Jan. 1824 (NAI, SOCP 1, 2618/18).

12. *DEP,* 27 Nov. 1821.

13. *DEP,* 3 Aug. 1822.

14. Daniel Mahony to Henry Goulburn, 22 Oct. 1822 (NAI, SOCP 1, 2349/65).

15. E.M. Kenny to Mahony, 30 Oct. 1822 (NAI, SOCP 1, 2347/24).

16. Among the signs of the times, this commentator glumly noted, were landlords without rent, laborers and tradesmen without employment, commerce "nearly at a stand," and no money in circulation (*DEP,* 10 Dec. 1822).

17. *DEP,* 5 June 1823.

18. Ibid. It should also be pointed out that numerous Irish landowners derived substantial income in more normal times from their activity as demesne farmers and graziers. But there was no profit, only losses, for those proprietors involved directly in agricultural pursuits in the years from 1819 to 1823. For the bitter experiences of Irish graziers in this period, see Dutton, *Galway,* pp. 118–19.

19. *DEP,* 4 Dec. 1821. For a few examples of slightly earlier abatements and other concessions, see *DEP,* 18, 25 May 1820. As Dickson points out, the agents of the Duke of Devonshire took the decision "to give a temporary across-the-board abatement in 1822 of 25 per cent on nearly all rents outstanding" on the duke's estates in Cork and Waterford (Dickson, *Old world colony,* p. 359), but this was far from the typical response.

20. Thomas Studdert to William Gregory, 27 May 1822 (NAI, SOCP 1, 2353/18).

21. In this instance the Rockites punished the new tenant by houghing an in-calf heifer (*DEP,* 30 Dec. 1823).

22. NAI, SOCP 1, 1824/2614/30.

23. *DEP,* 11 Sept. 1823.

24. *DEP,* 15 Dec. 1821. The Marquis of Lansdowne reportedly had not renewed a single lease during that time.

25. *DEP,* 24 Nov. 1821.

26. NAI, SOCP 1, 1824/2618/75.

27. *DEP,* 26 Jan. 1822.

28. Christopher Moore to Marquis of Waterford, 3 April 1822 (NAI, SOCP 1, 2357/25). Earlier instances of the same kind had occurred in County Waterford (NAI, SOCP 1, 1822/2357/6).

29. *DEP,* 20 Feb. 1823.

30. Capt. A. Warburton to Lt. Col. Mitchell, 1 Oct. 1822 (NAI, SOCP 1, 2347/6). See also NAI, SOCP 1, 1822/2347/1.

31. A. Woodley to Henry Goulburn, 16 Nov. 1822 (NAI, SOCP 1, 2347/35).

32. *DEP,* 14 Oct. 1823.

33. *DEP,* 25 Sept. 1823.

34. NAI, SOCP 1, 1823/2519/47; 1824/2606/6.

35. Maj. Samson Carter to Henry Goulburn, 10 Nov. 1823 (NAI, SOCP 1, 2516/7).

36. *DEP,* 14, 19 Sept. 1822. See also *DEP,* 25 March 1823.

37. *DEP*, 25 Sept. 1823.

38. *DEP*, 17 Jan. 1824.

39. Information of Michael Roche, 15 Nov. 1823 (NAI, SOCP 1, 2516/14). See also NAI, SOCP 1, 1823/2516/12. One of the keepers was killed at Killea near Mitchelstown, and the other keeper was murdered at Darragh More near Kilmallock in County Limerick. The authorities arrested the subtenants for these crimes—two brothers named Walsh (*DEP*, 20 Nov. 1823).

40. *DEP*, 15 Dec. 1821.

41. *DEP*, 5 Feb. 1822.

42. NAI, SOCP 1, 1823/2511/39.

43. *DEP*, 8 Jan. 1822.

44. Killaghy Castle itself was in County Tipperary. See NAI, SOCP 1, 1822/2639/17.

45. Capt. R. Mullen to [Military Secretary], 7 Aug. 1823 (NAI, SOCP 1, 2518/42).

46. *DEP*, 25 Sept. 1823.

47. Lt. Col. George Gauntlett to Lt. Col. J. Finch, 25 June 1824 (NAI, SOCP 1, 2621/42); *DEP*, 13 Nov. 1823.

48. Threatening notice, 25 Jan. 1822 (NAI, SOCP 1, 2343/47).

49. *DEP*, 7 Oct. 1823.

50. NAI, SOCP 1, 1823/2518/48.

51. *DEP*, 22 July 1824. Massey himself had been the target of an apparent attempt at assassination in January 1822 as he was returning home to Ballingarry from Newcastle West (NAI, SOCP 1, 1822/2350/42).

52. NAI, SOCP 1, 1822/2350/87. In the report just cited, Browne was described as a steward, but in a newspaper account he was termed a servant and was said to have been shot through the heart for resisting an arms raid (*DEP*, 28 Feb. 1822). The two reports are not necessarily incompatible.

53. *DEP*, 14 Jan. 1823.

54. *DEP*, 20 Dec. 1823.

55. NAI, SOCP 1, 1824/2621/27.

56. *DEP*, 23 Oct. 1823.

57. 56 Geo. 3, c. 88.

58. Burke, *History of Clonmel,* pp. 206–7.

59. Ibid., pp. 115, 116, 121, 128.

60. Ibid., p. 116.

61. Ibid., p. 142.

62. Threatening notice, n.d., but early 1822 (NAI, SOCP 1, 2369/5).

63. James Hickson to [Henry Goulburn], 18 Jan. 1822 (NAI, SOCP 1, 2348/15).

64. Lt. A. Morris to Goulburn, 9 March 1822 (NAI, SOCP 1, 2349/12).

65. Earl of Glengall to Goulburn, 5 Oct. 1822 (NAI, SOCP 1, 2356/61).

66. Lt. Col. Thomas Arbuthnot to Maj. J. Finch, 11 April 1823 (NAI, SOCP 1, 2512/16).

67. Threatening notice, 17 Feb. 1822, enclosed in Stephen Williams to Sir Edward S. Lees, 18 Feb. 1822 (NAI, SOCP 1, 2344/45). See also Gibbons, *Captain Rock,* p. 134.

68. Information of John Divine, 11 March 1823 (NAI, SOCP 1, 2519/47).

69. Gibbons, *Captain Rock,* p. 147.

70. One of the dairymen attacked, Daniel Connors, had come to the Doneraile district in the previous March from the barony of Orrery and Kilmore and was "considered an interloper by the disaffected." See Maj. Samson Carter to William Gregory, 25 May 1824 (NAI, SOCP 1, 2615/29). For other attacks on dairymen, see NAI, SOCP 1, 1824/2615/47, 50, 60; 1824/2614/72.

71. Rockites in County Kilkenny reportedly promulgated the regulation that "no person shall take ground or acquire property in any other place but that of his nativity" (*DEP*, 3 Sept. 1822).

72. Francis Despard to William Gregory, 3 April 1824 (NAI, SOCP 1, 2369/17).

73. Rev. Meade Dennis to [Henry Goulburn], 20 March 1822 (NAI, SOCP 1, 2349/21). The writer placed in his letter an extract from another letter sent to him by the agent for a Kerry estate.

74. John Flood to [Dublin Castle], 12 April 1822 (NAI, SOCP 1, 2369/27). Flood's view was almost universal among Kilkenny landlords in the spring of 1822. See Lt. Col. M. Lindsay to Col. Lord Greenock, 3 May 1822 (NAI, SOCP 1, 2369/41).

75. Thomas Studdert to William Gregory, 27 May 1822 (NAI, SOCP 1, 2353/18).

76. Andrew Batwell to Henry Goulburn, 7 April 1823 (NAI, SOCP 1, 2512/11).

77. Lt. Col. Thomas Arbuthnot to Maj. J. Finch, 11 April 1823 (NAI, SOCP 1, 2512/16).

78. NAI, SOCP 1, 1823/2511/55.

79. NAI, SOCP 1, 1823/2511/10.

80. Fortunately, those locked in the burning house (three men, two women, and four children) escaped the conflagration by breaking out after the Rockites had departed (NAI, SOCP 1, 1823/2511/53). In a later report on what was apparently the same incident, the victims were identified as a respectable farmer named Clanchy and his brother-in-law, both of whom had taken evicted farms at Aghacross in the Kildorrery district (NAI, SOCP 1, 1823/2512/10).

81. Lt. Col. Thomas Arbuthnot to Maj. J. Finch, 11 April 1823 (NAI, SOCP 1, 2512/16).

82. NAI, SOCP 1, 1822/2341/27. As noted in chapter 5, Rockites murdered Jackson's cattle driver and herd Michael Ryan in July 1823.

83. *DEP*, 15 April 1824.

84. NAI, SOCP 1, 1824/2613/20, 22, 29; *DEP*, 24 April 1824.

85. For evidence of the expression of this desire in threatening letters, see Gibbons, *Captain Rock*, pp. 118 (letter no. 164), 130 (letter no. 185), 147 (letter no. 222).

86. NAI, SOCP 1, 1824/2614/47.

87. For the activities of the dowager Countess of Kingston, see Bill Power, *White knights, dark earls: the rise and fall of an Anglo-Irish dynasty* (Cork, 2000), pp. 60–71. She became involved in sharp conflict with local Catholic priests and bishops, however, because some of her enterprises wore the appearance of Protestant proselytism (ibid., pp. 65–67).

88. Lt. Col. Thomas Arbuthnot to Maj. J. Finch, 12 Sept. 1823 (NAI, SOCP 1, 2515/13).

89. Ibid.

90. *DEP*, 16 Sept. 1823. See also NAI, SOCP 1, 1823/2515/9, 13.

91. Power, *White knights, dark earls*, pp. 60–61, 71, 72–94.

92. See note 67 above. See also *DEP*, 16 Sept. 1823. Some of the lands of Scart were held directly from the third Earl of Kingston. Among the alleged principals in the murder of the Frankses were Patrick Hogan and William Murphy. Though sometimes

described as tenants of Scart under Lord Kingston, they appear to have been the sons of direct tenants of the earl. After they were taken into custody for the crime, their fathers Michael Hogan and John Murphy appealed to Lord Kingston to intervene on their behalf. This Kingston did, though it was the disappearance of a key witness that secured their release. See Petition of Michael Hogan and John Murphy, Dec. 1823 (NAI, SOCP 1, 2516/59). See also NAI, SOCP 1, 1823/2515/47; Power, *White knights, black earls,* p. 82; Cal Hyland, "The Franks: murder for love or revenge," *Mallow Field Club Journal,* no. 4 (1986), pp. 141–43.

93. John Hyde to William Gregory, 25 Feb. 1822 (NAI, SOCP 1, 2344/60). Hyde may have exaggerated the licentiousness of the population of the Kingston estates; another observer claimed that it was largely free from agrarian crime of the Rockite types (*DEP,* 8 April 1823). But the Galtee mountains around Mitchelstown offered vast scope for illicit distillation, which seems to have been a valued occupation for many of the poorer tenants of the area. Thus the suspicion that he was an informer about private stills brought a savage reprisal against Patrick Drishane and his parents (the fury "seemed to be directed against the whole family") in January 1823. Drishane was murdered and his old parents were beaten so badly that at first they were not expected to survive (*DEP,* 21 Jan. 1823). The nine assailants who burst into Drishane's house declared that "they were Captain Rock's men," according to Drishane's father, who survived and successfully prosecuted three men (Andrew, John, and Patrick Donegan) for the killing of his son (*DEP,* 8 April 1823).

94. Maxwell Blacker to Henry Goulburn, 21 Sept. 1823 (NAI, SOCP 1, 2515/24).

95. If the third Earl of Kingston is to be believed, the murder of the Frankses eventually brought some relief to the tenants at Scart. Henry Franks, the earl told a select committee of the House of Lords in May 1825, had been merciless in extracting "the rent from them, as heavy a rent as he could, when it was due, never by chance paying his own, I am sorry to say (I was his landlord)." Although the earl sought to collect rent from these same tenants after the murders by distraining their crops or livestock as Henry Franks had done, he relented upon "finding that people had already paid their rent (to Franks) and that I must have taken absolutely the clothes of[f] their back to have got anything." For these quotations, see Power, *White knights, black earls,* p. 83.

96. Donnelly, "Journals of Sir John Benn-Walsh" [part 1], pp. 86–123. According to a parliamentary return of 1876, Benn-Walsh (or Baron Ormathwaite as he had by then become) owned about 8,900 acres in County Kerry and another 2,200 acres in County Cork (ibid., p. 86).

97. Ibid., p. 91.

98. Ibid., p. 93.

99. Ibid., p. 92.

100. Ibid.

101. Ibid., pp. 93–94.

102. Ibid., p. 94. In the end Julian did redeem Tullamore and retained the lands for years (ibid., pp. 98–99).

103. Ibid., p. 91.

104. Ibid., p. 116.

105. *DEP*, 1 March 1823. A member of the Hawkes family did hold another farm called Ballygroman in Desertmore parish in the same general area. A large holding controlled by a middleman and "belonging to one of the Hawkes's," Ballygroman fell into the hands of Benn-Walsh in about 1834 "by the death of the last life in the lease" (Donnelly, "Journals of Benn-Walsh," pp. 95).

106. Donnelly, "Journals of Benn-Walsh," p. 95.

107. The hired workers were said to have come from a neighboring village; they were not local laborers (*DEP*, 22 Nov. 1821).

108. *DEP*, 24 Nov. 1821.

109. Burke, *History of Clonmel*, pp. 209–10.

110. Quoted in *DEP*, 10 June 1824. For the ejectment of the Steeles from Rathpatrick, the acquisition of their lease by John Marum, and the restoration of the Steeles in June 1824, see Ó Macháin, *Six years in Galmoy*, pp. 101–3, 131–33, 138. As Ó Macháin has shown, this episode bore a strong resemblance to Marum's acquisition of another lease under similar circumstances elsewhere in the townland of Rathpatrick in 1819; in that earlier case Marum had evicted all the undertenants after the six-month period for the redemption of the lease had expired (ibid., pp. 33–35). See also chapter 8 below, pp. 259, 261.

111. *DEP*, 21 Feb. 1824.

112. *DEP*, 15 July 1824. See also NAI, SOCP 1, 1824/2621/46, 49, 50. The witnesses to the murders included a servant

girl and the two surviving Kinnealy brothers (Daniel and John). Soon after the crime it was said that the Rockite party that attacked the Kinnealys included eighteen men, and that one of the surviving brothers had recognized four of the assailants (*DEP*, 20 July 1824). The Reverend Bell served the grammar school in Clonmel with distinction from 1822 to 1841. At the time of the murders he was also the apparently nonresident rector of Outeragh parish and the custodian of its glebe house and glebe lands. He leased the tithes and employed a well-to-do local farmer as caretaker of the glebe (*DEP*, 21 Aug. 1824). For directing me to information about the Reverend Bell, I wish to thank Líam Ó Duibhir of Clonmel.

113. Maxwell Blacker to Henry Goulburn, 27 July 1824 (NAI, SOCP 1, 2621/49).

114. NAI, SOCP 1, 1822/2356/13; 1824/2621/22.

115. Maxwell Blacker to William Gregory, 27 Feb. 1824 (NAI, SOCP 1, 2621/22).

116. *DEP*, 31 July 1824.

117. *DEP*, 21 Aug. 1824.

118. Ibid.

119. I wish to thank John O'Gorman of the Tipperary Libraries for supplying information bearing on the ownership and letting of land in Outeragh parish and on the murder of the Kinnealys.

120. J.F. Fitzgerald (Knight of Glin) to Henry Goulburn, 30 Jan. 1822 (NAI, SOCP 1, 2350/35).

121. S. O'Grady to [?], 4 Feb. 1822 (NAI, SOCP 1, 2350/48).

122. Maj. J. Logan to Maj. J. Finch, 24 Feb. 1823 (NAI, SOCP 1, 2511/33).

123. Lord Lismore to Henry Goulburn, 4 Nov. 1822 (NAI, SOCP 1, 2356/74).

124. Lt. Col. Thomas Arbuthnot to Maj. J. Finch, 17 Jan. 1823 (NAI, SOCP 1, 2511/12).

125. Arbuthnot to Finch, 11 Oct. 1823 (NAI, SOCP 1, 2515/51).

126. Daniel Mahony to William Gregory, 5 June 1823 (NAI, SOCP 1, 2519/20). Mahony conceded that there was still "great distress here as elsewhere" and hence continuing discontent (ibid.).

127. Maj. Samson Carter to [Henry Goulburn], 13 Sept. 1823 (NAI, SOCP 1, 2515/12).

128. Col. Thomas Arbuthnot to Maj. J. Finch, 28 Feb. 1824 (NAI, SOCP 1, 2614/47).

129. John Waller to Francis Blackburne, 5 Feb. 1824 (NAI, SOCP 1, 2618/27).

130. Maj. George M. Drought to Henry Goulburn, 3 Oct. 1824 (NAI, SOCP 1, 2619/38). See also NAI, SOCP 1, 1824/2619/40. By a different observer the tranquility of the districts of Newcastle, Abbeyfeale, and Rathkeale, once the epicenter of "insurrection and outrage," was attributed in August 1824 to the employment afforded by local public works (DEP, 31 Aug. 1824).

131. Henry Croker to Henry Goulburn, 30 Jan. 1824 (NAI, SOCP 1, 2621/11).

132. DEP, 2 Nov. 1824.

133. This threatening notice was discovered "stuck in the thatch of a dairy-house" of a man named Connors, who had taken an evicted farm (quoted in Gibbons, Captain Rock, pp. 198–99).

Chapter Eight.
Patterns of Rockite Violence

1. Beames, Peasants and power, pp. 51–52, 72–74. For a highly informative discussion of "agrarian outrages" during and after the Great Famine, see W.A. Vaughan, Landlords and tenants in mid-Victorian Ireland (Oxford and New York, 1994), pp. 138–76.

2. NAI, SOCP 1, 1822/2350/72.

3. Donnelly, "Factions," passim.

4. DEP, 9 March 1822.

5. The sixteen Rockite-related murders in County Limerick during this particular period were those of John Corneal (or Corneille), James Buckley, Major Richard Going, John Walsh, John Ivis, Christopher and George Sparling, Thomas Murphy, Owen Cullinane, Michael Gorman, Jr., Major Thomas Hare, Frederick Petit, Denis Morrissy, Denis Browne, Henry Sheehan (a postboy), and Sheehan's unnamed dragoon escort. The sources for these murders are cited elsewhere in this chapter. In addition, Rockite violence may have been responsible for the slayings of a man named Fitzgerald, Thomas Foley, Ulick Burke, and Thomas Egan. For these murders, see DEP, 20 Oct. 1821, 22 Jan., 23 Feb., 2 March 1822; Leinster Journal, 27 Feb., 9, 27 March 1822; NAI, SOCP 1, 1821/2296/25; 1821/2354/45; 1822/2350/72.

6. DEP, 27 March 1823.

7. This was the killing of James Egan at Rathlogan near Urlingford in September 1822 (DEP, 14 Sept. 1822; Leinster Journal, 18 Sept. 1822; Ó Macháin, Six years in Galmoy, pp. 83–84). The murder of an unnamed policeman in October 1822 was almost

412

certainly a Rockite crime, and because it reportedly occurred "near Durrow," it has been assigned in the analysis below to County Kilkenny (*DEP*, 26 Oct. 1822).

8. *DEP*, 13 May 1824.

9. The victims in these six murders were Martin Quane, James Egan, John Neale, John Phelan (or Whelan), the opulent farmer John Marum, and Edmond (or Edward) Long. For Neale, a tithe proctor who was fatally beaten in Galmoy barony in April 1822, see Ó Macháin, *Six years in Galmoy*, pp. 60–61. The sources for the other agrarian killings in County Kilkenny are cited elsewhere in this chapter. The policeman mentioned in note 7 above might also have been included in this list.

10. The violent death of an unidentified member of the Palatine community in the Adare district early in December 1821 was reported without the attribution of a motive (NAI, SOCP 1, 1821/2354/45). Nevertheless, the targeting of other Palatines by the agrarian rebels made it seem likely that this should be considered a Rockite crime. Indeed, evidence discovered years after the author's original encounter with one truncated report of this killing shows that the victim was George Sparling, the brother of Christopher Sparling, whom a band of Rockites had murdered on 20 October 1821. George Sparling was the target of an arms raid and was beaten so badly that he died soon afterward (Memorial of Christopher Sparling to Marquis Wellesley, 27 April 1827, PRO, CO 384/12).

11. John Minton to William Gregory, 28 April 1822 (NAI, SOCP 1, 2345/67). See also NAI, SOCP 1, 1822/2343/15.

12. The victims in these eight cases were Fitzgerald, Thomas Foley, Ulick Burke, Thomas Egan, John Shea, Dennis Donoghue, Conway, and Thomas Kennedy. For the first four, see the sources cited in note 5 above; for the last four, see *DEP*, 23 Feb. 1822, 23 March, 11 May 1824; NAI, SOCP 1, 1822/2355/30; 1823/2518/49.

13. NAI, SOCP 1, 1821/2295/19.

14. *DEP*, 29 Nov. 1821; NAI, SOCP 1, 1821/2295/21.

15. *Leinster Journal*, 6 April 1822.

16. Ibid. See also *DEP*, 30 March 1822.

17. *DEP*, 12 Oct. 1822.

18. *DEP*, 23 March 1824; NAI, SOCP 1, 1824/2614/67–68.

19. *DEP*, 22 July 1823; NAI, SOCP 1, 1823/2513/56, 60, 68; 1823/2518/40.

20. *DEP*, 26 March 1822.

21. *DEP*, 12, 19, 24 Feb. 1824.

22. *DEP*, 10, 22 Jan., 5 Feb. 1822, 5, 15 July, 18 Oct. 1823.

23. *DEP*, 26 March, 5 Nov. 1822; *Leinster Journal*, 3 April 1822; NAI, SOCP 1, 1823/2518/1; Ó Macháin, *Six years in Galmoy*, pp. 60–61.

24. *Leinster Journal*, 25 Aug. 1824; NAI, SOCP 1, 1824/2621/33, 36.

25. *DEP*, 16 April 1822; *Leinster Journal*, 20 April 1822; NAI, SOCP 1, 1822/2369/28.

26. As indicated in note 7 above, the Kilkenny total includes a policeman murdered "near Durrow" in October 1822.

27. *DEP*, 27 Dec. 1821.

28. *DEP*, 26 Sept. 1822, 14 Jan., 20 Nov. 1823, 9, 14 Dec. 1824; *Leinster Journal*, 11 Dec. 1824; NAI, SOCP 1, 1823/2516/12; 1824/2618/53; 1824/2619/53, 56.

29. For Torrance, see *DEP*, 20 Dec. 1821. For Walsh, see *DEP*, 20, 23 Oct. 1821; NAI, SOCP 1, 1821/2296/25. For Murphy, see *DEP*, 6 Dec. 1821; NAI, SOCP 1, 1821/2354/45; 1822/2350/51. For Petit, see *DEP*, 21 Feb. 1822; *Leinster Journal*, 27 Feb. 1822; NAI, SOCP 1, 1822/2350/69, 71. For Neill, see *DEP*, 28 March 1822; *Leinster Journal*, 3, 27 April 1822; NAI, SOCP 1, 1822/2351/57, 60; 1822/2352/25; 1822/2353/32.

30. *DEP*, 21 Feb. 1822. One reason why the Rockites killed Petit was his close relationship to the murdered Palatine brothers and Adare yeomen Christopher and George Sparling (Petit's father-in-law). According to George Sparling's eldest son in 1827, Petit had been killed "merely for his alliance with the Sparlings." See Memorial of Christopher Sparling to Marquis Wellesley, 27 April 1827 (PRO, CO 384/12).

31. For Going, see *DEP*, 16, 20 Oct. 1821; NAI, SOCP 1, 1821/2296/21. For Gorman, see *DEP*, 5 Jan. 1822, 14 Aug. 1824; *Leinster Journal*, 9 Jan. 1822. For Hare, see *DEP*, 5 Feb. 1822, 10 Aug. 1824; *Leinster Journal*, 6 Feb. 1822. For Sheehan and his dragoon escort, see *DEP*, 5, 12 March 1822; *Leinster Journal*, 9 March 1822; NAI, SOCP 1, 1822/2351/5; 1822/2353/20; 1822/2354/9.

32. For Manning, see *DEP*, 23 Aug., 22 Dec. 1821. For Corneal, see *DEP*, 29 Sept., 20 Oct. 1821; *Leinster Journal*, 27 March 1822. For Ivis, see *DEP*, 20 Oct. 1821. In the two murders near Limerick city the victims were Denis Morrissy and an unnamed farm servant. For Morrissy, see *DEP*, 7 Aug. 1822, 26 Aug. 1823; NAI, SOCP 1, 1823/2517/62. For the farm servant, see *DEP*, 14 March 1822.

33. For Hoskins, see *DEP*, 6 Aug., 20 Sept. 1821. For Buckley, see *DEP*, 29 Sept., 20 Oct. 1821, 10 Sept. 1822. For Sparling, who was killed at Dromin near Newcastle, see *DEP*, 20 Oct., 6, 20 Dec. 1821; PRO, CO 384/14, pp. 558/1107–8. His brother George Sparling was fatally beaten early in December 1821 when Rockites raided his house near Adare for arms (PRO, CO 384/12). For Ambrose, see *DEP*, 14 March 1822.

34. In the murders near Ballingarry and Abbeyfeale the victims were Denis Browne and John Hartnett. For Browne, see *DEP*, 28 Feb. 1822; *Leinster Journal*, 9 March 1822; NAI, SOCP 1, 1822/2350/87. For Hartnett, see *DEP*, 14 Aug. 1824; NAI, SOCP 1, 1824/2618/7, 10, 12, 18.

35. *DEP*, 20 Dec. 1821.

36. *DEP*, 29 Sept. 1821.

37. *DEP*, 20 Oct. 1821.

38. Ibid.

39. For the murders of Manning, Going, and Ivis, see the relevant sources cited in notes 31 and 32 above.

40. For the burning at the Shea farm, see *DEP*, 22, 24, 27 Nov. 1821, 21 Aug. 1824.

41. *DEP*, 5 Feb. 1822; *Leinster Journal*, 6 Feb. 1822; NAI, SOCP 1, 1822/2344/16, 26, 33.

42. *DEP*, 26 Sept., 26 Oct. 1822, 17 April 1823; NAI, SOCP 1, 1824/2621/58.

43. The sources for eight of these murders have been cited in previous footnotes: Owen Cullinane (note 27), Susanna Torrance (note 29), Thomas

Murphy (note 29), Frederick Petit (notes 29 and 30), Thomas Hare (note 31), Denis Morrissy (note 32), George Sparling (note 33), and Denis Browne (note 34). For Thomas Max, see *DEP,* 21 Feb. 1822. For the wood-ranger (Phelan or Whelan), see *DEP,* 2 April 1822; *Leinster Journal,* 3 April 1822; NAI, SOCP 1, 1822/2355/67, 70; 1822/2621/38. For Richard Crofts, see *DEP,* 10 June 1823; NAI, SOCP 1, 1823/2513/56, 59–60, 68.

44. For the murders of Corneal, Murphy, and Petit, see the relevant sources cited in notes 29 and 36 above. For the triple murder of Thomas, Margaret, and Henry Franks, see above, chapter 7, pp. 234–36, and below, pp. 262–63, 320–21.

45. *DEP,* 29 Sept. 1821; NAI, SOCP 1, 1822/2354/25.

46. For the quotations in this paragraph, see *DEP,* 10 Sept. 1822.

47. For the murders of Jeremiah Scully, the laborer Stack, and John Sullivan, see *DEP,* 1 Dec. 1821 (Scully); *DEP,* 29 Dec. 1821, 5 Jan. 1822; *Leinster Journal,* 2 Jan. 1822 (Stack); and NAI, SOCP 1, 1822/2343/15; 1822/2345/67 (Sullivan).

48. For Murphy, see note 29 above; for the Rathcormack woman, see NAI, SOCP 1, 1822/2347/41.

49. *DEP,* 8 April 1823. See also *DEP,* 21 Jan. 1823.

50. Neill had lodged an information against twelve persons for participating in a Rockite attack on the postmaster's house at Shanagolden in September 1821 (NAI, SOCP 1, 1822/2352/25). The mere report that he had become an informer "caused half the male inhabitants of Ballyhahill in the mountains between Shanagolden and Glynn [*sic*] to leave their homes . . ." (Col. William Thornton to Lt. Col. Sorell, 26 March 1822, copy, NAI, SOCP 1, 2351/60). See also the relevant sources cited in note 29 above.

51. Thomas P. Vokes to William Gregory, 26 March 1822 (NAI, SOCP 1, 2351/57). Thirteen of the sixteen persons in the category of murdered informers or potential informers have been mentioned in the text. The other three were Daniel McAuliffe, slain near Cork city in February 1822 (*DEP,* 16 Feb. 1822); an unidentified "Scotch captain" killed in County Limerick in the spring of that year (*DEP,* 6 April 1822); and a prosecution witness named Ryan, murdered in County Clare in July 1823 (*DEP,* 19 July 1823).

52. Apart from the five victims in this category whose cases are discussed in the text, the following persons were murdered as land grabbers or their employees: Michael Gorman, Jr., killed near Rathkeale in late December 1821 (see note 31); a farm servant beaten to death near Limerick city in March 1822 (see note 32); Martin Quane, another farm servant, shot dead in daylight at Killamery, Co. Kilkenny, on 1 April 1822 (see note 25); Thomas Hill, a well-to-do farmer, slain near Clonakilty, Co. Cork, in October 1823 (*DEP,* 16 Oct. 1823, 2 Sept. 1824); an unidentified farmer waylaid and killed near Tipperary town in December 1823 (NAI, SOCP 1, 1823/2518/69); John Oakley, bludgeoned to death near Nenagh in the same county in April 1824 (*Leinster Journal,* 25 Aug. 1824; NAI, SOCP 1, 1824/2621/33, 36); and David Killigrew, beaten to death near Kilworth,

Co. Cork, in June 1824 (NAI, SOCP 1, 1824/2615/55).

53. *DEP*, 15, 20, 31 July, 21 Aug. 1824; NAI, SOCP 1, 1824/2621/46, 49–50. For an extended discussion of this episode, see chapter 7 above, p. 239.

54. Quoted in *DEP*, 1 Dec. 1821.

55. NAI, SOCP 1, 1824/2606/16, 19–21. See also *DEP*, 18 March, 26 Aug. 1824; *Leinster Journal*, 25 Aug. 1824. For a meticulous recent account of the murder of John Marum and of the events that preceded and followed it, see the model study by Ó Macháin, *Six years in Galmoy*, pp. 99–179.

56. Lt. Col. M. Lindsay to William Gregory, [?] March 1824 (NAI, SOCP 1, 2606/15).

57. Joseph Nicolson to Gregory, 18 March 1824 (NAI, SOCP 1, 2606/19).

58. Lt. Col. M. Lindsay to Gregory, [?] March 1824 (NAI, SOCP 1, 2606/15).

59. *DEP*, 10 June 1824. For a brief discussion of John Marum as a hated middleman, see chapter 7 above, pp. 239–40.

60. NAI, SOCP 1, 1824/2618/53; 1824/2619/56.

61. *DEP*, 14 Dec. 1824. See also NAI, SOCP 1, 1824/2619/53.

62. The victims in this category were the driver John Egan, shot to death by a defaulting tenant at Clonagoose, Co. Tipperary, in March 1822 (*DEP*, 26 March 1822; *Leinster Journal*, 3 April 1822); James Egan, stoned and pitchforked to death at Rathlogan, near Urlingford, Co. Kilkenny, in September 1822 while seeking to prevent the rescue of some distrained corn (*DEP*, 14 Sept. 1822; *Leinster Journal*,

18 Sept. 1822); William Green, waylaid and shot in the head near Tipperary town in August 1823 (NAI, SOCP 1, 1823/2518/42, 49); and the two keepers Laurence Lyons and Robert Vallance, both stoned to death on the same day in November 1823 within about two miles of each other, Lyons at Killeagh near Mitchelstown, Co. Cork, and Vallance at Darragh More near Kilmallock, Co. Limerick (*DEP*, 20 Nov. 1823; NAI, SOCP 1, 1823/2516/12, 14).

63. The bailiff Matthew Carey was slain at Lisnamuck near Clogheen, Co. Tipperary, in January 1823 while executing a legal decree (NAI, SOCP 1, 1823/2518/1); the process server (unidentified) was beaten to death near Bandon, Co. Cork, in May 1823 (*DEP*, 8 May 1823); and the driver Andrew Callaghan died in an affray in October 1823 while distraining lands in County Cork (*DEP*, 18 Oct. 1823).

64. For the murders of Hoskins and Ambrose, see the relevant sources cited in note 33 above.

65. *DEP*, 14 Aug. 1824; NAI, SOCP 1, 1824/2618/7, 10, 18. For the identity and social status of the three men who murdered Hartnett, see above, chapter 5, p. 158.

66. Thomas P. Vokes to Henry Goulburn, 20 Jan. 1824 (NAI, SOCP 1, 2618/12).

67. *DEP*, 14 Jan. 1823.

68. *DEP*, 16, 20 Dec. 1823.

69. *DEP*, 24 June 1824.

70. *DEP*, 16 Sept. 1823.

71. *DEP*, 15 April 1824.

72. NAI, SOCP 1, 1823/2515/8. See also NAI, SOCP 1, 1823/2515/19.

73. *DEP*, 16 Sept. 1823.

74. NAI, SOCP 1, 1823/2515/7, 9; *DEP*, 13, 16 Sept. 1823.

75. For an extended discussion of the conduct of Thomas Franks as a middleman under the dowager Countess of Kingston, see above, chapter 7, pp. 234–36.

76. Timothy Sheehan, a brother of Cornelius Sheehan, confessed to participation in the murders (NAI, SOCP 1, 1823/2515/25, 34). Sheehan escaped from the guardhouse at Doneraile early in December 1823 and was not recaptured until November 1824 (*DEP*, 4 Dec. 1823, 6 Jan., 26 Feb., 2 March 1824; NAI, SOCP 1, 1824/2617/35). Another accomplice to the crime, Edmond Magner, a laborer, made a confession in January 1824 and acted as an approver against three men convicted of the murders in the following April (NAI, SOCP 1, 1824/2511/15; *DEP*, 15 April 1824).

77. Maxwell Blacker to Henry Goulburn, 21 Sept. 1823 (NAI, SOCP 1, 2515/24).

78. See the source cited in note 73 above.

79. NAI, SOCP 1, 1823/2515/9.

80. For reports that at least some of this murderous band of Rockites wore women's clothes, see *DEP*, 13, 16 Sept. 1823.

81. The victims in these tithe conflicts were the proctor John Ivis, killed near Askeaton, Co. Limerick, in October 1821 (note 32); the driver Sughrue, slain near Cahersiveen, Co. Kerry, in January 1822 by an enraged crowd after he had shot dead a peasant boy (*DEP*, 10, 22 Jan. 1822); the process server Conway, stoned and bayoneted to death near Listowel, Co. Kerry, in the same month (*DEP*, 15 Jan. 1822; NAI, SOCP 1, 1822/2348/6, 18, 51); an unidentified proctor fatally shot near Thurles, Co. Tipperary, also in January 1822 (NAI, SOCP 1, 1822/2355/11); the driver Patrick Driscoll, killed by a crowd in an affray near Skibbereen, Co. Cork, in July 1823 (*DEP*, 5, 15 July 1823); and the process server Edmond (or Edward) Long, mortally wounded by the blow of a two-handled wattle at Rathealy near Tullaroan, Co. Kilkenny, in March 1824 (*DEP*, 23 March, 26 Aug. 1824; *Leinster Journal*, 25 Aug. 1824; Ó Macháin, *Six years in Galmoy*, p. 101).

82. For Going, see note 31 above.

83. For Walsh, see note 29 above.

84. *DEP*, 24 Jan. 1822; NAI, SOCP 1, 1822/2343/5.

85. For this incident, see the relevant sources cited in note 31 above.

86. *DEP*, 17 April 1823; NAI, SOCP 1, 1824/2621/58. Three soldiers also drowned in February 1822 as they tried to cross the flooding River Feale in a small boat. They were part of a detachment of troops sent from Listowel to a military post at Duagh, which Rockites had threatened to attack while it was manned temporarily by members of the Listowel yeomanry corps (*DEP*, 21 Feb. 1822; NAI, SOCP 1, 1822/2348/61).

87. For the attack on the Churchtown police barracks, see the sources cited in note 41 above.

88. *DEP*, 26 Sept., 26 Oct. 1822.

89. *DEP*, 23 Aug., 22 Dec. 1821, 5, 15 July 1823.

90. *DEP*, 5 Feb. 1822.

91. *DEP*, 5 Nov. 1882.

92. *DEP*, 13 Dec. 1823; NAI, SOCP 1, 1823/2516/38, 41, 48.

93. See, e.g., *DEP,* 20 Oct., 29 Nov. 1821, 14 Jan. 1823. Even if not life-threatening, some beatings left the victims maimed. A bailiff or driver named Harvey, searching for distrained goods that had been rescued illegally, lost an eye in a beating at Tubber fair in Clare in September 1823. Neville Payne Nunan, a gentleman executing a legal decree near Rathkeale in the same month, was attacked by a crowd, beaten, and hamstrung. A servant named Hanrahan, caring for a farm on the Earl of Clare's estate near Shanagolden, was attacked in April 1824 by a band of Rockites who broke three of his ribs and knocked out one of his eyes with the blow of a musket (*DEP,* 25 Sept. 1823, 20 April 1824).

94. *DEP,* 15 July 1824.

95. *DEP,* 17 April 1823.

96. NAI, SOCP 1, 1822/2348/51.

97. *DEP,* 19 Aug. 1824.

98. *DEP,* 24 Nov. 1821.

99. A. Crossley to Maj. Samson Carter, 8 Dec. 1823 (NAI, SOCP 1, 2516/38).

100. *Leinster Journal,* 3 April 1822.

101. Threatening notice, [?] Jan. 1822, enclosed in Clutterbuck Crone to Henry Goulburn, 15 Jan. 1822 (NAI, SOCP 1, 2342/34). For treatment of the popular rejoicing over the death and mutilation inflicted on the police involved in the 1831 "battle" of Carrickshock, see Gary Owens, "The Carrickshock incident, 1831: social memory and an Irish cause célèbre," *Cultural and Social History* 1, no. 1 (2004), pp. 40–42. This article is a scintillating example of both micro-history and the history of social memory.

102. Rudé, *Protest and punishment: the story of the social and political protesters transported to Australia, 1788–1868* (Oxford, 1978) p. 145.

103. See, e.g., *DEP,* 3 Oct. 1822, 19 April 1823.

104. NAI, SOCP 1, 1822/2346/52, 63; 1823/2511/13, 32.

105. NAI, SOCP 1, 1821/2295/8.

106. Lord Headley to Charles Grant, 22 Oct. 1821 (NAI, SOCP 1, 2295/13).

107. See, e.g., NAI, SOCP 1, 1822/2342/6.

108. NAI, SOCP 1, 1822/2343/44.

109. *DEP,* 27 Nov. 1821.

110. *DEP,* 2 March 1822; NAI, SOCP 1, 1822/2350/93.

111. For the burning of Knockane church, see *DEP,* 1 Dec. 1821; NAI, SOCP 1, 1822/2348/12. For Templenoe, see *DEP,* 12 Jan. 1822; NAI, SOCP 1, 1822/2348/3, 5. For Kilgarvan, see NAI, SOCP 1, 1822/2348/5. For Killeedy, see *DEP,* 12, 16 Feb. 1822; NAI, SOCP 1, 1822/2350/39, 53–54, 60; 1822/2344/22. For Ballybrood, see *DEP,* 26 Feb. 1822; NAI, SOCP 1, 1822/2350/73, 94. For Athlacca, see *DEP,* 2 March 1822; NAI, SOCP 1, 1822/2350/93.

112. NAI, SOCP 1, 1822/2345/62; 1822/2350/53.

113. See above, chapter 2, p. 79.

114. *DEP,* 1 Dec. 1821.

115. R.T. Herbert to Knight of Kerry, 9 Jan. 1822 (NAI, SOCP 1, 2348/5). The liberality of this comment was coupled with considerable naiveté about the attitudes of Catholic peasants. Neither priests nor Catholic gentlemen were to be found among Captain Rock's adher-

ents, whose sectarianism was otherwise made all too plain.

116. NAI, SOCP 1, 1822/2350/39, 60.

117. NAI, SOCP 1, 1822/2350/73.

118. NAI, SOCP 1, 1822/2348/3, 5, 12.

119. NAI, SOCP 1, 1822/2350/53, 73, 93.

120. NAI, SOCP 1, 1823/2515/21, 17, 30; *DEP*, 20 Sept. 1823. After the revival of the Rockite movement in the late summer of 1822, however, a few other Protestant churches were desecrated. See NAI, SOCP 1, 1823/2516/37, 40; *DEP*, 9 Dec. 1824.

121. *DEP*, 21, 23 Feb., 2, 5, 7, 9, 12, 30 March, 2, 9, 11 April 1822.

122. Edward Wilson to William Gregory, 6 April 1822 (NAI, SOCP 1, 2355/68).

123. Abraham Allen to [?], 6 March 1822 (NAI, SOCP 1, 2351/18).

124. *DEP*, 28 Feb., 2 March 1822.

125. *DEP*, 28 Feb. 1822.

126. *DEP*, 2 March 1822.

127. Ibid.

128. *DEP*, 23, 26 Feb., 2 March 1822.

129. See chapter 6 above, pp. 199–201. For vivid reports of the burning of tithe corn in north Cork in these months, see NAI, SOCP 1, 1822/2346/68; *DEP*, 1 Oct. 1822.

130. Robert Eames to Sir Edward S. Lees, 24 Sept. 1822 (NAI, SOCP 1, 2346/63). For the critical comment about the conduct of fearful landed gentlemen, see NAI, SOCP 1, 1822/2346/52.

131. Quoted in *DEP*, 1 Oct. 1822. This writer expressed surprise that this violent campaign was occurring in spite of over one hundred capital convictions in agrarian cases since the previous Christmas, leading to thirty or forty executions and the transportation of sixty to eighty other convicted prisoners (ibid.).

132. The first spate of burnings was concentrated around Mallow, Doneraile, and Buttevant; the fires soon extended to the districts of Youghal, Killeagh, and Ballynamona. Toward the end of 1822 some incendiary crimes took place in that portion of the barony of Condons and Clangibbon lying north of the River Blackwater. See *DEP*, 3 Oct. 1822; NAI, SOCP 1, 1822/2346/72; 1822/2347/49, 52. For the general absence of incendiary fires in Duhallow through the end of 1822, see NAI, SOCP 1, 1822/2347/5, 42, 43.

133. NAI, SOCP 1, 1823/2511/27.

134. "A statement of outrages that have occurred in the baronies of Orrery and Kilmore, Fermoy, Duhallow, Condons and Clongibbons [*sic*] from 6th January to 7th day of March 1823," 8 March 1823, by Maj. Samson Carter (NAI, SOCP 1, 2511/45).

135. NAI, SOCP 1, 1822/2347/38, 40.

136. In a notice sent to George Crofts in December 1822, the Rockites threatened to punish this gentleman "like the bloodhound Hill and like James Hill the robbing tyrant" (Threatening notice, [?] Dec. 1822, NAI, SOCP 1, 2347/63).

137. The damages suffered by Colonel Hill were estimated at a minimum of £500 (NAI, SOCP 1, 1822/2347/50, 53). Two possible motives were advanced for the attack, one of which was local opposition to tithes.

The grain destroyed was tithe corn from Caherduggan parish, where Hill was the "tithe farmer and his own proctor" (Rev. William O'Brien to Daniel Mahony, 24 Dec. 1822, NAI, SOCP 1, 2511/2). In addition, Hill was a scourge of local Rockites; he had "uniformly and openly manifested a determination to put down insurrection" (Col. Sir Hugh Gough to Maj. J. Finch, 13 Dec. 1822, copy, NAI, SOCP 1, 2347/53).

138. NAI, SOCP 1, 1822/2346/46, 50; 1822/2347/53.

139. NAI, SOCP 1, 1822/2347/61, 64.

140. Col. Sir Hugh Gough to William Gregory, 5 March 1823 (NAI, SOCP 1, 2511/55).

141. NAI, SOCP 1, 1823/2511/41, 43; *DEP*, 13 March 1823.

142. NAI, SOCP 1, 1823/2511/43; *DEP*, 11 March 1823.

143. Col. Sir Hugh Gough to Maj. J. Finch, 17 March 1823 (NAI, SOCP 1, 2511/55). For an earlier mention of the burning of Carker Lodge, see above, chapter 5, pp. 177-78.

144. Hickey's blunderbuss burst during the attack; his hand was so badly lacerated that it had to be amputated. The mishap led to his apprehension (NAI, SOCP 1, 1823/2511/43, 55).

145. A. Evans to Eliza Evans, 11 March 1823, copy (NAI, SOCP 1, 2511/50).

146. *DEP*, 6, 11, 20 Feb., 1 March 1823; NAI, SOCP 1, 1823/2511/4, 13, 27, 32; 1823/2518/5.

147. NAI, SOCP 1, 1823/2512/5. See also *DEP*, 6, 11, 13, 18, 20, 25, 27 March 1823; NAI, SOCP 1, 1823/2511/52-53; 1823/2512/10, 14; 1823/2517/17-18.

148. Maj. Gen. Sir John Lambert to Lieut. Gen. Lord Combermere, 1 April 1823, copy (NAI, SOCP 1, 2512/5).

149. Lt. Col. Sir Thomas Arbuthnot to Maj. J. Finch, 11 April 1823, copy (NAI, SOCP 1, 2512/16).

150. NAI, SOCP 1, 1823/2512/5.

151. Maxwell Blacker to William Gregory, 19 April 1823 (NAI, SOCP 1, 2518/19).

152. *DEP*, 29 April 1823.

153. *DEP*, 24, 26, 29 April, 1, 3, 6, 8, 13, 15, 17, 20, 24, 29, 31 May, 7, 17, 24 June, 1, 10, 12, 15, 22 July 1823.

154. *DEP*, Feb.-April 1823.

155. *DEP*, 20 March 1823.

156. Ibid.

157. NAI, SOCP 1, 1823/2512/21, 23; 1823/2513/7; *DEP*, 27 March, 3 May 1823.

158. NAI, SOCP 1, 1823/2512/21.

159. NAI, SOCP 1, 1823/2511/38; *DEP*, 18, 25, 29 March, 15 April 1823.

160. *DEP*, 1 March, 24 April 1823. For a discussion of the middleman John Hawkes, Sr., and his family, see above, chapter 7, pp. 238-39.

161. NAI, SOCP 1, 1823/2517/23.

162. NAI, SOCP 1, 1823/2517/6, 8, 17-18; *DEP*, 6 Feb., 1, 25, 27 March, 1, 3 April 1823.

163. NAI, SOCP 1, 1823/2517/27.

164. NAI, SOCP 1, 1823/2517/26.

165. See above, chapter 4, pp. 139-40.

166. NAI, SOCP 1, 1823/2517/25-26, 33; *DEP*, 3 May 1823.

167. NAI, SOCP 1, 1823/2517/27.

168. NAI, SOCP 1, 1823/2517/26.

169. The police eventually secured an information from an accomplice (William Thornhill) against twenty persons for participating in the attack on

Glenosheen. Thirteen were promptly arrested. Of the other seven, three had fled the country, two were already in jail at Cork for another crime, one was being held in "special custody," and one (Buckley) had been killed at Glenosheen. See Maj. Samson Carter to William Gregory, 31 July 1823 (NAI, SOCP 1, 2517/59).

170. *DEP*, 6 March 1823; NAI, SOCP 1, 1823/2519/4, 6, 8, 10.

171. NAI, SOCP 1, 1823/2519/8; *DEP*, 15 April 1823.

172. *DEP*, 15 April 1823.

173. *DEP*, 21 Jan., 1, 25 March 1823. See also *DEP*, 1, 6 Feb. 1823.

174. *DEP*, 3 April 1823.

175. *DEP*, 5, 7, 14 Aug. 1823.

176. For burnings in north Cork, see NAI, SOCP 1, 1823/1256/2, 57; 1824/2614/7, 9, 12, 55, 63, 72; 1824/2615/45; *DEP*, 4 Sept., 30 Oct., 6, 13, 15, 20 Nov., 9 Dec. 1823, 1 Jan., 6, 11, 24 March 1824. For burnings in west Limerick, including the Liberties of Limerick city, see NAI, SOCP 1, 1824/2618/63, 68; *DEP*, 18 Sept., 9, 16, 28, 30 Oct., 27 Nov., 2, 13 Dec. 1823, 6, 8, 15 Jan., 12, 19 Feb., 11 March, 8, 22, 29 April, 4, 13, 25, 27 May, 1 June 1824.

177. For burnings in Clare, see *DEP*, 3, 15, 24, 27 April, 4, 20, 27 May, 3, 10 June, 15, 29 July, 14 Aug. 1824.

178. A mother and her daughter were killed when the roof collapsed on them while they were trying to remove their belongings from a dwelling that some Rockites had set on fire.

179. NAI, SOCP 1, 1823/2512/10.

180. NAI, SOCP 1, 1823/2512/14.

181. NAI, SOCP 1, 1822/2351/18; 1823/2511/10.

182. *DEP*, 29 March, 24 July 1823.

183. *DEP*, 24 July 1823.

184. *DEP*, 29 March 1823.

185. 15 & 16 Geo. 3, c. 21, s. 8.

186. *DEP*, 29 March 1823.

187. NAI, SOCP 1, 1822/2351/18.

188. NAI, SOCP 1, 1822/2347/19.

189. Ibid.; *DEP*, 3 April 1823. See also above, chapter 5, p. 163.

190. See above, chapter 3, p. 184.

191. *DEP*, 6 March 1823.

192. NAI, SOCP 1, 1824/2617/18, 26.

193. Donnelly, "Irish agrarian rebellion," p. 310; idem, "Rightboy movement," p. 142.

194. Whooley, "Captain Rock's rebellion," p. 278.

195. *DEP*, 10 Nov. 1821.

196. See above, chapter 2, p. 79.

197. *DEP*, 4 Dec. 1821.

198. Whooley, "Captain Rock's rebellion," p. 278.

199. J. Lloyd to William Gregory, 30 Jan. 1822 (NAI, SOCP 1, 2350/36).

200. *DEP*, 17 Nov. 1821; NAI, SOCP 1, 1821–22/2345/32.

201. *DEP*, 10 Nov. 1821.

202. NAI, SOCP 1, 1823/2511/32.

203. *DEP*, 29 Jan. 1822.

204. Maj. Samson Carter to William Gregory, 1 March 1824 (NAI, SOCP 1, 2614/46).

205. Col. Sir Hugh Gough to Maj. J. Finch, 21 Feb. 1823 (NAI, SOCP 1, 2511/32); Andrew Batwell to William Gregory, 22 April 1823 (SOCP 1, 2512/32); Ensign Henry Bell to Col. Sir John Buchan, 26 July 1823, copy (NAI, SOCP 1, 2514/30); *DEP*, 14 June 1823.

206. Indeed, George Lidwill, a member of the Catholic Association who was at odds with O'Connell over the Tithe

Composition Act of 1823, pointed out correctly enough at one of its meetings in January 1824 that O'Connell had "defended more criminals for the perpetration of such offences, springing from the collection of tithes, than any other lawyer at the bar" (*DEP*, 13 Jan. 1824). See also chapter 4 above, p. 149.

207. Whooley, "Captain Rock's rebellion," p. 56.

208. See, e.g., chapter 1 above, pp. 47–48; chapter 5 above, p. 186.

209. NAI, SOCP 1, 1823/2517/81.

210. NAI, SOCP 1, 1822/2348/15.

211. NAI, SOCP 1, 1824/2614/58. See also NAI, SOCP 1, 1823/2513/27; 1824/2615/51–52.

212. NAI, SOCP 1, 1823/2513/67.

213. "Extracts of a letter from near Fermoy," [?] April 1823 (NAI, SOCP 1, 2512/1).

214. Whooley, "Captain Rock's rebellion," pp. 54–55.

215. For a case of attempted intimidation of this kind, see NAI, SOCP 1, 1823/2515/31.

216. *DEP*, 12 June 1823.

217. NAI, SOCP 1, 1823/2511/28.

218. Lt. Col. Thomas Arbuthnot to Maj. J. Finch, 11 April 1823, copy (NAI, SOCP 1, 2512/16).

219. Andrew Batwell to William Gregory, 22 April 1823 (NAI, SOCP 1, 2512/32).

220. For an example from the Rathkeale district, see NAI, SOCP 1, 1822/2350/19.

221. *DEP*, 17 Nov. 1821.

222. NAI, SOCP 1, 1823/2511/32.

223. NAI, SOCP 1, 1824/2615/ 47, 50.

224. Whooley, "Captain Rock's rebellion," p. 58.

225. *DEP*, 14 June 1823.

226. A notice posted on the chapel door at Killorglin, Co. Kerry, contained the following provision: "It is further resolved that no man whatever under the pain of first . . . a severe flogging and afterwards of decapitation shall take the liberty of demanding money in the name of Captain Rock, nothing being more contrary to our will" (Gibbons, *Captain Rock*, pp. 141–42 [no. 212]).

227. "John Rock" to Capt. Lindsey, 21 Nov. 1821, copy (NAI, SOCP 1, 2296/38).

228. NAI, SOCP 1, 1822/2349/5.

229. Whooley, "Captain Rock's rebellion," p. 60.

230. Col. Sir Hugh Gough to Lt. Col. J. Finch, 10 Jan. 1824, copy (NAI, SOCP 1, 2614/11).

231. NAI, SOCP 1, 1824/2617/18, 26, 33.

232. J.S. Donnelly, Jr., "Hearts of Oak, Hearts of Steel," *Studia Hibernica*, no. 21 (1981), p. 71.

233. Whooley, "Captain Rock's rebellion," pp. 61–62.

234. Ibid., p. 61.

Chapter Nine.
Repression of the Movement

1. Andrew Batwell to William Gregory, 22 Feb. 1823 (NAI, SOCP 1, 2511/13).

2. *DEP*, 13 May 1824.

3. NAI, SOCP 1, 1823/2511/6.

4. Maj. Gen. Sir John Lambert to Maj. J. Finch, 2 Jan. 1823, copy (NAI, SOCP 1, 2511/2).

5. Fr. William O'Brien to Daniel Mahony, 24 Dec. 1822, copy (ibid.).

6. James Crosbie to William Gregory, 26 Feb. 1822 (NAI, SOCP 1, 2348/63). See also NAI, SOCP 1, 1822/2349/38.

7. W.H.W. Newenham to Marquis of Londonderry, 4 Feb. 1822, copy (NAI, SOCP 1, 2344/39).

8. Thomas Studdert to William Gregory, 16 April 1822 (NAI, SOCP 1, 2352/23).

9. Broeker, *Rural disorder,* pp. 149–52; NAI, SOCP 1, 1822/2346/22; *DEP,* 8 July 1823.

10. Maj. Samson Carter to William Gregory, 16 July 1822 (NAI, SOCP 1, 2346/11).

11. Cornelius O'Leary to [Henry Goulburn], 8 June 1824 (NAI, SOCP 1, 2615/37). See also NAI, SOCP 1, 1824/2615/38; *DEP,* 19, 31 Aug. 1824.

12. Richard B. Osborne to [Goulburn], 7 Jan. 1822 (NAI, SOCP 1, 2355/5).

13. NAI, SOCP 1, 1822/2347/22.

14. Col. Sir Hugh Gough to Maj. J. Finch, 13 Dec. 1822, copy (NAI, SOCP 1, 2347/53). See also NAI, SOCP 1, 1822/2346/46, 50; 1822/2347/63.

15. NAI, SOCP 1, 1822/2347/61, 64.

16. Col. Sir Hugh Gough to Maj. J. Finch, 17 March 1823, copy (NAI, SOCP 1, 2511/55).

17. Maj. Samson Carter to William Gregory, 8 March 1823 (NAI, SOCP 1, 2511/43).

18. Confession of Edmond Magner, 28 Jan. 1824 (NAI, SOCP 1, 2511/15).

19. Donnelly, "Whiteboy movement," pp. 21, 27, 48, 51–52; idem, "Irish agrarian rebellion," pp. 305–6.

20. These figures relating to the murder of informers are based on reports appearing in the *Dublin Evening Post* and in the State of the Country Papers, Series 1.

21. Gerald Fitzgerald to William Gregory, 2 June 1822 (NAI, SOCP 1, 2356/24).

22. *DEP,* 16 Oct. 1823. See also NAI, SOCP 1, 1823/2517/78.

23. NAI, SOCP 1, 1824/2618/22, 24.

24. *DEP,* 12, 14 March, 25 April 1822.

25. Maj. Thomas P. Vokes to William Gregory, 26 March 1822 (NAI, SOCP 1, 2351/57). See also NAI, SOCP 1, 1822/2351/60.

26. 50 Geo. 3, c. 102, s. 5. See also 56 Geo. 3, c. 87, s. 3; NAI, SOCP 1, 1822/2351/72.

27. Maj. Richard Willcocks to William Gregory, 18 April 1822 (NAI, SOCP 1, 2352/25).

28. NAI, SOCP 1, 1823/2514/25.

29. Matthew Barrington to William Gregory, 5 April 1823 (NAI, SOCP 1, 2512/7).

30. *DEP,* 25 April 1822.

31. Col. Sir Hugh Gough to Lt. Col. J. Finch, 6 Feb. 1824, copy (NAI, SOCP 1, 2614/37).

32. Gough to Finch, 2 Nov. 1823, copy (NAI, SOCP 1, 2516/2).

33. Francis Despard to Henry Goulburn, 11 Dec. 1824 (NAI, SOCP 1, 2621/65).

34. Maj. George Warburton to William Gregory, 15 May 1823 (NAI, SOCP 1, 2513/28).

35. NAI, SOCP 1, 1824/2614/56.

36. See, e.g., NAI, SOCP 1, 1822/2353/37.

37. B. Herbert to Henry Goulburn, 21 Oct. 1822 (NAI, SOCP 1, 2349/64).

38. *DEP,* 27 May 1824.

39. NAI, SOCP 1, 1822/2351/60.

40. *DEP,* 19 Feb. 1822. According to another report a few days later, the county jail at Cork held over 240 persons charged with Whiteboy crimes (*DEP,* 23 Feb. 1822).

41. *DEP,* 21, 23, 26 Feb. 1822; NAI, SOCP 1, 1822/2344/53–54, 58. A figure of thirty-six capital convictions is given in some sources, but thirty-five appears to be the correct number for those sentenced to death.

42. Broeker, *Rural disorder,* pp. 135–37; Rudé, *Protest and punishment,* p. 76.

43. *DEP,* 28 Feb., 5, 9 March 1822. Raymond Gillespie has graphically described the awful sufferings of those executed by hanging in the early nineteenth century: "Before the scientific intervention by Samuel Haughton—the Trinity College professor of anatomy who introduced both the knot which broke the neck and the more precise calculation of the ratio of body weight and distance dropped—the victim strangled to death, a process that was far from instantaneous. The face would swell, especially the ears and lips, the eyelids would turn blue and the eyes red, and sometimes they were forced from their sockets. A bloody froth or mucus might escape from the mouth or nose and, very often, urine or faeces might be involuntarily expelled at the point of death. All this could happen even when the hangman was knowledgeable and efficient, but bungled executions were not unheard of. An incompetent hangman could either decapitate the victim or, alternatively, leave him or her dangling in the air for up to half an hour before death came." See Raymond Gillespie, "'And be

hanged by the neck until you are dead,'" Introduction, in Frank Sweeney, ed., *Hanging crimes* (Swords, Co. Dublin, 2005), p. 1.

44. NAI, SOCP 1, 1822/2344/67; 1822/2345/9; *DEP,* 28 Feb., 5, 7 March 1822.

45. *DEP,* 9 March 1822.

46. *DEP,* 21 Feb. 1822.

47. *DEP,* 26 Feb. 1822.

48. Ibid.

49. *Indictments and trials at assizes and special commissions: viz., returns of the number of persons indicted and tried at the several assizes and special commissions in Ireland during each of the last seven years, distinguishing between acquittals and convictions, and distinguishing the counties, cities, and towns,* pp. 1–14, H.C. 1823 (305), xvi, 625.

50. A.C. Macartney to William Gregory, 19 April 1822 (NAI, SOCP 1, 2349/38).

51. William Ponsonby to Gregory, 6 April 1822 (NAI, SOCP 1, 2349/3).

52. Francis Despard to Gregory, 12 Aug. 1822 (NAI, SOCP 1, 2356/52).

53. 3 Geo. IV, c. 1. This law, similar in its provisions to the Insurrection Acts of 1807 and 1814, was accompanied by another act (3 Geo. IV, c. 2) suspending habeas corpus until 1 August 1822. This was the first of many occasions on which habeas corpus was to be suspended to enable the government to fight agrarian crime in nineteenth-century Ireland. The act suspending habeas corpus was permitted to lapse after six months, but the Insurrection Act was renewed for a year in July 1822. See Virginia Crossman, "Emergency legislation and agrarian disorder in Ireland, 1821–41,"

Irish Historical Studies 27, no. 108 (Nov. 1991), pp. 309–32. Supplementing the Insurrection Act of February 1822 was the Irish Constabulary Act of August 1822 (3 Geo. IV, c. 102), which set up county police forces; these forces were centrally controlled by Dublin Castle, with the government now assuming half the costs. Though local magistrates still directed the operations of the police, the role of stipendiary magistrates was enhanced. For the best discussion of the implementation of the Insurrection Act of 1814 and of the Peace Preservation Force prior to this legislation of August 1822, see Stanley Palmer, *Police and protest in England and Ireland, 1780–1850* (Cambridge and New York, 1988), pp. 193–236. See also Palmer's treatment of the new constabulary during the years 1822–24 (*Police and protest*, pp. 237–72). In October 1822, soon after it was set up, the new police force numbered about 2,340 men, spread over sixteen counties, with 53 chief constables and 15 stipendiary police magistrates (*Police and protest*, p. 225).

54. Broeker, *Rural disorder,* pp. 138–41.

55. Ibid., p. 137; Beames, *Peasants and power*, p. 174.

56. *SOI evidence,* pp. 49–50, H.C. 1825 (20), vii, 1.

57. Ibid., p. 73.

58. Ibid., p. 49.

59. Ibid., p. 42.

60. Ibid., p. 53.

61. Ibid., p. 16.

62. Ibid., p. 42.

63. Ibid., p. 53.

64. Ibid., pp. 49–50.

65. 3 Geo. IV, c. 1.

66. *SOI evidence,* p. 50, H.C. 1825 (20), vii, 1.

67. Ibid., pp. 48–49.

68. Ibid., p. 71.

69. Ibid., pp. 53–54.

70. Rudé, *Protest and punishment,* pp. 77–79.

71. Ibid., p. 77.

72. *SOI evidence,* p. 72, H.C. 1825 (20), vii, 1.

73. Ibid., p. 52.

74. NAI, SOCP 1, 1823/2513/1.

75. *DEP,* 5, 19 Feb. 1824.

76. *DEP,* 1 Jan. 1824.

77. *DEP,* 8 Nov. 1823.

78. Maj. George M. Drought to Henry Goulburn, 28 March 1823 (NAI, SOCP 1, 2517/16).

79. *DEP,* 16 Dec. 1823.

80. NAI, SOCP 1, 1822/2353/15.

81. Maj. Samson Carter to [Henry Goulburn], 18 Jan. 1824 (NAI, SOCP 1, 2614/17).

82. Francis Blackburne to William Gregory, 17 Aug. 1823 (NAI, SOCP 1, 2517/67).

83. *DEP,* 15 Jan. 1824.

84. Maj. George M. Drought to Henry Goulburn, 28 March 1823 (NAI, SOCP 1, 2517/16).

85. *SOI evidence,* p. 74, H.C. 1825 (20), vii, 1.

86. Jemmett Browne to William Gregory, 8 May 1823 (NAI, SOCP 1, 2513/13).

87. W.H.W. Newenham to [William Gregory], 24 July 1823, copy (NAI, SOCP 1, 2514/25).

88. *SOI evidence,* p. 74, H.C. 1825 (20), vii, 1.

89. William and Denis Horan to William Burns, 27 April 1823 (NAI, SOCP 1, 2518/23).

90. *DEP*, 21 March 1822.

91. *DEP*, 1 Jan. 1824.

92. *DEP*, 27 May 1823.

93. *DEP*, 16 Oct. 1823.

94. *DEP*, 12 March 1822.

95. A.C. Macartney to William Gregory, 13 March 1822 (NAI, SOCP 1, 2349/15).

96. Rudé, *Protest and punishment*, p. 78.

97. *SOI evidence*, p. 16, H.C. 1825 (20), vii, 1.

98. Ibid., pp. 50–51.

99. Maj. Samson Carter to [Henry Goulburn], 29 Aug. 1823 (NAI, SOCP 1, 2519/30).

100. Ibid.

101. See, e.g., NAI, SOCP 1, 1822/2354/47.

102. NAI, SOCP 1, 1823/2515/62.

103. *DEP*, 4 Dec. 1823, 2 March 1824.

104. Maj. Thomas P. Vokes to Henry Goulburn, 11 Jan. 1824 (NAI, SOCP 1, 2618/4).

105. Ibid.

106. *DEP*, 17 April 1824.

107. Ibid.

108. *DEP*, 14 Aug. 1824.

109. *DEP*, 15 April 1824.

110. *DEP*, 10 Aug. 1824; NAI, SOCP 1, 1824/2619/12.

111. NAI, SOCP 1, 1824/2517/21.

112. Matthew Barrington to William Gregory, 8 April 1824 (NAI, SOCP 1, 2615/3).

113. *DEP*, 26 Aug. 1824.

114. Maj. Samson Carter to William Gregory, 24 June 1823 (NAI, SOCP 1, 2513/80).

115. Viscount Ennismore to Gregory, 25 June 1823 (NAI, SOCP 1, 2513/83).

116. Lt. Col. Thomas Arbuthnot to Maj. J. Finch, 23 June 1823, copy (NAI, SOCP 1, 2513/79).

117. Arbuthnot to Finch, 6 July 1823 (NAI, SOCP 1, 2514/8).

118. NAI, SOCP 1, 1823/2514/15.

119. Matthew Barrington to William Gregory, 1 Aug. 1822 (NAI, SOCP 1, 2354/11).

120. Lt. Col. Thomas Arbuthnot to Gregory, 27 June 1823 (NAI, SOCP 1, 2513/86).

121. *DEP*, 5 July 1823. See also NAI, SOCP 1, 1823/2514/4–5.

122. Maj. Samson Carter to William Gregory, 8 Aug. 1823 (NAI, SOCP 1, 2514/36).

123. *DEP*, 19 Aug. 1823; NAI, SOCP 1, 1823/2514/43.

124. Col. Sir Hugh Gough to William Gregory, 22 May 1824, copy (NAI, SOCP 1, 2615/28).

125. *DEP*, 31 Aug. 1824.

126. *DEP*, 24 Aug. 1824. Though newspaper reports of the burning of the Sheas identified the location as the townland of Gurtnapisha, a newspaper account of the trial of the Mahers gave the location of the crime as Tober, and it was reportedly from "a small holding at [Tober], on the slopes of Slievenamon," that Edmund Shea had evicted William Gorman prior to the burning. See *DEP*, 21 Aug. 1824; Burke, *History of Clonmel*, p. 209.

127. *DEP*, 31 Aug. 1824.

128. *DEP*, 7 Aug. 1823.

129. *DEP*, 6 Sept. 1823.

130. Maj. Thomas P. Vokes to Henry Goulburn, 10 Aug. 1824 (NAI, SOCP 1, 2619/13).

131. *DEP*, 8 April 1823.

132. Viscount Doneraile to Lieut.

Gen. Lord Combermere, [?] March 1823, copy (NAI, SOCP 1, 2512/3).

133. NAI, SOCP 1, 1823/2512/35.

134. *DEP*, 24 April 1823; *SOI evidence*, pp. 584–86, H.C. 1825 (129), viii, 455.

135. NAI, SOCP 1, 1823/2512/25.

136. W.H.N. Hodder to William Gregory, 18 April 1822 (NAI, SOCP 1, 2345/60).

137. Ibid.

138. *DEP*, 2 Sept. 1823.

139. NAI, SOCP 1, 1822/2352/11.

140. *DEP*, 18 March 1823.

141. *DEP*, 12 Aug. 1823.

142. NAI, SOCP 1, 1824/2615/5.

143. *SOI evidence*, pp. 580–81, H.C. 1825 (129), viii, 455.

144. Maj. Samson Carter to William Gregory, 12 April 1824 (NAI, SOCP 1, 2615/5).

145. Memorial of John, Patrick, and Maurice Cremins, n.d., but about April 1824 (NAI, SOCP 1, 2615/9).

146. *DEP*, 31 Aug. 1824.

147. Quoted ibid.

148. *DEP*, 21 Aug. 1823.

149. *DEP*, 26 Aug. 1823.

150. *DEP*, 10, 12 Aug. 1824.

151. *DEP*, 25 March, 15 April, 12, 26, 31 Aug. 1824.

152. Donnelly, "Whiteboy movement," p. 50; idem, "Irish agrarian rebellion," pp. 326–27.

153. See, e.g., *DEP*, 2 Sept. 1823.

154. *SOI evidence*, pp. 105–7, H.C. 1825 (20), vii, 1.

155. Major Richard Willcocks told a parliamentary committee in May 1824 that he knew of no cases "of late years" in which executed criminals had been buried "without the right of sepulture being allowed" (ibid., p. 105).

156. *DEP*, 31 Aug. 1824.

157. *DEP*, 2 March 1824.

158. *DEP*, 29 July 1824.

159. Ibid.

160. *DEP*, 12 Aug. 1824.

161. *SOI evidence*, pp. 48–49, 97, 145, 167–68, 178, 196, H.C. 1825 (20), vii, 1.

162. The estimate of about one hundred executions is based on a careful survey of reports in the *Dublin Evening Post* and in the State of the Country Papers, Series 1. The figure includes executions specifically reported and others for which the probability of their occurrence is very high, given the nature of the crimes and other circumstances.

163. Quoted in *DEP*, 8 July 1823.

164. John Keily to Francis Blackburne, 8 July 1823 (NAI, SOCP 1, 2517/46).

165. Maxwell Blacker to [William Gregory], 7 Aug. 1823 (NAI, SOCP 1, 2514/35).

166. Godfrey Massey to [Henry Goulburn], 14 Aug. 1823 (NAI, SOCP 1, 2517/65).

167. Maj. Thomas P. Vokes to William Gregory, 21 March 1824 (NAI, SOCP 1, 2618/49).

168. NAI, SOCP 1, 1822/2345/59.

169. *DEP*, 21 Aug. 1824.

170. *DEP*, 24 June, 12 Aug. 1824.

171. Maj. George M. Drought to William Gregory, 27 June 1824 (NAI, SOCP 1, 2621/42).

172. *DEP*, 21 Aug. 1824.

173. Ibid. See also *DEP*, 19 Aug. 1824.

174. Maj. George M. Drought to Henry Goulburn, 18 Aug. 1824 (NAI, SOCP 1, 2621/52).

175. *DEP,* 19 Aug. 1824.

176. *DEP,* 29 July 1824.

177. *DEP,* 17 Aug. 1824.

178. *DEP,* 19 Aug. 1824.

179. *DEP,* 26, 28, 31 Aug., 2 Sept. 1824.

180. Six members of this original group (Kingston, Mount Cashell, Wrixon-Becher, Ennismore, Jephson, and Doneraile) would be joined by two other north Cork landowners ("Captain Roberts" of Charleville and Ennismore's son-in-law Richard Aldworth of Newmarket House) in providing from their estates the great majority of the emigrants of 1825. See Donald MacKay, *Flight from famine: the coming of the Irish to Canada* (Plattsburgh, N.Y., 2002; originally published 1990), pp. 52, 82. The person identified by Donald MacKay as Captain Roberts may have been the Charleville magistrate Jonathan Bruce Roberts. Whether he himself was a substantial landowner in the district is doubtful, though he may have acted as a land agent for another member of his family or for some other local proprietor.

181. Kingston to Bathurst, 20 June 1822, quoted ibid., p. 53.

182. Ibid., pp. 53–54. See also R.D.C. Black, *Economic thought and the Irish question, 1817–1870* (Cambridge, 1960), pp. 206–7; W.F. Adams, *Ireland and Irish emigration to the new world from 1815 to the famine* (reprint ed., Baltimore, 1980; originally published 1932), pp. 275–77; *SOI evidence,* p. 11, H.C. 1825 (129), viii, 1.

183. Peter Robinson's memorandum to Sir Robert Wilmot Horton, n.d., but early 1825, p. 2, Sir Robert Wilmot Horton miscellaneous correspondence (Microfilm 7–2167), Public Archives of Canada, Ottawa. I am very grateful to Kerby Miller for providing me with copies of his typed transcripts from, and copious notes on, the contents of Microfilm 7–2167.

184. Robinson's memorandum to Horton, n.d., but early 1825, pp. 3–4, 6 (see note 183 above). Robinson paid special tribute to the enthusiastic support that he had received from the parish priests of Mallow and Newmarket. In fact, when asked before a parliamentary committee in February 1825 whether his plan "was likely to have succeeded if you had not met with that cordial cooperation on the part of the Roman Catholic clergy," he replied, "I should think not; the effect their influence might have had if exerted against me upon the minds of the people, who were still suspicious that all was not right, was evident. . . ." See *SOI evidence,* p. 22, H.C. 1825 (129), viii, 1.

185. Robinson's memorandum to Horton, n.d., but early 1825, p. 4 (see note 183 above). Besides combating widespread anxieties about the arduous conditions of the transatlantic journey and relieving common fears about the frontier environment of Upper Canada, Robinson had to counter the assumption among some of his hearers that this assisted-emigration scheme was simply a subterfuge for the equivalent of penal transportation to Australia. Or as one farmer complained to him, "The government found they could not get rid of them fast enough by the Insurrection Act and were trying another plan" (quoted in MacKay, *Flight from famine,* p. 59). See also *SOI evidence,* p. 20, H.C. 1825 (129), viii, 1.

186. Robinson's memorandum to Horton, n.d., but early 1825, pp. 7–8 (see note 183 above). See also ibid., pp. 6–7.

187. Speaking of the 1823 emigrants as a group, Robinson remarked, "I found them much more intelligent than I expected; most of them could write and calculate their allowance of rations to the eighth part of an ounce; in that way they were quite intelligent." See *SOI evidence,* p. 23, H.C. 1825 (129), viii, 1. See also MacKay, *Flight from famine,* p. 60.

188. Robinson's memorandum to Horton, n.d., but early 1825, p. 8 (see note 183 above).

189. For MacKay's evaluation, see his *Flight from famine,* p. 63. Robinson winnowed lists compiled by landowners or their agents in order to arrive at his final choices. As he told a parliamentary committee in February 1825, "The emigrants I took [in 1823] were selected from the persons who were recommended to me by the principal noblemen and gentlemen of the country as being absolutely paupers, and such as it was particularly desirous to get rid of." See *SOI evidence,* p. 21, H.C. 1825 (129), viii, 1. Pressed as to what exactly he meant by "paupers," he answered initially that they were persons "who held no land," but he promptly amended that characterization to say that "the greater proportion had been partially employed at home, living on one acre or two acres, and among these were found weavers, blacksmiths, and other tradesmen who cultivated a few acres each." Robinson also observed that he had especially "endeavoured to get small farmers who had been dispossessed of their lands," as well as "persons without employment, absolute paupers, and I was always assured by the gentlemen recommending them that they were such" (ibid., pp. 20, 22). R.D.C. Black has asserted, however, that in spite of Robinson's public insistence that the 1823 project was mainly designed to relieve paupers, "the private correspondence between Robinson, Ennismore, and Horton, preserved in CO 384/12, shows clearly that the primary object was to get rid of 'troublesome characters'" (Black, *Economic thought,* p. 207, fn. 3).

190. After arriving at Quebec city, the emigrants traveled partly by land and partly by river to Montreal, Lachine, and Prescott in Upper Canada (a distance of about three hundred miles), before trekking another sixty miles across the country to the Bathurst District. Robinson then proceeded to inspect and allot lands to a total of 182 families—82 in the township of Ramsay, 34 in Huntley, 29 in Pakenham, 26 in Goulburn, and the remaining 11 in Bathurst, Beckwith, Darling, and Lanark. The dry and warm autumn weather permitted Robinson to have log cabins built for all of these families before the beginning of November: "To do this I was obliged to go to some additional expense, as the settlers were not sufficiently acquainted with the use of the axe to put up log buildings themselves." For details of the embarkation, the voyage, and the path to settlement in Upper Canada, see Robinson's memorandum to Horton, n.d., but early 1825 (and note 183 above). See also *SOI evidence,* pp. 20–26, H.C. 1825 (129), viii, 1; MacKay, *Flight from famine,* pp. 63–69.

191. Sir Robert Wilmot Horton, memorandum entitled "Emigration to Canada," 30 Dec. 1824, Sir Robert Wilmot Horton miscellaneous correspondence (Microfilm 7-2167), Public Archives of Canada, Ottawa; *SOI evidence*, pp. 7-9, H.C. 1825 (129), viii, 1.

192. Viscount Doneraile to Peter Robinson, 23 Nov. 1824, Sir Robert Wilmot Horton miscellaneous correspondence (Microfilm 7-2167, n.p.), Public Archives of Canada, Ottawa.

193. Earl Mount Cashell to Robinson, 20 Oct. 1824 (ibid., n.p.). The lists accompanying this letter identified the parishes from which the prospective emigrants were drawn. The adjoining north Cork parishes of Kilworth, Macroney, Leitrim, and Kilcrumper supplied 299 persons (46 families), while the six parishes of Fermoy, Rathcormack, Gortroe, Middleton, "Curriglass" (perhaps the townland of Curryglass in Shandrum parish), and Tallow (in County Waterford) furnished another 263 persons (37 families). In this latter group the overwhelming majority came from Fermoy and Rathcormack parishes. In general, the overlap with areas of Rockite militancy in northeast Cork is very striking. For an interesting portrait of the third Earl Mount Cashell, who owned the town of Kilworth and about 6,000 acres around it (mostly in the hands of middlemen on long leases and full of cottiers), along with a demesne of about 1,000 acres around his Georgian mansion of Moore Park, see C.A. Wilson, *A new lease on life: landlords, tenants, and immigrants in Ireland and Canada* (Montreal, Buffalo, and London, 1994), pp. 13-91. His estates were by no means confined to

County Cork. Chased by creditors for much of his life, he died almost penniless in England in 1883, having lost possession of all of his once extensive Irish and Canadian properties.

194. Quoted in MacKay, *Flight from famine*, pp. 77-78.

195. Earl Mount Cashell to Peter Robinson, 20 Oct. 1824, Sir Robert Wilmot Horton miscellaneous correspondence (Microfilm 7-2167, n.p.), Public Archives of Canada, Ottawa.

196. Adams, *Ireland and Irish emigration*, pp. 277-78; MacKay, *Flight from famine*, p. 79.

197. MacKay, *Flight from famine*, pp. 81-82. At Westminster, however, politicians showed much less enthusiasm for a repetition of the assisted-emigration scheme of 1823, which was often criticized as extravagant. Though the government prevailed in the Commons' vote on a public subsidy, the chancellor of the exchequer promised a future parliamentary committee of inquiry on emigration—"an announcement greeted with cheers." See Adams, *Ireland and Irish emigration*, pp. 278-79; Black, *Economic thought*, p. 208.

198. William Wrixon-Becher to Peter Robinson, 3 Oct. 1824, Sir Robert Wilmot Horton miscellaneous correspondence (Microfilm 7-2167, n.p.), Public Archives of Canada, Ottawa. Ballygiblin House, Wrixon-Becher's residence, lay in the northeast Cork parish of Brigown, adjacent to several of those parishes from which Earl Mount Cashell wished to send out paupers under Robinson's scheme.

199. MacKay, *Flight from famine*, pp. 82-83.

200. The government grant of £30,000 for 1825 was twice that for 1823, but of the £15,000 provided in 1823, only £10,000 had supposedly been designated for the assisted emigration to Upper Canada; the other £5,000 had been earmarked for Cape Colony at the southern tip of Africa. See *SOI evidence,* p. 11, H.C. 1825 (129), viii, 1. See also MacKay, *Flight from famine,* pp. 87-88.

201. MacKay, *Flight from famine,* pp. 82, 87 (quotation). Robinson insisted to Horton that the emigrants of 1825 constituted a "better description of people than those taken out in 1823, although they are wretchedly poor." Again, MacKay challenges the claim: "In fact, they were much like the earlier group" (ibid., p. 89).

202. The assisted emigrants of 1823 included 186 children under the age of fourteen, or 32 percent of total; those of 1825 included 851 such children, or 42 percent of the much larger number of those departing. See *SOI evidence,* p. 6, H.C. 1825 (129), viii, 1; MacKay, *Flight from famine,* pp. 87, 89.

203. MacKay, *Flight from famine,* pp. 88-91 (quotation on p. 89).

204. Ibid., pp. 91-102 (quotation on p. 99). The largest group of settlers (142 families) was concentrated in the township of North Emily, but five other townships attracted significant numbers: Ennismore (67), Douro (60), Otonobee (51), Smith (36), and Asphodel (36). A small number (15 families) went to the outlying township of Ops (ibid., p. 99). See also J.J. Mannion, *Irish settlements in eastern Canada: a study of cultural transfer and adaptation* (Toronto and Buffalo, 1974), pp. 16-17, 21.

205. MacKay, *Flight from famine,* p. 110.

206. Mannion, *Irish settlements,* pp. 24, 27, 182 (quotation on p. 24). Donald Akenson has emphasized the high degree of long-term continuity that distinguished the Irish settlers of North Emily township and their economic progress as rural immigrants: "Despite initial disadvantages, by 1861 the Irish Catholics had the same levels of capital accumulation and were as firmly committed to mixed commercial agriculture as their non-Catholic and non-Irish neighbours." See D.H. Akenson, *The Irish in Ontario: a study in rural history* (Kingston and Montreal, 1984), pp. 40-41. The central point here is developed much more extensively in *Small differences: Irish Catholics and Irish Protestants, 1815-1922: an international perspective,* a book that Akenson published in 1988 to challenge what he perceived as certain erroneous claims about Catholic-Protestant differences made by Kerby Miller in his prize-winning work *Emigrants and exiles* of 1985.

207. Horton's enthusiasm for assisted emigration was at times almost boundless, and his public statements of his views frightened all those who saw the government's proper role in this area as much more limited. When reviewing the scale and costs of the 1823 scheme before a parliamentary committee in February 1825, Horton declared: "I see no reason why 100,000, 200,000, or 500,000 are not to be located at the same rate of expense [he had earlier cited a figure of

£22 per head]. I think it is a material point to establish that fact; in Upper Canada alone . . . there is the most distinct evidence that 160,000 persons can be received." Nor was this all. Horton added that the Gaspé district of Lower Canada "alone would absorb a population . . . of 500,000 persons" with "the greatest facility." See *SOI evidence,* pp. 10–11, H.C. 1825 (129), viii, 1.

208. Miller, *Emigrants and exiles,* pp. 195–97. Especially important was the repeal of all restrictions on emigration from Britain and Ireland in 1827, which resulted in lower fares.

209. Adams, *Ireland and Irish emigration,* p. 282. As R.D.C. Black pointed out, the Colonial Office "continued to receive applications for assistance to emigrate to North America from Ireland years after the schemes had been terminated" (Black, *Economic thought,* p. 209, fn. 1). See Petitions to emigrate, CO, 384/14 (1826); 384/16 (1827); 384/19 (1828); 384/21 (1829); 384/23 (1830); 384/24 (1831); 384/25 (1832); 384/30 (also 1832); 384/31 (1833); 384/34 (1834). Petitions arrived at the Colonial Office in subsequent years, though in diminished numbers. I am very grateful to Kerby Miller for drawing my attention to this long series of documents.

210. Adams, *Ireland and Irish emigration,* p. 283. Not everyone agreed that "partial removal" would be ineffective. Horton certainly did not (Black, *Economic thought,* p. 215, fn. 1).

211. For the best treatment of the debates surrounding the hearings of Horton's two committees on emigration, see Black, *Economic thought,* pp.

209–15. See also Adams, *Ireland and Irish emigration,* pp. 281–92.

212. 7 Geo. IV, c. 28. This law provided that no leaseholder was to sublet or otherwise alienate his holding after June 1826 without the written consent of the landowner or lessor. If this statute had been strictly enforced, it would have prevented many new acts of subletting and encouraged proprietors to consolidate small holdings into larger farms through eviction. But there is little evidence to demonstrate that the law increased evictions substantially or prevented subletting. It was repealed in 1832.

213. Black, *Economic thought,* pp. 209–11.

214. Ibid., p. 213.

215. Ibid., pp. 213–14. Under Horton's proposals Irish landlords were in effect to shoulder the costs of the scheme, though they could borrow the money from the local authorities. The repayment period for these government loans was set at thirty years. Irish tenants who had been dispossessed, or who were about to be evicted, were to have a prior claim on these government funds (Adams, *Ireland and Irish emigration,* p. 291).

216. Black, *Economic thought,* pp. 214–15 (quotations from Peel on p. 215). See also Adams, *Ireland and Irish emigration,* pp. 289–95.

Conclusion

1. For middlemen, see David Dickson, "Middlemen," in Thomas Bartlett and D.W. Hayton, eds., *Penal era and golden age: essays in Irish history, 1690–1800* (Belfast, 1979), pp. 162–85; Lyons,

"Vicissitudes of a middleman," pp. 300–18; Dickson, *Old world colony*, pp. 329–37, 345–46.

2. Landlords and tenants in Ulster did not escape the economic adversities that battered their counterparts in Munster and Connacht. For a discussion of the impact of the agricultural depression of 1813–16 (compounded by the effects of bad weather, food and fuel shortages, and epidemic disease in 1816–18) on the Earl of Gosford's estates in County Armagh, see F.M.L. Thompson and D. Tierney, eds., *General report on the Gosford estates in County Armagh, 1821, by William Greig* (Belfast, 1976), pp. 112–21. Greig explained to Lord Gosford how his tenants had been devastated by this combination of calamities: "In the course of two years (1816–17) the gains which had been saved during the years of prosperity were almost entirely exhausted in the payment of rent without an adequate return of produce, or perhaps more generally in the purchase of food" (ibid., p. 116).

3. For the Palatines in this region of west Limerick, see P.J. O'Connor, *All Ireland is in and around Rathkeale* (Newcastle West, 1996).

4. For an instructive graph (derived from the painstaking work of Peter Solar) showing "Irish agricultural exports measured in real and current prices, 1810–1825," see Ó Gráda, *Ireland*, p. 161. The conclusion that Ó Gráda draws from the statistical evidence appears more than a little too sanguine. For the most sophisticated attempt yet made to graph Irish agricultural prices from the mid-eighteenth century to the early twentieth, see Kennedy and Solar, *Irish agriculture*, pp. 92–93 (figs. 5.1 and 5.2). For a study of prefamine subsistence crises and of public and private efforts to relieve the resulting distress, see O'Neill, "The state, poverty, and distress in Ireland."

5. As James Patterson has recently pointed out, Whiteboys in northeast Cork in 1798–99 also called for the complete abolition of tithes, and other examples of such an escalation in demands during the tumultuous 1790s might also be cited. See Patterson, "'Educated Whiteboyism,'" p. 27.

6. On the subjects of sectarianism and interdenominational relations before, during, and after the 1820s, see Bowen, *Protestant crusade*; David Hempton, "The Methodist crusade in Ireland, 1795–1845," *Irish Historical Studies* 22, no. 85 (March 1980), pp. 33–48; Ian d'Alton, *Protestant society and politics in Cork, 1812–1844* (Cork, 1980); S.J. Connolly, *Religion and society in nineteenth-century Ireland* (Dundalk, 1985); Fergus O'Ferrall, *Catholic emancipation: Daniel O'Connell and the birth of Irish democracy, 1820–30* (Dublin, 1985); David Hempton, *Religion and political culture in Britain and Ireland: from the Glorious Revolution to the decline of empire* (Cambridge and New York, 1996); Proinnsíos Ó Duigneáin, *The priest and the Protestant woman: the trial of Rev. Thomas Maguire, P.P., Dec. 1827* (Dublin and Portland, Ore., 1997); Jacqueline Hill, *From patriots to unionists: Dublin civic politics and Irish Protestant patriotism, 1660–1840* (Oxford and New York, 1997); Thomas McGrath, *Politics, interdenominational relations, and education in the public ministry of Bishop*

James Doyle of Kildare and Leighlin, 1786-1834 (Dublin, 1999); McGrath, *Religious renewal*; Whelan, *Bible war.*

7. For an excellent study of the yeomanry in the late eighteenth and early nineteenth centuries, see Blackstock, *Ascendancy army*, esp. pp. 232-68; for the development of the police in Ireland over the same period, see the classic study by Palmer, *Police and protest,* and Broeker, *Rural disorder;* for the role of the regular army, see Virginia Crossman, "The army and law and order in the nineteenth century," in Thomas Bartlett and Keith Jeffery, eds., *A military history of Ireland* (Cambridge and New York, 1996), pp. 358-78.

8. When marking the shift toward extreme violence in the 1790s, Bartlett insisted that Ireland "for most of the eighteenth century was not a particularly violent country, certainly not by the standards of the continent and almost certainly not by English standards either." As Bartlett showed, the harsh repression of the Irish antimilitia riots of 1793, with about 230 lives lost in barely more than eight weeks of disturbances, served as a prelude to "the massive bloodletting of the 1798 rebellion" and its aftermath. See Bartlett, "End to moral economy," pp. 193-94. Recently, Neal Garnham has argued that the rate of murder in Ireland in the eighteenth century, though not high by continental standards, was far above the corresponding rate in England (perhaps as much as four times higher). See Neal Garnham, "How violent was eighteenth-century Ireland?" *Irish Historical Studies* 30, no. 119 (May 1997), pp. 377-92, quotation on p. 92.

9. Familiarity with weapons and military discipline reached unprecedented levels in Ireland in the early nineteenth century. More than ninety thousand Irish recruits joined the British army during the fifteen years before the Battle of Waterloo, and by 1830 Irishmen (over forty thousand were then serving) actually outnumbered Englishmen in this branch of the military. See E.M. Spiers, "Army organisation and society in the nineteenth century," in Bartlett and Jeffery, *Military history,* pp. 335-37. See also Dickson, *Old world colony,* p. 490.

10. Estimates of military and civilian casualties in the 1798 rebellion and its immediate aftermath vary considerably. In an authoritative essay Thomas Bartlett has indicated that by September 1798, "perhaps 25,000 rebels (including a high proportion of noncombatants) and some hundreds of soldiers had been slain, and large areas of the country had been effectively laid waste." See Bartlett, "Defence, counter-insurgency, and rebellion: Ireland, 1793-1803," in Bartlett and Jeffery, *Military history,* p. 287.

11. Long ago George O'Brien drew attention to the malign significance of the Ejectment Act of 1816 in his book *The economic history of Ireland from the union to the famine* (London and New York, 1921), pp. 157-59. Under its terms landlords could now evict a tenant within two months at a cost of only forty shillings, "whereas a similar process by an English landlord would take at least twelve months and cost eighteen pounds." The same law also authorized landlords to distrain and sell the growing crops of defaulting tenants

and to impose the costs of this exercise on the tenant. The powers granted under the 1816 act were strengthened by further legislation in 1818 and 1820 (56 Geo. III, c. 88; 58 Geo. III, c. 39; and 1 Geo. IV, c. 87). As O'Brien points out, "Daniel O'Connell pronounced the power of distraining growing crops as 'the fruitful source of murder and outrage,' and stated before the select committee of 1824–5 that the power was responsible for some of the worst agrarian outrages of the south" (*Economic history*, pp. 157–58).

12. In "The Catholic rent," one of his poems in Irish as translated by Douglas Hyde, the poet Anthony Raftery (Antaine Raiftearaí) sought rhetorically to strip Protestants of their dignity and power (Hyde, *Songs ascribed to Raftery*, p. 115):

> On observing the signs, I see fear for
> the fanatics
> Who fast not on Fridays and jeer at the
> Catholics;
> Success is denied them, defeat shall be
> absolute.
> As Peter and Jesus have spoken.
> Wrote Pastoreeni [Pastorini], you'll
> see it made manifest,
> A rascally meeting each month in each
> hamlet. But
> Clonmel shall make pieces of New
> Lights and Orangemen,
> And Loughrea shall defeat them and
> beat their rascality;
> We have lost our good Clayton, but
> Daly's as bad for them;
> Their bible's mendacious, we'll shame
> them and sadden them,
> We'll give them ('twill please us) a
> token.

13. Lewis, *Local disturbances* (Tower Books ed.), p. 192.

14. See esp. Palmer, *Police and protest*, pp. 237–62.

15. The currency of the term "Botany Bay" derived from the destination of the first fleet carrying Irish convicts in 1791. Since the area around that bay was considered unsuitable for settlement, a colony was established at Sydney Cove instead. Subsequently, however, "Botany Bay" became in common Irish parlance synonymous with the principal penal colony in New South Wales. In the years immediately before the Rockite movement, the London government was putting pressure on the managers of this penal colony to ensure that it remained "an object of real terror to all classes of the community" everywhere in the United Kingdom; the accents were to be on "the strict discipline, the unremitting labour, [and] the severe but wholesome privations" to which prisoners had been condemned. See the letter of January 1819 by Earl Bathurst, the colonial secretary, quoted in T.J. Kiernan, *The Irish exiles in Australia* (Dublin and London, 1954), pp. 11–12. See also Con Costello, *Botany Bay: the story of the convicts transported from Ireland to Australia, 1791–1853* (Cork and Dublin, 1987), p. 9.

16. For the Insurrection Act of 1822 and its counterparts in other periods of disturbance, see Crossman, "Emergency legislation," pp. 309–23. Stanley Palmer states that almost four thousand persons were arrested under the Insurrection Act in the years 1822–24, and that about four hundred were transported (Palmer, *Police and protest*, p. 237). But as noted earlier in this book (see chap. 9) and below, many

others were also transported for Rockite offenses.

17. Rudé, *Protest and punishment,* pp. 77-78.

18. Con Costello puts the total number of Irish convicts transported to "Botany Bay" at about 45,000 over the whole period 1787-1868, including perhaps 6,000 apprehended in England and transported from there. Of this total, some 40,000 were "ordinary criminals," another 4,300 were "social or agrarian offenders," and perhaps 600 were "political" convicts. See Costello, *Botany Bay,* pp. 161-62.

19. Finbarr Whooley explored the brigandage that followed in the wake of the Rockite movement in County Cork; he showed how farmers were now often ready to support and even to join local bodies established to stamp out such brigandage ("Captain Rock's rebellion," pp. 54-60).

20. The need is great for other guides to surviving estate records such as that recently produced for the period of the Great Famine and the years immediately preceding and following it. See Andrés Eiríksson and Cormac Ó Gráda, *Estate records of the Irish famine: a second guide to famine archives, 1840-1855* (Dublin, 1995). See also Terence Dooley, *Sources for the history of landed estates in Ireland* (Dublin and Portland, Ore., 2000).

21. For members of educated society in Britain and Ireland the significance of the Rockite movement was interpreted in an exceptional way by the writer Thomas Moore, who published his satirical *Memoirs of Captain Rock, the celebrated chieftain, with some account of his ancestors, written by him-*

self, in April 1824. Fergus O'Ferrall has best summed up the nature, purpose, and impact of this work: "Under the transparent disguise of tracing the history of the Rock family, the leaders of popular disturbance in Ireland, Moore reviewed with bitter, ironic power the history of English misrule in Ireland. He attacked the proselytising educational societies, tithes, and the Anglican church 'as by law (and constables) established in Ireland!' His work, a most readable and comprehensive arraignment of misgovernment in Ireland, created an immediate sensation and rapidly went into a number of editions; long quotations were carried in *The Times* and the *Morning Chronicle.* This angry indictment of the penal laws made a tremendous impact on educated English opinion, all the more so because of Moore's pre-eminent position in literary and liberal circles." See O'Ferrall, *Catholic emancipation,* p. 80. The most thorough scholarly discussion of this famous work is that of Emer Nolan. See Thomas Moore, *Memoirs of Captain Rock, the celebrated Irish chieftain, with some account of his ancestors, written by himself,* edited and introduced by Emer Nolan, with annotations by Seamus Deane (Dublin, 2008), pp. xi-li. See also Tadhg O'Sullivan, "'The violence of a servile war': three narratives of Irish rural insurgency post-1798," in Geary, *Rebellion and remembrance,* pp. 73-92.

22. For examples of the use of the "illustrious name" of Captain Rock in later threatening letters or notices and in Irish counties outside those strongly affected by the movement of 1821-24, see Gibbons, *Captain Rock,* pp. 118,

124, 129, 135, 144–47, 150, 152–53, 157, 161–62, 164–65, 172, 178–79, 184–87, 189–90, 193–94, 200–201, 203–5, and passim.

23. Even when the 1823 act was amended in 1824 (5 Geo. IV, c. 63), the vote in the parish vestries was still limited to the twenty-five landholders paying the highest sums in county cess or grand-jury rates, and the system of weighted voting prescribed in 1823 was continued. Nevertheless, a combination of factors, including popular pressure and a desire on the part of many Anglican clergymen to avoid annual battles over tithes if they could achieve an acceptable monetary settlement, led to the spread of tithe composition during the remainder of the decade. In fact, more than half of the 2,450 parishes in Ireland had become subject to compositions by 1830. Where adopted, compositions lowered the tax burden and usually reduced the tithe income of Anglican clergymen. See Macintyre, *Liberator,* pp. 170–72.

24. For the tithe war of the 1830s, see the articles by Patrick O'Donoghue: "Causes of opposition to tithes, 1800–38," *Studia Hibernica,* no. 5 (1965), pp. 7–28; "Opposition to tithe payments in 1830–31," *Studia Hibernica,* no. 6 (1966), pp. 69–98; "Opposition to tithe payment in 1832–3," *Studia Hibernica,* no. 12 (1972), pp. 77–108. See also O'Donoghue, "The tithe war, 1830–33" (M.A. thesis, University College Dublin, 1961); Macintyre, *Liberator,* pp. 167–200.

25. M.R. Beames, "Rural conflict," pp. 264–81. "One of the main characteristics distinguishing rural violence in Ireland from that in the rest of the

British Isles in the first half of the nineteenth century," observes Beames, "was the willingness and the ability of the Irish Whiteboy movements to resort to assassination to achieve their objectives" (ibid., p. 264).

26. Vaughan, *Landlords and tenants,* pp. 24–26; J.S. Donnelly, Jr., "Mass eviction and the great famine," in Cathal Póirtéir, ed., *The great Irish famine* (Cork and Dublin, 1995), pp. 155–73; J.S. Donnelly, Jr., *The great Irish potato famine* (Stroud, Gloucestershire, 2001), pp. 138–62.

27. Some clearances did undoubtedly occur before 1845, along with many cases of piecemeal consolidation of holdings, but the desire among landowners and agents to engineer consolidation through large-scale evictions was typically much stronger than their willingness to brave the consequences. See the section on consolidation of farms in J.P. Kennedy, ed., *Digest of evidence taken before her majesty's commissioners of inquiry into the state of the law and practice in respect to the occupation of land in Ireland,* 2 vols. (Dublin, 1847), 1:451–72. See also Black, *Economic thought,* pp. 20–21; Dickson, *Old world colony,* p. 353.

28. For O'Connell's election manifesto, see Norman Gash ed., *The age of Peel* (London and New York, 1968), pp. 22–24.

29. This issue is carefully explored in John Cornelius, "The legal career of Daniel O'Connell" (M.A. thesis, University of Wisconsin–Madison, 2004). For O'Connell's heroic status in Irish folklore, arising in large part from his fame as a brilliant barrister, see Dáithí Ó hÓgáin, *The hero in Irish folk history*

(Dublin and New York 1985), pp. 99–119.

30. In a sparkling essay in 1975, Gearóid Ó Tuathaigh argued that "the language of politics" for Daniel O'Connell, as for the Irish speakers in his audiences, "was apocalyptic, hyperbolic, and rhetorical. They were both inheritors, if only in part, of the same [Gaelic] tradition." As Ó Tuathaigh noted, "it is remarkable how the language of O'Connellite 'grievance' speeches and the millennial language of popular poetry both tend towards the same end, the inflation of popular expectations from political exertion." See Ó Tuathaigh, "Gaelic Ireland, popular politics, and Daniel O'Connell," *Journal of the Galway Archaeological and Historical Society* 34 (1974–75), pp. 21–34 (quotations on p. 33). For illuminating comments on O'Connell's messianic role by another distinguished historian writing more recently, see O'Ferrall, *Catholic emancipation,* p. 73.

Bibliography

Contemporary Sources

Manuscript Material in Public Repositories

National Archives of Ireland, Dublin, State of the Country Papers, Series 1

Clare: 1822—2341/1–55; 1823—2510/1–80; 1824—2613/1–69.

Cork: 1821—2293/1–15; 1822—2342/1–66; 2343/1–47; 2344/1–68; 2345/1–89; 2346/1–72; 2347/1–64; 1823—2511/1–62; 2512/1–42; 2513/ 1–90; 2514/1–57; 2515/1–64; 2516/ 1–59; 1824—2614/1–73; 2615/1–66; 2616/ 1–48; 2617/1–56.

Kerry: 1821—2295/1–26; 1822—2348/1–69; 2349/1–78; 1823—2519/1–37; 1824—2620/1–9.

Kilkenny, 1821—2292/11–14; 1822—2369/1–63; 1823—2506/1–37; 1824— 2606/1–75.

Limerick: 1821—2296/1–56; 1822—2350/1–94; 2351/1–72; 2352/1–38; 2353/1–38; 2354/1–48; 1823—2517/1–91; 1824—2618/1–88; 2619/ 1–60.

Tipperary: 1821—2297/1–24; 1822—2355/1–91; 2356/1–84; 1823—2518/1–71; 1824—2621/1–66.

Waterford: 1821—2294/1–5; 1822—2357/1–34; 1823—2519/38–47.

National Library of Ireland, Dublin

Dublin Corn Market prices, 1785–1839 (NLI MS 4168)

Cork Archives Institute, Cork

Anstis, Eady, and Hewston letters

Public Record Office of Northern Ireland, Belfast

Redesdale papers (T.3030/C.34/13/2, 3)

Public Record Office, Kew (now the National Archives)

Colonial Office papers (CO 384)

Public Archives of Canada, Ottawa

Sir Robert Wilmot Horton miscellaneous correspondence (Microfilm 7–2167)

Manuscript Material in Private Possession

John Cussen papers

Printed advertisement for auction on 23 July 1808: "Sketch of the title of the Right Honorable William, Lord Viscount Courtenay, to lands in the manor of Meane and Bewly in the county of Limerick. Advertized to be sold by auction at the Royal Exchange Coffee-House on Saturday, the 23[r]d of July, 1808. [Issued by]

Furlong and Chambers, solicitors for Lord Courtenay and his trustees, 60, Aungier Street, Dublin. Printed by William Porter, Grafton Street [Dublin]." Three pages. In the possession of Robert Cussen and Son, Solicitors, Bridge Street, Newcastle West, Co. Limerick.

Printed advertisement for auction on 1 Dec. 1817: "To be sold by auction at the Commercial Buildings, Dame Street, Dublin, on Monday, the 1st December, 1817, and the following days, in twenty-five lots, according to the within particulars and conditions, sundry valuable freehold estates, part of the domain of the Right Hon. Lord Viscount Courtenay, desirably situate[d] contiguous to the post and market towns of Newcastle, Rathkeale, and Charleville . . . ," 6 pages. In the possession of Robert Cussen and Son, Solicitors, Bridge Street, Newcastle West, Co. Limerick.

Printed poster dated 26 Dec. 1820: "Outrage and Reward—At a meeting of the undersigned magistrates held at Newcastle in the county of Limerick on the 26th day of December 1820 to take into consideration the present state of the country, and particularly the barbarous attempt of assassination on Alexander Hoskins, Esq., on the 16th day of December instant. . . . Printed at the Chronicle Office, 2, Rutland Street, Limerick." In the possession of Robert Cussen and Son, Solicitors, Bridge Street, Newcastle West, Co. Limerick.

"Rental of the Irish estates of the Right Honorable William Lord Viscount Courtenay situate[d] in the baronies of Upper and Lower Connelloe and county of Limerick for one year from Lady Day 1827 to Lady Day 1828, containing Mr. Furlong's sixth general account with Lord Viscount Courtenay for that period, being [the] second rental stated in the imperial measurement and currency." In the possession of Robert Cussen and Son, Solicitors, Bridge Street, Newcastle West, Co. Limerick.

Michael Dore to John Cussen, 22 July 1986. In the possession of Robert Cussen and Son, Solicitors, Bridge Street, Newcastle West, Co. Limerick.

Statutes and Parliamentary Debates

A collection of the general public statutes passed in the forty-first year of the reign of . . . George the Third, in the first session of the first parliament of the United Kingdom of Great Britain and Ireland [etc.], 1801–69, 74 vols. London, 1801–69.

The parliamentary debates . . . , published under the superintendence of T.C. Hansard, new series, 1820–29, vols. i–xxii. London, 1820–29.

Parliamentary Papers (in Chronological Order)

A return of the number of troops or corps of effective yeomanry . . . , so far as relates to Ireland, H.C. 1821 (306), xix, 177.

Papers presented by his majesty's command relative to the disturbed state of Ireland, H.C. 1822 (2), xiv, 741.

Indictments and trials at assizes and special commissions: viz., returns of the number of persons indicted and tried at the several assizes and special commissions in Ireland during each of the last seven years, distinguishing between acquittals and convictions, and distinguishing the counties, cities, and towns, H.C. 1823 (305), xvi, 625.

Abstract of returns relative to magistrates, constables, and sub-constables appointed under the constables' act for Ireland, H.C. 1824 (257), xxii, 405.

Minutes of evidence taken before the select committee appointed to inquire into the disturbances in Ireland, in the last session of parliament, 13 May–18 June 1824, H.C. 1825 (20), vii, 1.

Minutes of evidence taken before the select committee of the House of Lords appointed to examine into the nature and extent of the disturbances which have prevailed in those districts of Ireland which are now subject to the provisions of the insurrection act, and to report to the House, 18 May–23 June 1824, H.C. 1825 (200), vii, 501.

Report from the select committee on the state of Ireland, ordered to be printed 30 June 1825, with the four reports of minutes of evidence, H.C. 1825 (129), viii, 1, 173, 293, 455.

Minutes of evidence taken before the select committee of the House of Lords appointed to inquire into the state of Ireland, more particularly with reference to the circumstances which may have led to the disturbances in that part of the United Kingdom, 18 February–21 March 1825, H.C. 1825 (181), ix, 1.

Minutes of evidence taken before the select committee of the House of Lords appointed to inquire into the state of Ireland, more particularly with reference to the circumstances which may have led to the disturbances in that part of the United Kingdom (24 March–22 June 1825), brought from the Lords, 5 July 1825, H.C. 1825 (521), ix, 249.

Report from the select committee on the butter trade of Ireland, H.C. 1826 (406), v, 135.

Report from the select committee on agriculture, with the minutes of evidence taken before them, and an appendix and index, H.C. 1833 (612), v, 1.

Census of Ireland for the year 1851, pt. v: tables of deaths, vol. i, containing the report, tables of pestilences, and analysis of the tables of deaths, [2087-I], H.C. 1856, xxix, 261.

Newspapers

Dublin Evening Post, 1819–25
Ennis Chronicle and Clare Advertiser
Leinster Journal, 1822–24
Limerick Evening Post

Other Contemporary Publications

Anon. An authentic report of the discussion which took place by agreement at Carrick-on-Shannon on the 9th November 1824 between three Roman Catholic priests and three clergymen of the established church, accompanied by the certificates of the reporters appointed by each party, and by that of the committee of gentlemen authorized to publish a report of the proceedings. Dublin, 1824.

Anon. Commutation of tythe by an acreable charge upon land, calculated to prevent the necessity of tythe proctors, &c., &c., by a beneficed clergyman of the established church. London, 1816.

Anon. *Commutation of tythes in Ireland injurious not only to the church establishment but to the poor, addressed, without permission, to the gentry of Kerry, Galway, and Tipperary.* London, 1808.

Anon. *Considerations addressed to the landed proprietors of the county of Clare.* Limerick, [1831].

Anon. *An enquiry into the history of tithe, its influence upon the agriculture, population, and morals of Ireland, with a plan for modifying that system and providing an adequate maintenance for the Catholic and Presbyterian clergy.* Dublin, 1808.

Anon. *A full and authentic report of all the debates that have taken place on the Irish tithe question in the last session of parliament.* Dublin, 1833.

Anon. *A full report of the proceedings at the late special commission holden [sic] in Cork from Thursday, the 22nd, to Friday, the 30th October 1829, before the Hon. Baron Pennefather and the Hon. Justice Torrens, being a detailed account of the several trials for the Doneraile conspiracy, with the luminous speeches of the solicitor-general and the able defence of Counsellor O'Connell.* Cork, 1829.

Anon. *Lachrymae Hibernicae, or the grievances of the peasantry of Ireland, especially in the western counties, by a resident native.* Dublin, 1822.

Anon. *The life and adventures of James Freney, together with an account of the actions of several other noted highwaymen.* Dublin, n.d.

Anon. *Old Bailey solicitor.* n.p., [1822?].

Anon. *Pastorini proved to be a bad prophet and a worse divine, in an address to the Roman Catholics of Ireland, earnestly recommended to their serious perusal, by Pastor Fido.* Dublin, 1823.

Anon. *The present state of Tipperary as regards agrarian outrages, their nature, origin, and increase considered, with suggestions for remedial measures, respectfully submitted to the Rt. Hon. Lord Eliot, M.P., chief secretary to the lord lieutenant, by a magistrate of the county.* Dublin, 1842.

Anon. *Remarks on a letter from Lord Cloncurry to the Duke of Leinster on the police and present state of Ireland.* Dublin, 1822.

Anon. *A report of the trial of Edward Browne and others for administering, and of Laurence Woods for taking, an illegal oath.* Dublin, 1822.

Anon. *A report of the trial of Michael Keenan for administering an unlawful oath.* Dublin, 1822.

Anon. *Report on the present state of the disturbed district in the south of Ireland, with an enquiry into the causes of the distresses of the peasantry and farmers.* Dublin, 1822.

Anon. *State of Ireland considered, with an inquiry into the history and operation of tithe and a plan for modifying that system and providing an adequate maintenance for the Catholic and Presbyterian clergy . . . , with an appendix containing the Rev. Mr. Hewlett's plan of commutation and a proposition for taxing absentees.* 2nd ed., Dublin, 1810.

Anon. *A summary digest of the most material portions of the evidence taken before "The select committee of the House of Lords on the state of Ireland in respect of crime," as relating to the Ribbon society, with an appendix* containing verbatim extracts from the evidence. London, 1839.

Anon. *The surprising life and adventures of the gentleman-robber Redmond O'Hanlon, generally called the captain general of the Irish robbers, protector of the rights and properties of his benefactors and redresser of the wrongs of the poor and distressed.* Glasgow, n.d.

Atkinson, A. *The Irish tourist, in a series of picturesque views, travelling incidents, and observations, statistical, political, and moral, on the character and aspect of the Irish nation.* Dublin, 1815.

Barker, F., and J. Cheyne. *An account of the rise, progress, and decline of the fever lately epidemical in Ireland, together with communications from physicians in the provinces and various official documents.* 2 vols. Dublin, 1821.

Carleton, William. "The Irish prophecy man," in *Irish Penny Journal* 1, no. 50 (12 June 1841), pp. 393–96.

——. *Tales and stories illustrating the character, usages, traditions, sports, and pastimes of the Irish peasantry.* Dublin, 1854; originally published 1845.

——. *Traits and stories of the Irish peasantry.* London, n.d.

——. *The works of William Carleton.* 3 vols. Collier's unabridged ed., New York, 1881.

Carr, John. *The stranger in Ireland, or a tour in the southern and western parts of that country in the year 1805.* London, 1806.

Croker, T.C. *Researches in the south of Ireland illustrative of the scenery, architectural remains, and the manners and superstitions of the peasantry.* London, 1824.

Curwen, J.C. *Observations on the state of Ireland, principally directed to its agriculture and rural population, in a series of letters written on a tour through the country.* 2 vols. London, 1818; originally published 1812.

D'Alton, John. *The history of tithes, church lands, and other ecclesiastical benefices, with a plan for the abolition of the former and the better distribution of the latter, in accordance with the trusts for which they were originally given.* Dublin, 1832.

[Doyle, James]. *Letters on the state of Ireland addressed by J.K.L. to a friend in England.* Dublin, 1825.

Doyle, James. *The pastoral address of the Right Rev. Dr. Doyle, Roman Catholic bishop of Kildare and Leighlin, against the illegal associations of Ribbonmen.* Dublin, 1822.

Dutton, Hely. *A statistical and agricultural survey of the county of Galway, with observations on the means of improvement, drawn up for the consideration and by the direction of the Royal Dublin Society.* Dublin, 1824.

[Emerson, J.S.]. *One year of the administration of his excellency the Marquess of Wellesley in Ireland.* London and Dublin, 1823.

The evidence taken before the select committee of the Houses of Lords and Commons appointed in the sessions of 1824 and 1825 to inquire into the state of Ireland. London, 1825.

Finlay, John. *A treatise on the law of tithes in Ireland and ecclesiastical law connected therewith.* Dublin, 1828.

Gorman, Peter, comp. *A report of the proceedings under a special commis-*

sion of oyer and terminer in the counties of Limerick & Clare in the months of May and June 1831, including the proceedings at the adjourned commission in Ennis, taken in short-hand. Limerick, 1831.

Hay, Edward. History of the Irish insurrection of 1798, giving an authentic account of the various battles fought between the insurgents and the king's army, and a genuine history of transactions preceding that event, with a valuable appendix. New ed., Boston, n.d.; originally published 1803.

Index to acts in force extending and relating to Ireland, passed in the parliaments of the United Kingdom from the union, 41 Geo. III, A.D. 1801, to the end of the session 3 & 4 William IV, A.D. 1833. Dublin, 1834.

Kennedy, J.P., ed. Digest of evidence taken before her majesty's commissioners of inquiry into the state of the law and practice in respect to the occupation of land in Ireland. 2 vols. Dublin, 1847.

Kernan, Randall, comp. A report of the trials of the Caravats and Shanavests at the special commission for the several counties of Tipperary, Waterford, and Kilkenny, before the Right Hon. Lord Norbury and the Right Hon. S. O'Gready [sic], commencing at Clonmel on Monday, February 4th, 1811, taken in short-hand by Randall Kernan, esq., barrister-at-law. Dublin, 1811.

Lewis, George Cornewall. Local disturbances in Ireland. Tower Books ed., Cork, 1977; originally published 1836.

Lewis, Samuel. A topographical dictionary of Ireland, comprising the several counties, cities, boroughs, corporate, market, and post towns, parishes, and villages, with historical and statistical descriptions. . . . 2 vols. Genealogical Publishing Company ed., 1984; originally published 1837.

Mac Nevin, Thomas. A letter to the Rt. Hon. the Earl of Roden, K.P., on the nature and causes of crime in Ireland. London, 1838.

McSkimin, Samuel. Annals of Ulster [from 1790 to 1798], ed. E.J. McCrum. New ed., Belfast, 1906; originally published 1849.

Moore, Thomas. Memoirs of Captain Rock, the celebrated Irish chieftain, with some account of his ancestors, written by himself. London, 1824.

Musgrave, Sir Richard. Memoirs of the different rebellions in Ireland from the arrival of the English; also a particular detail of that which broke out the 23d of May, 1798; with the history of the conspiracy which preceded it, ed. S.W. Myers and D.E. McKnight. 4th ed., Fort Wayne, Ind., 1995; 3rd ed. published 1802.

[O'Beirne, T.L.]. A letter from an Irish dignitary to an English clergyman on the subject of tithes in Ireland, written during the administration of the Duke of Bedford, with the addition of some observations and notes suggested by the present state of this momentous question. Dublin, 1822.

O'Connell. John, ed. The select speeches of Daniel O'Connell, M.P., edited with historical notices by his son John O'Connell, esq. 2 vols. in one. Dublin, n.d.; originally published 1854.

O'Driscol, John. Views of Ireland, moral, political, and religious. 2 vols. London, 1823.

O'Leary, Joseph. *The law of statutable composition for tithes in Ireland, with an appendix containing the necessary parts of the tithe acts and some precedents of pleadings.* Dublin, Belfast, Limerick, and Cork, 1834.

[O'Sullivan, Mortimer]. *Captain Rock detected, or the origin and character of the recent disturbances, and the causes, both moral and political, of the present alarming condition of the south and west of Ireland, fully and fairly considered and exposed by a Munster farmer.* London, 1824.

Oulton, A.N., comp. *Index to the statutes at present in force in, or affecting, Ireland from the year 1310 to 1835 inclusive.* Dublin, 1836.

Phelan, William, and Mortimer O'Sullivan. *A digest of the evidence taken before select committees of the two houses of parliament appointed to inquire into the state of Ireland, 1824–1825, with notes historical and explanatory, and a copious index.* 2 vols. London, 1826.

Reade, John. *Observations upon tythes and rents addressed to the clergy and lay impropriators of Ireland; dedicated to the Rt. Hon. Robert Peel, with a supplementary section, a postscript on the road acts, &c.* 2nd ed., Dublin, 1818.

Reid, Thomas. *Travels in Ireland in the year 1822, exhibiting brief sketches of the moral, physical, and political state of the country, with reflections on the best means of improving its condition.* London, 1823.

Ridgeway, William, comp. *A report of the proceedings under a special commission of oyer and terminer and gaol delivery for the counties of Sligo,* Mayo, Leitrim, Longford, and Cavan in the month of December 1806. Dublin, 1807.

Ryan, Richard. *Directions for proceeding under the tithe act, with observations and an appendix containing precedents of applications, notices, certificates, &c., required under the act.* Dublin, 1823.

Taylor, George. *A history of the rise, progress, and suppression of the rebellion in the county of Wexford in the year 1798, to which is annexed the author's account of his captivity and merciful deliverance.* Reprint ed., Dublin, 1907; originally published 1800.

Teeling, C.H. *History of the Irish rebellion of 1798: a personal narrative.* London and Glasgow, n.d.; originally published 1828.

The treble almanack for the year 1824, containing (I) John Watson Stewart's almanack; (II) the English court registry; (III) Wilson's Dublin directory, with a new, correct plan of the city; forming the most complete lists published of the present civil, military, and naval establishments of Great Britain & Ireland. Dublin, 1824.

Trimmer, J.K. *A brief enquiry into the present state of agriculture in the southern part of Ireland, and its influence on the manners and condition of the lower classes of the people, with some considerations on the ecclesiastical establishment of that country.* London, 1809.

Trimmer, J.K. *Further observations on the present state of agriculture and condition of the lower classes of the people in the southern parts of Ireland, with an estimate of the agricultural resources of that country, and a plan for carrying into effect a commutation for tithe and a project for poor laws.* London, 1812.

Wakefield, Edward. *An account of Ireland, statistical and political.* 2 vols. London, 1812.

[Walmesley, Charles]. *The general history of the Christian church from her birth to her final triumphant state in heaven, chiefly deduced from the Apocalypse of St. John the apostle, by Sig[nor] Pastorini.* Dublin, 1790.

———. *The general history of the Christian church....* 4th ed., Dublin, 1805.

———. *The general history of the Christian church....* 6th ed., Belfast, 1816.

———. *The general history of the Christian church....* 6th ed., Cork, 1820.

Woodward, Richard. *The present state of the Church of Ireland, containing a description of its precarious situation and the consequent danger to the public; recommended to the serious consideration of the friends of the Protestant interest; to which are subjoined some reflections on the impracticability of a proper commutation for tythes, and a general account of the origin and progress of the insurrections in Munster.* New ed., Dublin, 1808; originally published 1787.

Young, Arthur. *Arthur Young's tour in Ireland (1776–1779),* ed. A.W. Hutton. 2 vols. London and New York, 1892.

Later Works

Aalen, F.H.A., Kevin Whelan, and Matthew Stout, eds. *Atlas of Irish rural landscape.* Cork, 1997.

Adams, J.R.R. *The printed word and the common man: popular culture in Ulster, 1700–1900.* Belfast, 1987.

———. "Swine-tax and eat-him-all-Magee: the hedge schools and popular education in Ireland," in Donnelly and Miller, *Irish popular culture,* pp. 97–117.

Adams, W.F. *Ireland and Irish emigration to the new world from 1815 to the famine.* Baltimore, 1980; originally published 1932.

Adas, Michael. *Prophets of rebellion: millenarian protest movements against the European colonial order.* Chapel Hill, N.C., 1979.

Akenson, D.H. *The Church of Ireland: ecclesiastical reform and revolution, 1800–1885.* New Haven, Conn., and London, 1971.

———. *The Irish education experiment: the national system of education in the nineteenth century.* London and Toronto, 1970.

———. *The Irish in Ontario: a study in rural history.* Kingston, ON, and Montreal, 1984.

———. *Small differences: Irish Catholics and Irish Protestants, 1815–1922: an international perspective.* Kingston, ON, and Montreal, 1988.

Bardon, Jonathan. *A history of Ulster.* Belfast, 1992.

Bartlett, Thomas. "Defence, counter-insurgency and rebellion: Ireland, 1793–1803," in Bartlett and Jeffery, *A military history of Ireland,* pp. 247–93.

———. "Defenders and Defenderism in 1795," *Irish Historical Studies* 24, no. 95 (May 1985), pp. 373–94.

———. "An end to moral economy: the Irish militia disturbances of 1793," in Philpin, *Nationalism and popular protest,* pp. 191–218.

———. *The fall and rise of the Irish na-*

tion: the Catholic question, 1690–1830. Dublin and Savage, Md., 1992.

——, ed. Life of Theobald Wolfe Tone, compiled and arranged by William Theobald Wolfe Tone. Dublin, 1998.

Bartlett, Thomas, Kevin Dawson, and Dáire Keogh. Rebellion: a television history of 1798. Dublin, 1998.

Bartlett, Thomas, David Dickson, Dáire Keogh, and Kevin Whelan, eds. 1798: a bicentenary perspective. Dublin, 2003.

Bartlett, Thomas, and D.W. Hayton, eds. Penal era and golden age: essays in Irish history, 1690–1800. Belfast, 1979.

Bartlett, Thomas, and Keith Jeffery, eds. A military history of Ireland. Cambridge and New York, 1996.

Beames, M.R. Peasants and power: the Whiteboy movements and their control in pre-famine Ireland. Brighton, Sussex, and New York, 1983.

——. "The Ribbon societies: lower-class nationalism in pre-famine Ireland," in Philpin, Nationalism and popular protest, pp. 245–63.

——. "Rural conflict in pre-famine Ireland: peasant assassinations in Tipperary," in Philpin, Nationalism and popular protest, pp. 264–83.

Beattie, J.M. Crime and the courts in England, 1660–1800. Oxford, 1986.

Beatty, J.D., ed. Protestant women's narratives of the Irish rebellion of 1798. Dublin and Portland, Ore., 2001.

Beiner, Guy. Remembering the year of the French: Irish folk history and social memory. Madison, Wis., 2007.

Black, R.D.C. Economic thought and the Irish question, 1817–1870. Cambridge and New York, 1960.

Blackstock, Allan. An ascendancy army: the Irish yeomanry, 1796–1834. Dublin and Portland, Ore., 1998.

Bourke, P.M.A. 'The visitation of God'? The potato and the great Irish famine, ed. Jacqueline Hill and Cormac Ó Gráda. Dublin, 1993.

Bowen, Desmond. The Protestant crusade in Ireland, 1800–70: a study in Protestant-Catholic relations between the act of union and disestablishment. Dublin and Montreal, 1978.

Breathnach, Breandán. Folk music and dances of Ireland. Rev. ed., Cork and Dublin, 1977; originally published 1971.

Brennan, Brian. Máire Bhuí Ní Laoire: a poet of her people. Cork, 2000.

Bric, M.J. "Priests, parsons, and politics: the Rightboy protest in County Cork, 1785–1788," in Philpin, Nationalism and popular protest, pp. 163–90.

——. "The tithe system in eighteenth-century Ireland," Proceedings of the Royal Irish Academy 86, sec. C (1986), pp. 271–88.

Broeker, Galen. Rural disorder and police reform in Ireland, 1812–36. London and Toronto, 1970.

Buchanan, R.H., Emrys Jones, and Desmond McCourt, eds. Man and his habitat: essays presented to Emyr Estyn Evans. London and New York, 1971.

Burke, W.P. History of Clonmel. Waterford, 1907.

Casey, D.J. "Wildgoose Lodge: the evidence and the lore," Journal of the County Louth Archaeological and Historical Society 18, no. 2 (1974), pp. 140–61; 18, no. 3 (1975), pp. 211–31.

Casey, D.J., and R.E. Rhodes, eds. *Views of the Irish peasantry, 1800–1916*. Hamden, Conn., 1977.

Clark, Samuel. *Social origins of the Irish land war*. Princeton, N.J., 1979.

Clark, Samuel, and J.S. Donnelly, Jr., eds. *Irish peasants: violence and political unrest, 1780–1914*. Madison, Wis., and Manchester, 1983.

Clear, Caitríona. *Nuns in nineteenth-century Ireland*. Dublin and Washington, D.C., 1987.

Clifford, Brendan, ed. *Billy Bluff and the squire (a satire on Irish aristocracy) and other writings by Rev. James Porter, who was hanged in the course of the United Irish rebellion of 1798*. Belfast, 1991.

Cohn, Norman. "Medieval millenarianism: its bearing on the comparative study of millenarian movements," in Thrupp, *Millennial dreams*, pp. 31–43.

——. *The pursuit of the millennium: revolutionary millenarians and mystical anarchists of the middle ages*. Rev. ed., New York, 1970.

Connell, K.H. *Irish peasant society: four historical essays*. Oxford, 1968.

——. *The population of Ireland, 1750–1845*. Westport, Conn., 1975; originally published 1950.

——. *The population of Ireland, 1750–1845*. Oxford, 1950.

Connolly, S.J. "'Ag déanamh *commanding*': elite responses to popular culture, 1660–1850," in Donnelly and Miller, *Irish popular culture*, pp. 1–29.

——. "The 'blessed turf': cholera and popular panic in Ireland, June

1832," *Irish Historical Studies* 23, no. 91 (May 1983), pp. 214–32.

——. "The Houghers: agrarian protest in early eighteenth-century Connacht," in Philpin, *Nationalism and popular protest*, pp. 139–62.

——. *Priests and people in pre-famine Ireland, 1780–1845*. Dublin, 1982.

——. *Religion and society in nineteenth-century Ireland*. Dundalk, 1985.

——. *Religion, law, and power: the making of Protestant Ireland, 1660–1760*. Oxford, 1992.

——. "Violence and order in the eighteenth century," in O'Flanagan, Ferguson, and Whelan, *Rural Ireland*, pp. 42–61.

Corish, Patrick. *The Irish Catholic experience: a historical survey*. Dublin, 1985.

——, ed. *Radicals, rebels, and establishments (Historical Studies 15)*. Belfast, 1985.

Costello, Con. *Botany Bay: the story of the convicts transported from Ireland to Australia, 1791–1853*. Cork and Dublin, 1987.

Crawford, W.H., and Brian Trainor, eds. *Aspects of Irish social history, 1750–1800*. Belfast, 1969.

Cronin, Maura. "Memory, story, and balladry: 1798 and its place in popular memory in pre-famine Ireland," in Geary, *Rebellion and remembrance in modern Ireland*, pp. 112–34.

Crossman, Virginia. "The army and law and order in the nineteenth century," in Bartlett and Jeffery, *A military history of Ireland*, pp. 358–78.

——. "Emergency legislation and agrarian disorder in Ireland, 1821–41," *Irish Historical Studies* 27, no. 108 (Nov. 1991), pp. 309–23.

Crotty, R.D. *Irish agricultural production: its volume and structure.* Cork, 1966.

Cullen, L.M. *An economic history of Ireland since 1660.* London, 1972.

———. *The emergence of modern Ireland.* London, 1981.

———. "Irish history without the potato," in Philpin, *Nationalism and popular protest,* pp. 126–38.

Curtin, Gerard. *West Limerick: crime, popular protest, and society, 1820–1845.* Ballyhahill, Co. Limerick, 2008.

Curtin, N.J. *The United Irishmen: popular politics in Ulster and Dublin, 1791–1798.* Oxford, 1994.

d'Alton, Ian. *Protestant society and politics in Cork, 1812–1844.* Cork, 1980.

Daly, M.E. *Social and economic history of Ireland since 1800.* Dublin, 1981.

Daly, M.E., and David Dickson, eds. *The origins of popular literacy in Ireland: language change and educational development, 1700–1900.* Dublin, 1990.

Danaher, Kevin. *Gentle places and simple things.* Cork, 1964.

———. *In Ireland long ago.* Cork, 1962.

———. *Irish country people.* Cork, 1966.

———. *The pleasant land of Ireland.* Cork, 1970.

———. *The year in Ireland.* 4th ed., Cork and Dublin, and St. Paul, Minn., 1972.

Davis, N.Z. "Women on top," in N.Z. Davis, *Society and culture in early modern France.* Stanford, Calif., 1975.

de Nie, Michael. *The eternal Paddy: Irish identity and the British press, 1798–1882.* Madison, Wis., 2004.

Dickson, Charles. *The Wexford rising in 1798: its causes and its course.* Tralee, n.d.

Dickson, David, *Arctic Ireland: the extraordinary story of the great frost and forgotten famine of 1740–41.* Belfast, 1997.

———. "Middlemen," in Bartlett and Hayton, *Penal era and golden age,* pp. 162–85.

———. *Old world colony: Cork and South Munster, 1630–1830.* Madison, Wis., and Cork, 2005.

———. "The other great Irish famine," in Póirtéir, *The great Irish famine,* pp. 50–59.

———. "Taxation and disaffection in late eighteenth-century Ireland," in Clark and Donnelly, *Irish peasants,* pp. 37–63.

Dickson, David, Dáire Keogh, and Kevin Whelan, eds. *The United Irishmen: republicanism, radicalism, and rebellion.* Dublin, 1993.

Dickson, David, Cormac Ó Gráda, and Stuart Daultrey. "Hearth tax, household size, and Irish population change, 1672–1821," *Proceedings of the Royal Irish Academy* 82, sec. C, no. 6 (Dec. 1982), pp. 125–81.

Donnelly, J.S., Jr. "Captain Rock: ideology and organization in the Irish agrarian rebellion of 1821–24," *Éire-Ireland* 42, nos. 3–4 (Fall-Winter 2007), pp. 60–103.

———. "Captain Rock: the origins of the Irish agrarian rebellion of 1821–24," *New Hibernia Review* 11, no. 4 (Winter 2007), pp. 47–72.

———, ed. "A contemporary account of the Rightboy movement: the John Barter Bennett manuscript," *Jour-*

nal of the Cork Historical and Archaeological Society 88, no. 247 (Jan.–Dec. 1983), pp. 1–50.

———. "Factions in prefamine Ireland," in Audrey Eyler and Robert Garratt, eds., The uses of the past: essays on Irish culture. Newark, Del., and London, 1988), pp. 113–30.

———. The great Irish potato famine. Stroud, Gloucestershire, 2001.

———. "Hearts of Oak, Hearts of Steel," Studia Hibernica, no. 21 (1981), pp. 7–73.

———. "Irish agrarian rebellion: the Whiteboys of 1769–76," Proceedings of the Royal Irish Academy 83, sec. C, no. 12 (Dec. 1983), pp. 293–331.

———, ed. "The journals of Sir John Benn-Walsh relating to the management of his Irish estates, 1823–64 [part 1]," Journal of the Cork Historical and Archaeological Society 80, no. 230 (July–Dec. 1974), pp. 86–123.

———, ed. "The journals of Sir John Benn-Walsh relating to the management of his Irish estates, 1823–64 [part 2]," Journal of the Cork Historical and Archaeological Society 81, no. 231 (Jan.–June 1975), pp. 15–42.

———. The land and the people of nineteenth-century Cork: the rural economy and the land question. London and Boston, 1975.

———. Landlord and tenant in nineteenth-century Ireland. Dublin, 1973.

———. "Mass eviction and the great famine," in Póirtéir, The great Irish famine, pp. 155–73.

———. "Propagating the cause of the United Irishmen," Studies: An Irish Quarterly Review 69, no. 273 (Spring 1980), pp. 5–23.

———. "Republicanism and reaction in the 1790s," Irish Economic and Social History 11 (1984), pp. 94–100.

———. "The Rightboy movement, 1785–8," Studia Hibernica, nos. 17–18 (1977–78), pp. 120–202.

———. "Sectarianism in 1798 and in Catholic nationalist memory," in Geary, Rebellion and remembrance in modern Ireland, pp. 15–37.

———. "The social composition of agrarian rebellions in early nineteenth-century Ireland: the case of the Carders and Caravats, 1813–16," in Corish, Radicals, rebels, and establishments, pp. 151–69.

———. "The Terry Alt movement, 1829–31," History Ireland 2, no. 4 (Winter 1994), pp. 30–35.

———. "The Whiteboy movement, 1761–5," Irish Historical Studies 21, no. 81 (March 1978), pp. 20–54.

Donnelly, J.S., Jr., and K.A. Miller, eds. Irish popular culture, 1650–1850. Dublin and Portland, Ore., 1998.

Dooley, Terence. The decline of the big house in Ireland: a study of Irish landed families, 1860–1960. Dublin, 2001.

———. The murders at Wildgoose Lodge: agrarian crime and punishment in pre-famine Ireland. Dublin and Portland, Ore., 2007.

———. Sources for the history of landed estates in Ireland. Dublin and Portland, Ore., 2000.

Dore, Michael. "The murder of Thomas Hoskins," Annual Observer (June 1983), p. 40.

Dorian, Hugh. The outer edge of Ulster: a memoir of social life in nineteenth-

century Donegal, ed. Breandán Mac Suibhne and David Dickson. Dublin, 2000.

Dowling, Martin. *Tenant right and agrarian society in Ulster, 1600–1870.* Dublin and Portland, Ore., 1999.

Dowling, P.J. *The hedge schools of Ireland.* Rev. ed., Cork, 1968; originally published 1935.

Doyle, Danny, and Terence Folan. *The gold sun of Irish freedom: 1798 in song and story.* Cork and Boulder, Col., 1998.

Dunbabin, J.P.D. *Rural discontent in nineteenth-century Britain.* London, 1974.

Dunne, Tom. *"The Installation of Captain Rock,"* in Murray, *Daniel Maclise,* pp. 100–105.

——. "'Tá Gaedhil bhocht cráidhte': memory, tradition, and the politics of the poor in Gaelic poetry and song," in Geary, *Rebellion and remembrance in modern Ireland,* pp. 93–111.

——. *Rebellions: memoir, memory, and 1798.* Dublin, 2004.

Eiríksson, Andrés, and Cormac Ó Gráda. *Estate records of the Irish famine: a second guide to famine archives, 1840–1855.* Dublin, 1995.

Elliott, B.S. *Irish migrants in the Canadas: a new approach.* Kingston, ON, and Montreal, and Belfast, 1988.

Elliott, Marianne. *The Catholics of Ulster: a history.* New York, 2001.

——. *Partners in revolution: the United Irishmen and France.* New Haven, Conn., and London, 1982.

——. *Wolfe Tone: prophet of Irish independence.* New Haven, Conn., and London, 1989.

Evans, E.E. *Irish folk ways.* London, 1957.

——. *The personality of Ireland: habitat, heritage, and history.* Cambridge and New York, 1973.

Evans, E.J. "Some reasons for the growth of English anti-clericalism, *c.*1750–*c.*1830," *Past & Present,* no. 66 (Feb. 1975), pp. 84–109.

Farrell, Sean. *Rituals and riots: sectarian violence and political culture in Ulster, 1784–1886.* Lexington, Ky., 2000.

Fegan, Melissa. *Literature and the Irish famine, 1845–1919.* Oxford and New York, 2002.

Fitzpatrick, David. "Class, family, and rural unrest in nineteenth-century Ireland, in P.J. Drudy, ed., *Irish studies 2: land, politics, and people.* Cambridge, 1982, pp. 37–75.

——. "The disappearance of the Irish agricultural labourer, 1841–1912," *Irish Economic and Social History* 7 (1980), pp. 66–92.

——. "The modernisation of the Irish female," in O'Flanagan, Ferguson, and Whelan, *Rural Ireland,* pp. 162–80.

——. *Oceans of consolation: personal accounts of Irish emigration to Australia.* Ithaca, N.Y., and London, 1994.

Fitz-Patrick, W.J. *The life, times, and correspondence of the Right Rev. Dr. Doyle, bishop of Kildare and Leighlin.* 2 vols. New ed., Dublin, 1880.

Flanagan, Thomas. *The Irish novelists, 1800–1850.* New York and London, 1959.

Foley, Tadhg, and Seán Ryder, eds. *Ideology and Ireland in the nineteenth century*. Dublin and Portland, Ore., 1998.

Freeman, T.W. *Ireland: its physical, historical, social, and economic geography*. London and New York, 1950.

———. *Pre-famine Ireland: a study in historical geography*. Manchester, 1957.

Freyer, Grattan. *Bishop Stock's "Narrative" of the Year of the French: 1798*. Ballina, Co. Mayo, 1982; reprint of 2nd ed., Dublin, 1800.

Furlong, Nicholas. *Fr. John Murphy of Boolavogue, 1753–1798*. Dublin, 1991.

Gahan, D.J. *The people's rising: Wexford, 1798*. Dublin, 1995.

———. *Rebellion: Ireland in 1798*. Dublin, 1997.

Gailey, Alan. *Irish folk drama*. Cork, 1969.

Garnham, Neal. *The courts, crime, and the criminal law in Ireland, 1692–1760*. Dublin, 1996.

———. "How violent was eighteenth-century Ireland?" *Irish Historical Studies* 30, no. 119 (May 1997), pp. 377–92.

Garvin, Tom. "Defenders, Ribbonmen, and others: underground political networks in pre-famine Ireland," in Philpin, *Nationalism and popular protest*, pp. 219–44.

Gash, Norman, ed. *The age of Peel*. London and New York, 1968.

———. *Mr. Secretary Peel: the life of Sir Robert Peel to 1830*. Cambridge, Mass., and London, 1961.

Gattrell, V.A.C. *The hanging tree: execution and the English people, 1700–1868*. Oxford, 1994.

Geary, L.M. "John Milner Barry and public health in early nineteenth-century Cork," *Journal of the Cork Historical and Archaeological Society* 106 (2001), pp. 131–42.

———. *Medicine and charity in Ireland, 1718–1851*. Dublin, 2004.

———, ed. *Rebellion and remembrance in modern Ireland*. Dublin and Portland, Ore., 2001.

Gibbon, Peter. *The origins of Ulster unionism: the formation of popular Protestant politics and ideology in nineteenth-century Ireland*. Manchester, 1975.

Gibbons, Luke. "Between Captain Rock and a hard place: art and agrarian insurgency," in Foley and Ryder, *Ideology and Ireland*, pp. 23–44.

———. "Identity without a centre: allegory, history, and Irish nationalism," in Luke Gibbons, *Transformations in Irish culture* (Cork, 1996), pp. 134–47.

Gibbons, S.R., ed. *Captain Rock, night errant: the threatening letters of pre-famine Ireland, 1801–1845*. Dublin and Portland, Ore., 2004.

Gillespie, Raymond. "'And be hanged by the neck until you are dead,'" introduction, in Sweeney, *Hanging crimes*, pp. 1–9.

Gillespie, Raymond, and Gerard Moran, eds. *"A various country": essays in Mayo history, 1500–1900*. Westport, Conn., 1987.

Gough, Hugh, and David Dickson, eds. *Ireland and the French Revolution*. Blackrock, Co. Dublin, 1990.

Guha, Ranajit. "The prose of counter-insurgency," in Ranajit Guha and G.C. Spivak, eds., *Selected subaltern studies* (New York and Oxford, 1988), pp. 45–88.

Harrison, J.F.C. *The second coming: popular millenarianism, 1780–1850.* London and New Brunswick, N.J., 1979.

Hay, Douglas, et al. *Albion's fatal tree: crime and society in eighteenth-century England.* London, 1975.

Hayley, Barbara, ed. *A bibliography of the writings of William Carleton.* Gerrards Cross, Bucks., 1985.

———. *Carleton's "Traits and stories" and the 19th century Anglo-Irish tradition.* Gerrards Cross, Bucks., and Totowa, N.J., 1983.

Hempton, David. *Methodism and politics in British society, 1750–1850.* London and Stanford, Calif., 1984.

———. "The Methodist crusade in Ireland, 1795–1845," *Irish Historical Studies* 22, no. 85 (March 1980), pp. 33–48.

———. *Religion and political culture in Britain and Ireland: from the Glorious Revolution to the decline of empire.* Cambridge and New York, 1996.

Henry, Brian. *Dublin hanged: crime, law enforcement, and punishment in late eighteenth-century Dublin.* Blackrock, Co. Dublin, 1994.

Hill, Jacqueline. *From patriots to unionists: Dublin civic politics and Irish Protestant patriotism, 1660–1840.* Oxford and New York, 1997.

———. "The legal profession and the defence of the ancien regime in Ireland, 1790–1840," in Dáire Hogan and W.N. Osborough, eds., *Brehons, serjeants, and attorneys* (Dublin, 1990), pp. 181–210.

———. "National festivals, the state, and 'Protestant ascendancy' in Ireland, 1790–1829," *Irish Historical*

Studies 24, no. 93 (May 1984), pp. 30–51.

Hobsbawn, E.J. *Primitive rebels: studies in archaic forms of social movement in the 19th and 20th centuries.* Manchester, 1959.

Hobsbawn, E.J., and George Rudé. *Captain Swing.* London, 1968.

Holmes, R.F. *Henry Cooke.* Belfast and Ottawa, 1981.

Hoppen, K.T. *Elections, politics, and society in Ireland, 1832–1885.* Oxford and New York, 1984.

———. *Ireland since 1800: conflict and conformity.* London and New York, 1989.

Houston, C.J., and W.J. Smyth. *Irish emigration and Canadian settlement: patterns, links, and letters.* Toronto and Belfast, 1990.

Huggins, Michael. *Social conflict in pre-famine Ireland: the case of County Roscommon.* Dublin and Portland, Ore., 2007.

Hyde, Douglas. *Abhráin atá leagtha ar an Reachtúire,* [or] *Songs ascribed to Raftery, being the fifth chapter of "The songs of Connacht."* Shannon, 1973; originally published 1903.

Hyland, Cal. "The Franks: murder for love or revenge," *Mallow Field Club Journal,* no. 4 (1986), pp. 141–43.

Inglis, Brian. *The freedom of the press in Ireland, 1784–1841.* Westport, Conn., 1975; originally published 1954.

Joyce, John. *General Thomas Cloney: a Wexford rebel of 1798.* Dublin, 1988.

Jupp, Peter, and Eoin Magennis, eds. *Crowds in Ireland, c.1720–1920.* London, 2000.

Katsuta, Shunsuke. "The Rockite

movement in County Cork in the early 1820s," *Irish Historical Studies* 33, no. 131 (May 2003), pp. 278–96.

Kavanagh, P.F. *A popular history of the insurrection of 1798, derived from every available record and reliable tradition.* 2nd centenary ed., Cork, 1898; originally published 1870.

Kavanagh, Patrick, ed. *The autobiography of William Carleton.* Rev. ed., London, 1968; originally published 1896.

Kee, Robert. *The green flag: the turbulent history of the Irish national movement.* London, 1972.

Keenan, D.J. *The Catholic church in nineteenth-century Ireland: a sociological study.* Dublin and Totowa, N.J., 1983.

Kelly, James, ed. *Gallows speeches from eighteenth-century Ireland.* Dublin and Portland, Ore., 2001.

——. "The genesis of 'Protestant ascendancy': the Rightboy disturbances of the 1780s and their impact on Protestant opinion," in Gerard O'Brien, ed., *Parliament, politics, and people: essays in eighteenth-century Irish history* (Dublin, 1989), pp. 93–127.

Kennedy, Liam. *Colonialism, religion, and nationalism in Ireland.* Belfast, 1996.

Kennedy, Liam, and Martin Dowling. "Prices and wages in Ireland, 1700–1850," *Irish Economic and Social History* 24 (1997), pp. 62–104.

Kennedy, Liam, and P.M. Solar. *Irish agriculture: a price history from the mid-eighteenth century to the eve of the First World War.* Dublin, 2007.

Kenny, Kevin. *Making sense of the Molly Maguires.* New York and Oxford, 1998.

Keogh, Dáire. *"The French Disease": The Catholic church and radicalism in Ireland, 1790–1800.* Blackrock, Co. Dublin, 1993.

Keogh, Dáire, and Nicholas Furlong, eds. *The mighty wave: the 1798 rebellion in Wexford.* Dublin and Portland, Ore., 1996.

——, eds. *The women of 1798.* Dublin and Portland, Ore., 1998.

Kerr, D.A. "Priests, pikes, and patriots: the Irish Catholic church and political violence from the Whiteboys to the Fenians," in S.J. Brown and D.W. Miller, eds., *Piety and power in Ireland, 1760–1960: essays in honour of Emmet Larkin.* Belfast and Notre Dame, Ind., 2000, pp. 16–42.

Kiely, Benedict. *Poor scholar: a study of the works and days of William Carleton (1794–1869).* London, 1948.

Kiernan, T.J. *The Irish exiles in Australia.* Dublin and London, 1954.

Killen, John, ed. *The decade of the United Irishmen: contemporary accounts, 1791–1801.* Belfast, 1997.

King-Harman, R.D. *The Kings, Earls of Kingston.* Cambridge, n.d. [1959].

Kselman, T.A. *Miracles and prophecies in nineteenth-century France.* New Brunswick, N.J., 1983.

Large, David. "The wealth of the greater Irish landowners, 1750–1815," *Irish Historical Studies* 15, no. 57 (March 1966), pp. 21–47.

Larkin, Emmet. *The pastoral role of the Roman Catholic church in pre-famine Ireland, 1750–1850.* Dublin and Washington, D.C, 2006.

Lecky, W.E.H. *A history of Ireland in the eighteenth century.* 5 vols. London, 1892.

Lee, Joseph. *The modernisation of Irish society, 1848–1918.* Dublin, 1973.

——. "The Ribbonmen," in Williams, *Secret societies in Ireland,* pp. 26–35.

Lenihan, Maurice. *Limerick: its history and antiquities.* Dublin, 1866.

Logan, Patrick. *Fair day: the story of Irish fairs and markets.* Belfast, 1986.

Luddy, Maria. *Women and philanthropy in nineteenth-century Ireland.* Cambridge and New York, 1995.

——. *Women in Ireland, 1800–1918: a documentary history.* Cork, 1995.

Lyons, F.S.L. "The vicissitudes of a middleman in County Leitrim, 1810–27," *Irish Historical Studies* 9, no. 35 (March 1955), pp. 300–318.

McAnally, Sir Henry. *The Irish militia, 1793–1816.* Dublin and London, 1949.

McAuliffe, John. "Richard Griffith and road-making," *Journal of the Newcastle West Historical Society,* no. 1 (1990), pp. 20–22.

McBride, Ian. "Review article: Reclaiming the rebellion: 1798 in 1998," *Irish Historical Studies* 31, no. 123 (May 1999), pp. 395–410.

——. *Scripture politics: Ulster Presbyterians and Irish radicalism in the late eighteenth century.* Oxford, 1998.

——. *The siege of Derry in Ulster Protestant mythology.* Dublin and Portland, Ore., 1997.

McCormack, W.J. *The Dublin paper war of 1786–1788.* Dublin, 1993.

MacDonagh, Oliver. *The emancipist: Daniel O'Connell, 1830–47.* London, 1989.

——. *The hereditary bondsman: Daniel O'Connell, 1775–1829.* London, 1988.

McDowell, R.B. *Ireland in the age of imperialism and revolution, 1760–1801.* Oxford, 1979.

——. *The Irish administration, 1801–1914.* London and Toronto, 1964.

——. *Irish public opinion, 1750–1800.* London, 1944.

——. *Public opinion and government policy in Ireland, 1801–1846.* Westport, Conn., 1975; originally published 1952.

——, ed. *Social life in Ireland, 1800–45.* Dublin, 1957.

McEvoy, Frank, ed. *Life & adventures of James Freney, written by himself.* Kilkenny, 1988; originally published 1754.

McGrath, Mícheál, ed. and trans. *Cinnlae Amhlaoibh Uí Súilleabháin (The diary of Humphrey O'Sullivan).* 4 vols. London, 1936–37; originally published 1928–31.

McGrath, Thomas. *Politics, interdenominational relations, and education in the public ministry of Bishop James Doyle of Kildare and Leighlin, 1786–1834.* Dublin, 1999.

——. *Religious renewal and reform in the pastoral ministry of Bishop James Doyle of Kildare and Leighlin, 1786–1834.* Dublin, 1999.

McHugh, R.J., ed. *Carlow in 98: the autobiography of William Farrell of Carlow.* Dublin, 1949.

McMahon, Seán, and Jo O'Donoghue, eds. *Taisce duan: a treasury of Irish poems with translations in English.* Swords, Co. Dublin, 1992.

Macintyre, Angus. *The liberator: Daniel O'Connell and the Irish party, 1830–1847.* London, 1965.

MacKay, Donald. *Flight from famine: the coming of the Irish to Canada.* Plattsburgh, N.Y., 2002; originally published 1990.

Madden, R.R. *The United Irishmen, their lives and times, with numerous original portraits and additional authentic documents; the whole matter newly arranged and revised,* ser. 1–4. 2nd ed., 1857–60.

Magennis, E.F. "A 'Presbyterian insurrection'?: reconsidering the Hearts of Oak disturbances of July 1763," *Irish Historical Studies* 31, no. 122 (Nov. 1998), pp. 165–87.

Magray, Mary Peckham. *The transforming power of the nuns: women, religion, and cultural change in Ireland, 1750–1900.* New York, 1998.

Maguire, W.A. *The Downshire estates in Ireland, 1801–1845: the management of Irish landed estates in the early nineteenth century.* Oxford, 1972.

——, ed. *Letters of a great Irish landlord: a selection from the estate correspondence of the third Marquess of Downshire, 1809–45.* Belfast, 1974.

Malcolm, Elizabeth. "The rise of the pub: a study in the disciplining of popular culture," in Donnelly and Miller, *Irish popular culture,* pp. 50–77.

Malcomson, A.P.W. *John Foster: the politics of the Anglo-Irish Ascendancy.* Oxford, 1978.

Mannion, J.J. *Irish settlements in eastern Canada: a study of cultural transfer and adaptation.* Toronto and Buffalo, N.Y., 1974.

Marnane, D.G. "Land and violence in nineteenth-century Tipperary," *Tipperary Historical Journal* (1988), pp. 53–89.

Maxwell, Constantia. *Country and town in Ireland under the Georges.* Rev. ed., Dundalk, 1949; originally published 1940.

——. *Dublin under the Georges, 1714–1830.* London and Dublin, 1946.

——. *A history of Trinity College, 1591–1892.* Dublin, 1946.

——. *The stranger in Ireland from the reign of Elizabeth to the great famine.* London, 1954.

Maxwell, W.H. *History of the Irish rebellion in 1798, with memoirs of the union and Emmett's insurrection in 1803.* London, 1887; originally published 1845.

Miller, D.W. "The Armagh troubles, 1784–95," in Clark and Donnelly, *Irish peasants,* pp. 155–91.

——, ed. *Peep O'Day Boys and Defenders: selected documents on the County Armagh disturbances, 1784–96.* Belfast, 1990.

——. "Presbyterianism and 'modernization' in Ulster," *Past & Present,* no. 80 (Aug. 1978), pp. 66–90.

Miller, K.A. *Emigrants and exiles: Ireland and the Irish exodus to North America.* New York and Oxford, 1985.

——. "The lost world of Andrew Johnston: sectarianism, social conflict, and cultural change in southern Ireland during the pre-famine era," in Donnelly and Miller, *Irish popular culture,* pp. 222–41.

Miller, K.A., Arnold Schrier, B.D. Boling, and D.N. Doyle, eds. *Irish emigrants in the land of Canaan: letters and memoirs from colonial and revolutionary America, 1675–1815.* Oxford and New York, 2003.

Mokyr, Joel. *Why Ireland starved: a quantitative and analytical history of the Irish economy, 1800–1850.* London and Boston, 1983.

Moody, T.W., and W.E. Vaughan, eds. *A new history of Ireland,* vol. 4: *eighteenth-century Ireland, 1691–1800.* Oxford and New York, 1986.

Moore, Thomas. *The life and death of Lord Edward Fitzgerald.* London and Manchester, n.d.; originally published 1832.

———. *Memoirs of Captain Rock, the celebrated Irish chieftain, with some account of his ancestors, written by himself,* edited and introduced by Emer Nolan, with annotations by Seamus Deane. Dublin, 2008.

Moran, Gerard, ed. *Radical Irish priests, 1660–1970.* Dublin, 1998.

———. *Sending out Ireland's poor: assisted emigration to North America in the nineteenth century.* Dublin and Portland, Ore., 2004.

Murphy, J.A., ed. *The French are in the bay: the expedition to Bantry Bay, 1796.* Cork, 1997.

———. "The support of the Catholic clergy in Ireland, 1750–1850," in J.L. McCracken, ed. *Historical Studies 5* (Philadelphia, 1965), pp. 103–21.

Murphy, Maura. "The ballad singer and the role of the seditious ballad in nineteenth-century Ireland," *Ulster Folklife* 25 (1979), pp. 79–102.

Murphy, Sean. "Irish Jacobinism and freemasonry," *Eighteenth-Century Ireland* 9 (1994), pp. 79–82.

Murray, A.C. "Agrarian violence and nationalism in nineteenth-century Ireland: the myth of Ribbonism," *Irish Economic and Social History* 13 (1986), pp. 56–73.

Murray, Peter, ed. *Daniel Maclise, 1806–1870: romancing the past.* Cork and Kinsale, 2008.

Nolan, Emer. "Irish melodies and discordant politics: Thomas Moore's *Memoirs of Captain Rock* (1824)," *Field Day Review,* no. 2 (2006), pp. 41–53.

Nowlan, K.B. "Agrarian unrest in Ireland, 1800–1845," *University Review* 2, no. 6 (1959), pp. 7–16.

O'Brien, George. *The economic history of Ireland from the union to the famine.* London and New York, 1921.

O'Brien, R.B., ed. *The autobiography of Theobald Wolfe Tone.* 2 vols. Dublin, Cork, and Belfast, n.d..

O'Carroll, Gerald, ed. *Robert Day (1746–1841): the diaries and the addresses to grand juries, 1793–1829.* Tralee, 2004.

Ó Catháin, Séamas. *Irish life and lore.* Dublin and Cork, 1982.

Ó Ceallaigh, Tadhg. "Peel and police reform in Ireland, 1814–18," *Studia Hibernica,* no. 6 (1966), pp. 25–48.

Ó Ciardha, Éamonn. *Ireland and the Jacobite cause, 1685–1766: a fatal attachment.* Dublin and Portland, Ore., 2002.

Ó Ciosáin, Niall. "The Irish rogues," in Donnelly and Miller, *Irish popular culture,* pp. 78–96.

———. *Print and popular culture in Ireland, 1750–1850.* London, 1996.

O'Connell, M.R., ed. *The correspondence of Daniel O'Connell.* 8 vols. Shannon and Tallaght, Co. Dublin, 1972–80.

O'Connor, P.J. *All Ireland is in and around Rathkeale.* Newcastle West, 1996.

———. "Ireland's last best west: evidence from the Courtenay/Devon estate," *Journal of the Newcastle West Historical Society,* no. 1 (1990), pp. 23–29.

Ó Crualaoich, Gearóid. "The 'merry wake,'" in Donnelly and Miller, *Irish popular culture,* pp. 173–200.

O'Donnell, Patrick. *The Irish faction fighters of the 19th century.* Dublin, 1975.

O'Donnell, Ruán. *Aftermath: post-rebellion insurgency in Wicklow, 1799–1803.* Dublin and Portland, Ore., 2000.

———. *The rebellion in Wicklow, 1798.* Dublin and Portland, Ore., 1998.

O'Donoghue, Patrick. "Causes of opposition to tithes, 1800–38," *Studia Hibernica,* no. 5 (1965), pp. 7–28.

———. "Opposition to tithe payments in 1830–31," *Studia Hibernica,* no. 6 (1966), pp. 69–98.

———. "Opposition to tithe payment in 1832–3," *Studia Hibernica,* no. 12 (1972), pp. 77–108.

O'Donovan, John. *The economic history of livestock in Ireland.* Dublin and Cork, 1940.

Ó Duigneáin, Proinnsíos. *The priest and the Protestant woman: the trial of Rev. Thomas Maguire, P.P., Dec. 1827.* Dublin and Portland, Ore., 1997.

O'Faolain, Sean. *King of the beggars: a life of Daniel O'Connell.* Dublin, 1986; originally published 1938.

O'Farrell, Patrick. "Millen[n]ialism, messianism, and utopianism in Irish history," in *Anglo-Irish Studies* 2 (1976), pp. 45–68.

O'Ferrall, Fergus. *Catholic emancipa-*
tion: Daniel O'Connell and the birth of Irish democracy, 1820–30.* Dublin and Atlantic Highlands, N.J., 1985.

O'Flanagan, Patrick, and C.G. Buttimer, eds. *Cork: history and society.* Dublin, 1993.

O'Flanagan, Patrick, Paul Ferguson, and Kevin Whelan, eds. *Rural Ireland: modernisation and change, 1600–1900.* Cork, 1987.

Ó Giolláin, Diarmuid. "The pattern," in Donnelly and Miller, *Irish popular culture,* pp. 201–21.

Ó Gráda, Cormac. *Black '47 and beyond: the great Irish famine in history, economy, and memory.* Princeton, N.J., and Chichester, West Sussex, 1999.

———. *Ireland: a new economic history, 1780–1939.* Oxford and New York, 1994.

———. *Ireland before and after the famine: explorations in economic history.* 2nd ed., Manchester, 1993.

O'Higgins, Paul, ed. *A bibliography of Irish trials and other legal proceedings.* Abingdon, Oxon., 1986.

Ó hÓgáin, Dáithí. *The hero in Irish folk history.* Dublin and New York, 1985.

Ó Macháin, Pádraig. *Six years in Galmoy: rural unrest in County Kilkenny, 1819–1824.* Dublin, 2004.

Ó Muimhneacháin, Aindrias, ed. and trans. *Stories from the Tailor.* Cork and Dublin, n.d.; originally published 1978.

Ó Muireadhaigh, Sailbheastar. "Na Carabhait agus na Sean-Bheisteanna," *Galvia* 8 (1961), pp. 4–20.

———. "Buachaillí na Carraige, 1820–25," *Galvia* 9 (1962), pp. 4–13.

O'Neill, Kevin. *Family and farm in pre-famine Ireland: the parish of Killashandra.* Madison, Wis., and London, 1984.

O'Shaughnessy, Peter, ed. *Rebellion in Wicklow: General Joseph Holt's personal account of 1798.* Dublin and Portland, Ore., 1998.

Ó Súilleabháin, Seán. *Irish folk custom and belief.* Dublin, n.d.

———. *Irish wake amusements.* Cork, 1967.

O'Sullivan, Tadhg. "'The violence of a servile war': three narratives of Irish insurgency post-1798," in Geary, *Rebellion and remembrance in modern Ireland,* pp. 73–92.

Ó Tuathaigh, Gearóid. "Gaelic Ireland, popular politics, and Daniel O'Connell," *Journal of the Galway Archaeological and Historical Society* 34 (1974–75), pp. 21–34.

———. *Ireland before the famine, 1798–1848.* Dublin, 1972.

Owens, Gary. "The Carrickshock incident, 1831: social memory and an Irish cause célèbre," *Cultural and Social History* 1, no. 1 (2004), pp. 36–64.

———. "'A moral insurrection': faction fighters, public demonstrations, and the O'Connellite campaign, 1828," *Irish Historical Studies* 30, no. 120 (Nov. 1997), pp. 513–39.

———. "Nationalism without words: symbolism and ritual behaviour in the repeal 'monster meetings' of 1843," in Donnelly and Miller, *Irish popular culture,* pp. 242–69.

Pakenham, Thomas. *The year of liberty: the story of the great Irish rebellion of 1798.* London, 1969.

Palmer, S.H. *Police and protest in England and Ireland, 1780–1850.* Cambridge and New York, 1988.

Patterson, J.G. "'Educated Whiteboyism': the Cork tithe war, 1798–9," *History Ireland* 12, no. 4 (Winter 2004), pp. 25–29.

———. "Republicanism, agrarianism, and banditry in the west of Ireland, 1798–1803," *Irish Historical Studies* 35, no. 137 (May 2006), pp. 17–39.

Peacock, A.J. *Bread or blood: a study of the agrarian riots in East Anglia in 1816.* London, 1965.

Philpin, C.H.E., ed. *Nationalism and popular protest in Ireland.* Cambridge, 1987.

Póirtéir, Cathal, ed. *The great Irish famine.* Cork and Dublin, 1995.

———, ed. *The great rebellion of 1798.* Cork and Boulder, Col., 1998.

Post, J.D. *The last great subsistence crisis in the western world.* Baltimore and London, 1977.

Power, Bill. *White knights, dark earls: the rise and fall of an Anglo-Irish dynasty.* Cork, 2000.

Power, Thomas. "Father Nicholas Sheehy (*c.*1728–1766)," in Moran, *Radical Irish priests,* pp. 62–78.

———. *Land, politics, and society in eighteenth-century Tipperary.* Oxford, 1993.

Reynolds, J.A. *The Catholic emancipation crisis in Ireland, 1823–1829.* Westport, Conn., 1970; originally published 1954.

Roberts, P.E.W. "Caravats and Shanavests: Whiteboyism and faction fighting in East Munster, 1802–11," in Clark and Donnelly, *Irish peasants,* pp. 64–101.

Ronan, M.V., ed. *Insurgent Wicklow, 1798: the story as written by Rev. Bro. Luke Cullen, O.D.C. (1793–1859), with additional material from other Mss.* Dublin, 1948.

Rudé, George. *Protest and punishment: the story of the social and political protesters transported to Australia, 1788–1868.* Oxford, 1978.

Salaman, R.N. *The history and social influence of the potato, with a chapter on industrial uses by W.G. Burton,* ed. J.G. Hawkes. Cambridge and New York, 1985; originally published 1949.

Savage, John. *98 and 48: the modern revolutionary history and literature of Ireland.* New York, 1882.

Senior, Hereward. *Orangeism in Ireland and Britain, 1795–1836.* London and Toronto, 1966.

Shepperson, George. "The comparative study of millenarian movements," in Thrupp, *Millennial dreams,* pp. 44–52.

Smyth, Jim. *The men of no property: Irish radicals and popular politics in the late eighteenth century.* Dublin, 1992.

Smyth, W.J. "Social, economic, and landscape transformations in County Cork from the mid-eighteenth to the mid-nineteenth century," in O'Flanagan and Buttimer, *Cork,* pp. 655–98.

Smyth, W.J, and Kevin Whelan, eds. *Common ground: essays on the historical geography of Ireland presented to T. Jones Hughes.* Cork, 1988.

Spiers, E.M. "Army organisation and society in the nineteenth century," in Bartlett and Jeffery, *A military history of Ireland,* pp. 335–57.

Stevenson, John. *Popular disturbances in England, 1700–1832.* 2nd ed., London and New York, 1992.

Stewart, A.T.Q. *A deeper silence: the hidden origins of the United Irishmen.* London and Boston, 1993.

———. *The narrow ground: aspects of Ulster, 1609–1969.* London, 1977.

———. *The summer soldiers: the 1798 rebellion in Antrim and Down.* Belfast, 1995.

Sweeney, Frank, ed. *Hanging crimes.* Cork, 2005.

Thompson, F.M.L., and D. Tierney, eds. *General report on the Gosford estates in County Armagh, 1821, by William Greig.* Belfast, 1976.

Thrupp, S.L., ed. *Millennial dreams in action: studies in revolutionary religious movements.* Schocken Books paperback ed., New York, 1970.

Thuente, Mary Helen. *The harp restrung: the United Irishmen and the rise of Irish literary nationalism.* Syracuse, N.Y., 1994.

Tilly, Charles. "Collective violence in European perspective," in H.D. Graham and T.R. Gurr, eds., *The history of violence in America: historical and comparative perspectives* (New York and London, 1969), pp. 4–45.

Tillyard, Stella. *Citizen lord: Edward Fitzgerald, 1763–1798.* London, 1997.

Trench, C.C. *Grace's card: Irish Catholic landlords, 1690–1800.* Cork, 1997.

Vaughan, W.E. *Landlords and tenants in mid-Victorian Ireland.* Oxford and New York, 1994.

———, ed. *A new history of Ireland,* vol. 5: *Ireland under the union, part 1: 1800–70.* Oxford and New York, 1989.

Vaughan, W.E., and A.J. Fitzpatrick,

eds. *Irish historical statistics: population, 1821-1971*. Dublin, 1978.

Wall, Maureen. "The Whiteboys," in Williams, *Secret societies in Ireland*, pp. 13-25.

Wall, Thomas. *The sign of Doctor Hay's head, being some account of the hazards and fortunes of Catholic printers and publishers in Dublin from the later penal times to the present day*. Dublin, 1958.

Walsh, J.E. *Ireland one hundred and twenty years ago, being a new and revised edition of Ireland sixty years ago*, ed. Dillon Cosgrave. Dublin, 1911.

Weber, Eugen. *Apocalypse: prophecies, cults, and millennial beliefs through the ages*. Cambridge, Mass., 1999.

Wells, Roger. "The Irish famine of 1799-1801," in Adrian Randall and Andrew Charlesworth, eds., *Markets, market culture, and popular protest in eighteenth-century Britain and Ireland*. Liverpool, 1996, pp. 163-93.

Westropp, T.J. *The Desmond Castle, Newcastle West, Co. Limerick*. Newcastle West, 1983; originally published 1910.

Whelan, Irene. *The bible war in Ireland: the "Second Reformation" and the polarization of Protestant-Catholic relations, 1800-1840*. Dublin and Madison, Wis., 2005.

Whelan, Kevin. *Fellowship of freedom: the United Irishmen and the 1798 rebellion*. Cork, 1998.

———. *The tree of liberty: radicalism, Catholicism, and the construction of Irish identity, 1760-1830*. Cork, 1996.

———. "An underground gentry? Catholic middlemen in eighteenth-century Ireland," in Donnelly and Miller, *Irish popular culture*, pp. 118-72.

Whelan, Kevin, and T.P. Power, eds. *Endurance and emergence: Catholics in Ireland in the eighteenth century*. Dublin, 1990.

Wilde, W.R. *Irish popular superstitions*. Dublin, 1979; originally published 1852.

Williams, David. *The Rebecca riots: a study in agrarian discontent*. Cardiff, Wales, 1955.

Williams, T.D., ed. *Secret societies in Ireland*. Dublin and New York, 1973.

Wilson, C.A. *A new lease on life: landlords, tenants, and immigrants in Ireland and Canada*. Montreal, Buffalo, N.Y., and London, 1994.

Wilson, D.A. *United Irishmen, United States: immigrant radicals in the early republic*. Dublin and Portland, Ore., 1998.

Woods, C.J., ed. *Journals and memoirs of Thomas Russell, 1791-5*. Dublin and Belfast, 1991.

Worsley, Peter. *The trumpet shall sound: a study of "cargo" cults in Melanesia*. 2nd ed., New York, 1968.

Yates, Nigel. *The religious condition of Ireland, 1770-1850*. Oxford and New York, 2006.

Zimmermann, Georges-Denis. *The Irish storyteller*. Dublin, 2001.

———. *Songs of Irish rebellion: political street ballads and rebel songs, 1780-1900*. Dublin, 1967.

Master's Theses and Doctoral Dissertations

Cornelius, John. "The legal career of Daniel O'Connell." M.A. thesis, University of Wisconsin-Madison, 2004.

Curtin, N.J. "The origins of Irish republicanism: the United Irishmen in Dublin and Ulster, 1791–1798." Ph.D. dissertation, University of Wisconsin–Madison, 1988.

de Nie, Michael. "The eternal Paddy: Irish identity and the British press, 1798–1882." Ph.D. dissertation, University of Wisconsin–Madison, 2001.

Dickson, David. "An economic history of the Cork region in the eighteenth century." Ph.D. dissertation, University of Dublin, 1977.

Dowling, Martin. "The Abercorn estate: economy and society in northwest Ulster, 1745–1800." M.A. thesis, University of Wisconsin–Madison, 1986.

Farrell, Sean. "Conflicting visions: sectarian violence in Ulster, 1784–1886." Ph.D. dissertation, University of Wisconsin–Madison, 1996.

Hogan, P.M. "Civil unrest in the province of Connacht, 1793–98: the role of the landed gentry in maintaining order." M.Ed. thesis, University College Galway, 1976.

Lenahan, David. "Ribbonmen of the West? The Connacht outrages of 1819–20." M.A. thesis, University of Wisconsin–Madison, 2003.

Lynch, Matthew. "The mass evictions in the Kilrush poor law union, 1847–1852." M.A. thesis, University of Limerick, 2000.

O'Donoghue, Patrick. "The tithe war, 1830–33." M.A. thesis, University College Dublin, 1961.

O'Neill, J.W. "Popular culture and peasant rebellion in pre-famine Ireland." Ph.D. dissertation, University of Minnesota, 1984.

O'Neill, Timothy P. "The state, poverty, and distress in Ireland, 1815–45." Ph.D. dissertation, University College Dublin, 1971.

Peckham, Mary. "Catholic female congregations and religious change in Ireland, 1770–1870." Ph.D. dissertation, University of Wisconsin–Madison, 1993.

Tally, Patrick. "The growth of the Dublin weekly press and the development of Irish nationalism, 1810–1879." Ph.D. dissertation, University of Wisconsin–Madison, 1993.

Whelan, Irene. "Evangelical religion and the polarization of Protestant-Catholic relations in Ireland, 1780–1840." Ph.D. dissertation, University of Wisconsin–Madison, 1994.

Whooley, Finbarr. "Captain Rock's rebellion: Rockites and Whiteboys in County Cork, 1820–25." M.A. thesis, University College Cork, 1986.

Wolf, Nicholas. "Language change and the evolution of religion, community, and culture in Ireland, 1800–1900." Ph.D. dissertation, University of Wisconsin–Madison, 2008.

Internet Web Sites

Beckford, William (1759–1844): see the entry for him at http://www.heureka.clara.net/art/beckford.htm (accessed 6 Jan. 2007).

Courtenay, third Viscount: see the note about him at http://www.powderham.co.uk (accessed 4 Jan. 2007). (This site is dedicated to the history and preservation of Powderham Castle.)

Kanellis, Monica, "The Gothic labyrinth," at http://www.omni.sytes.net/~monica/mysterious.htm (accessed 6 Jan. 2007).

Norton, Rictor, "A visit to Fonthill," in "Gay history and literature: essays by Rictor Norton," at http:// www.infopt.demon.co.uk/beckfor3.htm (accessed 6 Jan. 2007).

Index

226; arms raids in, 60–61, 74, 105, 106, 248–49; Benn-Walsh's estates in, 236–39, 410n96; calm in, 80; collecting from defaulting tenants in, 222–23; conacre lettings in, 166; corn sales in, 171–72; cottier's cabin in, *161*; crowd-yeomanry clash in, 139; decline of Rockite movement in, 244, 412n126; evictions in, 237–39; executions in, 323; funds collected in, 285, 287, 422n226; incendiarism in, 136, 237, 238–39, 268, 269, 270, 271, 279; Insurrection Act in, 302; localism in, 89, 91, 172; mail coaches attacked in, 70; middlemen-farmers attacked in, 242; migrant workers from, 107, 167; murders and other violence in, mapped, *250–51*; murders in, motivations for, 417n81; murders in, summarized, 253; outside activists' attacks in, 109–10; Pastorini's influence in, 125, 126, 134; police of, 140; rent issues in, 219–20; Ribbonism and Rockite movement linked in, 102; Rockite leadership in, 180, 183, 185; Rockite mobilization in, 60–62; Rockite notices in, 47, 89, 91, 92, 95, 166, 171, 229, 395n95; social character of unrest in, 154, 156; tenant attacked in, 229; tithe collections in, 196–97; tithe commutation supported in, 405n127; tithe proctors and other tithe agents murdered in, 194, 196; withdrawn from Insurrection Act, 323; yeomanry in, 138, 139. *See also* assizes, Kerry; *specific baronies, districts, towns, and parishes*
Kiladysert: incendiarism in, 279
Kilbarry: Rockites' attack on, 70. *See also* Barry, James

Kilbehy: affray over distrained cattle at, 226
Kilcaragh: Benn-Walsh's estate in, 237
Kilcash: arms raid at, 168
Kilchreest: evictions in, 234
Kilcrumper: emigrants from, 430n193
Kildare, county: Insurrection Act in, 302; livestock fair in, 54
Kildare Place Society, 343
Kildimo: potato sales in, 172; Rockite leadership in, 180; Rockite notices at or near, 93, 95, 96
Kildorrery: absentee property around, 236; faction fighting in, 114, 243–44; funds collected in, 286; incendiarism in, 199, 235, 280; localism and workers in, 173; military wives raped near, 144–45; millenarian songs heard at, 134; murder near, 249; Rockites' efficacy in controlling land in, 233
Kilfinnane: funds collected in, 286; Pastorini's prophecies in, 391n157
Kilfinny: Rockite schoolmaster of, 180
Kilflyn: livestock pound destroyed near, 223; Rockite notice in, 47; sectarian incident in, 126
Kilgarvan: church burned in, 269, 270
Kilkenny city: affray over distrained cattle near, 226; arms raids near, 396–97n136; executions near, 318; murder near, 262; Ribbonism in, 174; Rockite allegiance around, 76; Rockites of, 174; Templemartin gang in or near, 175–77; tithe issues in or near, 189, 203; wheat seized and rescued near, 225
Kilkenny city, Liberties: arms raid in, 177
Kilkenny, county: affray over distrained cattle in, 226; arms raids in, 60–61, 74, 105, 106, 248–49;